Boundary-Value Problems for Transonic Flow

Boundary-Value Problems for Transonic Flow

Alexander G. Kuz'min

St. Petersburg State University, Russia

JOHN WILEY & SONS, LTD

Phone (+44) 1243 779777

Email (for orders and customer service enquiries): cs-books@wiley.co.uk
Visit our Home Page on www.wiley.co.uk or www.wiley.com

Other Wiley Editorial Offices

John Wiley & Sons, Inc., 605 Third Avenue, New York, NY 10158-0012, USA

Jossey-Bass, 989 Market Street, San Francisco, CA 94103-1741, USA

Wiley-VCH Verlag GmbH, Pappelallee 3, D-69469 Weinheim, Germany

John Wiley & Sons Australia, Ltd, 33 Park Road, Milton, Queensland 4064, Australia

John Wiley & Sons (Asia) Pte Ltd, 2 Clementi Loop #02-01, Jin Xing Distripark, Singapore 129809

John Wiley & Sons Canada, Ltd, 22 Worcester Road, Etobicoke, Ontario, Canada M9W 1L1

British Library Cataloguing in Publication Data

A catalogue record for this book is available from the British Library

ISBN 0-471-48641-8

Produced from LaTeX files supplied by the author, typeset by T&T Productions Ltd, London.
Printed and bound in Great Britain by Antony Rowe, Chippenham, Wiltshire.
This book is printed on acid-free paper responsibly manufactured from sustainable forestry in which at least two trees are planted for each one used for paper production.

To D.K., an amazing person who greatly assisted me in research work for years

Contents

Preface

There are two fundamental difficulties in the mathematical study of partial differential equations which govern steady inviscid transonic flow. The first one is associated with the mixed type of the equations and their nonlinearity in the physical coordinates. The nonlinearity leads to the variable position of sonic lines, which separate subsonic regions from supersonic ones, so that they finally depend on the geometry of an airfoil/channel and on the prescribed boundary conditions. The second difficulty is concerned with the formation of shock waves, which introduces additional complications into the mathematical analysis.

Since the late 1970s, remarkable progress has been made in the theory of second-order equations of mixed type. New techniques for obtaining *a priori* estimates for solutions in Sobolev spaces have been developed, and conditions for the coefficients of the equation which provide the unique and Fredholm solvabilities of boundary-value problems have been found.

The results obtained in PDE theory have made it possible to study perturbation problems for *shock-free* transonic flows. The existence and uniqueness of smooth solutions were proved, and the nonlinear stability of transonic flow at certain conditions was demonstrated. Shock-free flows occur in practice, e.g. in the case of accelerating flow through a Laval nozzle. Flow over an airfoil with a local supersonic region may be shock-free if a damping boundary condition, which prevents formation of shock waves, is prescribed on a portion of the airfoil. The above-mentioned results concerned with the solvability of boundary-value problems are presented in Chapters 1–3 of this book.

Transonic flow *with shock waves* still remains a difficult matter to investigate mathematically. Therefore, in Chapter 4 and in a few sections of Chapter 3 we employ asymptotic and high-resolution numerical methods in order to analyse flow patterns with shock waves. The goal of this analysis is to gain an insight into the structure of solutions and the physical phenomena which take place in transonic flow.

In this book, some effort has therefore been made to present results of significance for the theory of boundary-value problems associated with transonic flow. Considerable attention is also paid to practical problems, such as airfoil design, flow control, and the high sensitivity of transonic flow to a small perturbation of the boundary conditions. At the same time, the book does not pretend to be a complete survey of the theory and its applications.

The author believes that the problems under discussion will be of interest to both pure and applied mathematicians because of the close relationship between the points related to PDE theory and theoretical aerodynamics. The book will possibly attract the attention of CFD researchers who deal with transonic flow and wish to learn basic concepts and the underlying mathematical theory.

A significant part of the manuscript was prepared during the author's visit to Karlsruhe University, in the summer of 2001, which was funded by the Deutscher Akademischer Austauschdienst; the author expresses deep appreciation to Professor M. Schneider and Professor R. Scherer for valuable discussions and a stimulating environment for work. The author is obliged to Dr A. Kotov for providing computer code for solving the Euler equations. Many figures for this book have been thoroughly prepared by Ms A. Ivanova, to whom the author addresses his sincere gratitude. Support of the work by the Russian Foundation for Basic Research is also acknowledged.

Alexander Kuz'min
St. Petersburg, December 2001

Introduction

For a long time, the theory of mixed-type equations has lagged behind the theory of, for example, equations of elliptic type. The early studies of transonic flow problems concentrated on the construction of particular solutions to the Tricomi and Chaplygin equations governing inviscid transonic flow in the hodograph plane, where the independent variables are the velocity magnitude and the flow deflection angle.

In the late 1970s and in the 1980s, a breakthrough was achieved in the theory of boundary-value problems for the linear equation of mixed type

$$k(x, y)u_{xx} + [a(x, y)u_y]_y - \alpha(x, y)u_x + c(x, y)u = f(x, y), \tag{1}$$

where

$$a(x, y) > 0 \quad \text{in a rectangle } \overline{G},$$

while the coefficient $k(x, y)$ may change its sign arbitrarily in G. Thus, Equation (1) is of elliptic type where $k(x, y) > 0$ and of hyperbolic type where $k(x, y) < 0$. The set of the parabolic degeneracy $k(x, y) = 0$ may be of an essentially general structure. Basic results obtained by Karatoprakliev (1977) and Vragov (1977), who studied in fact a more general equation as compared with (1), were followed by developments of many other researchers. In Chapter 1 we present principal points of the theory. Solvability of a basic boundary-value problem for Equation (1) is established under the important condition

$$2\alpha \pm k_x > 0 \quad \text{in } \overline{G}, \tag{2}$$

which implies that the coefficient $\alpha(x, y)$ must be large enough. In the basic problem, we assume that

$$k(0, y) > 0, \qquad k(l, y) > 0,$$

i.e. the type of Equation (1) is elliptic on the front and exit sides of the rectangle G, while the coefficient $k(x, y)$ may change its sign in G and on the upper and lower sides of G arbitrarily. Analysis of this problem enables one to study the equation with negative k on the exit side $x = l$ and the equation with a local hyperbolic subdomain in which condition (2) is violated (Figure 1).

The theory developed has made it possible to study the solvability of transonic flow problems directly in the physical plane (x, y) or in the streamline coordinates (φ, ψ) ignoring the hodograph variables (Larkin 1983, 1991; Kuz'min 1984, 1986a, 1990). Thus, the study of problems for Equation (1) in cases of interest for aerodynamics

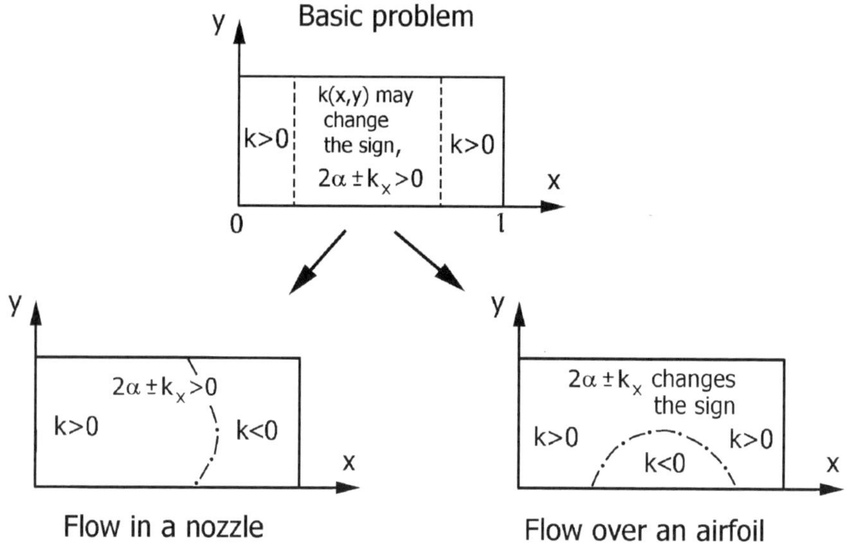

Figure 1 A strategy of the analysis of boundary-value problems.

turned out to be easier than that for the Tricomi equation $yu_{xx} + u_{yy} = 0$, which looks simple only at first sight.

In Chapter 2 we consider a perturbation problem for transonic flow that accelerates from subsonic to supersonic velocities in a channel. The perturbation $u(x, y)$ of a given velocity potential $\varphi(x, y)$ in the physical plane is governed by the equation

$$(1 - \varphi_x + u_x)u_{xx} + \frac{1}{1+\gamma}u_{yy} - \varphi_{xx}u_x = 0. \tag{3}$$

It easily follows from the von Kármán equation $(\gamma + 1)(1 - \varphi_x)\varphi_{xx} + \varphi_{yy} = 0$, which is valid if the flow velocity is close to the speed of sound and the flow deflection angle $\theta = \arctan(\varphi_y/\varphi_x)$ is small.

For the linearized version of Equation (3), the principal condition of solvability (2) reduces to

$$\varphi_{xx}(x, y) > 0, \tag{4}$$

because $\alpha = \varphi_{xx}$, $k = 1 - \varphi_x$, and

$$2\alpha + k_x = \varphi_{xx}, \qquad 2\alpha - k_x = 3\varphi_{xx}. \tag{5}$$

Since φ_x is the velocity component in the x-direction, φ_{xx} represents the flow acceleration in this direction. The condition of positive acceleration (4) typically holds for the flow in a Laval nozzle. This condition provides the existence and uniqueness of a smooth solution to the perturbation problem due to the results of Chapter 1.

The perturbation approach has been extended to encompass the Chaplygin equation in the streamline coordinates (φ, ψ),

$$K(\Phi_\varphi)\Phi_{\varphi\varphi} + \Phi_{\psi\psi} = 0, \tag{6}$$

where ψ is the stream function, φ is the velocity potential. The function $K(\Phi_\varphi)$ is negative in the subdomain of supersonic flow and positive in the subdomain of subsonic flow. Equation (6) governs inviscid flow which can substantially deviate from the sonic conditions. It is equivalent to the full-potential equation

$$(a^2 - \varphi_x^2)\varphi_{xx} - 2\varphi_x\varphi_y\varphi_{xy} + (a^2 - \varphi_y^2)\varphi_{yy} = 0,$$

where $a^2 = [\gamma + 1 - (\gamma - 1)(\varphi_x^2 + \varphi_y^2)]/2$ is the squared local speed of sound and γ is the ratio of specific heats. The full-potential equation in its turn is equivalent to the system of the continuity and Euler equations if the flow is shock free and irrotational.

For the analysis of finite-element discretizations of transonic boundary-value problems, we employ a weak formulation of the problems at hand. It consists of finding a finite-element solution u_h which satisfies an integral relation that follows from (3) or (6). This can be reformulated as a minimization problem for a functional $\mathcal{J}_h(u_h)$ as suggested by Glowinski and Pironneau (1978). Developments by Mandel and Nečas (1987) and Berger (1989) have made it possible to prove the convergence of a subsequence of solutions u_h as $h \to 0$ under the assumption that there exists a unique solution $u(x, y)$ to the continuous problem. The latter is true if condition (4) holds as discussed above.

Chapter 3 deals with boundary-value problems for flow over an airfoil with a local supersonic region. In well-known papers, Morawetz (1956, 1957) proved the nonexistence of a smooth solution to the perturbation problem for flow with a local supersonic region over a solid airfoil. From the viewpoint of the theory of Equations (1), (3), a flow with a local supersonic region is inevitably associated with the violation of condition (4). Indeed, the flow first accelerates and then decelerates in the supersonic region. Therefore, φ_{xx} is of variable sign. Hence, the expression $2\alpha \pm k_x$ changes its sign due to (5).

The nonexistence of smooth solutions demonstrates that shock-free flow with a local supersonic region is unstable to perturbation of boundary conditions, so that weak shock waves arise under any perturbation of the shockless airfoil along its supersonic arc. Consequently, for the analysis of transonic flow over a given solid airfoil, one should employ a class of functions admitting jumps in the gradient of the velocity potential and incorporate the entropy condition to be satisfied across the shock waves. Unfortunately, this is associated with formidable mathematical difficulties.

Developments in the theory of boundary-value problems for Equation (1) have shown that violation of the condition of positive acceleration (4) results in the arising of a singularity of the solution to the linearized problem at the downstream endpoint B of the sonic line. Moreover, it is the violation of (4) at the point B which plays a crucial role in the appearance of the singularity. Therefore, one can try a modification of the slip condition on the airfoil in order to provide a smooth flow field, and in this way to separate mathematical complications caused by the mixed type of Equation (3) from those concerned with the presence of shock waves.

That is why, to prevent the appearance of shock waves in the flow with a local supersonic region, we employ an oblique derivative condition, which is a kind of damping boundary condition, on a portion of the airfoil embracing point B. From the physical point of view, this means employment of a permeable (porous or perforated) surface of the airfoil. The modification of boundary conditions enables one to study the problem in spite of violation of the principal condition (2). However, this requires an advanced technique for obtaining *a priori* estimates in Sobolev spaces. Success in analysis of the problem with the damping boundary condition has been achieved due to the Fredholm alternative established for Equation (3) and a technique for analysis of the uniqueness worked out by Morawetz (1956) and Cook (1978).

In our mathematical study, we therefore concentrate on the perturbation of a given smooth solution, not on 'global' ones. This approach is justified by the fact that examples of exact solutions, which illustrate transonic flow in many situations of practical interest, were elaborated by the 1990s with the theory of self-similar solutions and other techniques. Hence, in order to study mathematically the direct problem for a given geometry of the airfoil/channel, one can refer to an available exact solution with a similar shape of streamlines, then consider it as an initial guess, and concentrate on solving the perturbation problem under the given small variation of the shape of the boundary.

The obtained results on the solvability of the steady perturbation problems resemble the well-known results for the unsteady compressible Euler equations, which assert that if the initial Cauchy data with respect to the flow density and velocity components are smooth enough, then there exists a maximum time interval on which there exists a unique smooth solution to the system of the continuity and Euler equations.

The study of shock-free transonic flow appears to be quite a natural step in the development of the theory and a prerequisite for the treatment of increasingly complex flows with shock waves. The latter still remains a difficult point to investigate mathematically because of the intricacy introduced by the entropy condition to be satisfied across the shock waves. These complications can hardly be overcome without a clear understanding of the physical phenomena which are associated with the formation of shock waves.

The manner in which shock waves arise in transonic flow is often difficult to predict, but it influences the structure of the flow field, development of instabilities in the boundary layer, the laminar–turbulent transition, and, as a consequence, the aerodynamic lift and drag. Fundamentals of the arising of shock waves were established by Nikolsky and Taganov (1946), Frankl (1947), Busemann (1949) and Guderley (1953). Following their ideas, we employ asymptotic methods in order to analyse the principal concepts of shock waves formation and propagation in the local supersonic region.

In Chapter 4, we discuss transonic flow patterns revealed by numerical simulation. The emphasis is laid on the structure of transonic flow at nearly shock-free conditions. The observed behaviour of compression and shock waves in a local supersonic region contributes to the understanding of the nature of the great sensitivity of transonic flow to perturbation of the airfoil shape and of the freestream Mach number.

Throughout the book, we concentrate on direct problems in which the domain and boundary conditions are given, while the solution of a differential equation is to be found. A brilliant elucidation of the relation between direct and inverse approaches was given by Ferrari and Tricomi (1968, p. 106):

> In the inverse method, which has been the approach most commonly discussed in the literature, one examines the nature of specific particular solutions or of specific families of such particular solutions of the governing differential equations in hope of being able to discover one that holds promise for practical use. In the direct method, one looks for a solution that satisfies the boundary conditions that have been imposed *a priori*, as dictated by the needs of the technical problem. Of course, this latter way of proceeding is more difficult, because, as mentioned earlier, even the curves in the hodograph plane along which the boundary conditions are given cannot be presumed known beforehand.

1

Linear Partial Differential Equations of Mixed Type

Consider the equation

$$Lu := k(x, y)u_{xx} + [a(x, y)u_y]_y - \alpha(x, y)u_x + c(x, y)u = f(x, y), \qquad (1.1)$$

where the coefficients are supposed to be smooth enough: $k, a \in C^3(\overline{G})$; $\alpha, c \in C^2(\overline{G})$, and

$$a(x, y) \geqslant \delta > 0 \qquad (1.2)$$

in the rectangle

$$G = \{(x, y) \in \mathbb{R}^2 : 0 < x < l, \ -1 < y < 1\}.$$

Formulation of boundary-value problems for Equation (1.1) depends on the signs of the coefficient $k(x, y)$ at $x = 0$ and $x = l$. In a *basic problem* we suppose that the type of equation is elliptic on the front and exit sides $x = 0$, $x = l$ of the rectangle G, and prescribe boundary conditions on the whole perimeter of G. Moreover, we assume that the inequality $2\alpha \pm k_x \geqslant \delta > 0$ with δ an arbitrary small positive constant holds in the whole domain G. In order to prove the solvability, we employ a method of singular perturbations, which is similar to the streamline diffusion technique well-known in the theory of convection–diffusion equations.

To obtain a basic *a priori* estimate, we multiply Equation (1.1) by u_x and integrate the result over G as in the theory of hyperbolic equations. Then special assumptions with respect to the coefficients of Equation (1.1) are applied which do not involve the sign of the coefficient k in the domain G.

Even though the analysis of the basic problem requires some effort, it paves the way for the study of a number of other problems with different signs of k on the front and exit sides of G. By and large, the theory worked out is similar to the classical theory of elliptic equations. At the same time, in a special case when the type of the equation is hyperbolic in the given domain, the obtained results reduce to the well-known unique solvability of the initial–boundary-value problem for the purely hyperbolic equation.

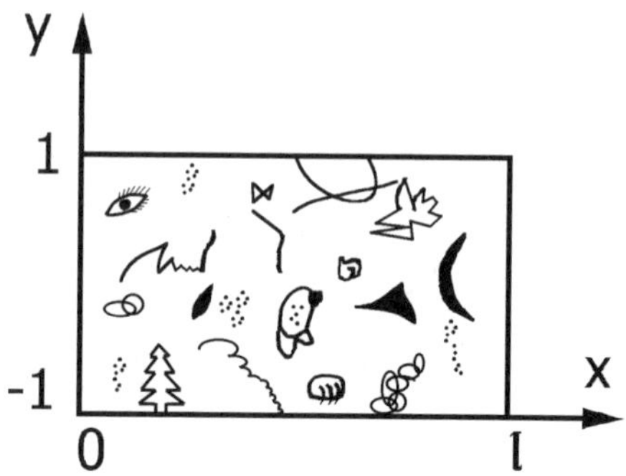

Figure 1.1 A set at which the coefficient $k(x, y)$ may vanish.

1.1 Solvability of a Basic Boundary-Value Problem

In this section, we suppose that

$$k(0, y) > 0 \quad \text{and} \quad k(l, y) > 0 \quad \text{at} \; -1 \leqslant y \leqslant 1, \tag{1.3}$$

while the coefficient $k(x, y)$ may change its sign arbitrarily inside the domain G and on the upper and lower sides $y = \pm 1$, $0 < x < l$ (Figure 1.1). Under the assumption (1.3), we prescribe the mixed Dirichlet–Neumann boundary conditions on the boundary ∂G:

$$\left.\begin{array}{lll} u(0, y) = 0 & \text{at } x = 0, & -1 < y < 1, \\ u_x(0, y) = 0 & \text{at } x = l, & -1 < y < 1, \\ u_y(x, \pm 1) = 0 & \text{at } y = \pm 1, & 0 < x < l. \end{array}\right\} \tag{1.4}$$

Theorem 1.1. *Let inequalities (1.2), (1.3) hold, and*

$$2\alpha \pm k_x \geqslant \delta > 0, \tag{1.5}$$

$$a_x \leqslant 0, \quad c \leqslant 0, \quad c_x \geqslant 0 \tag{1.6}$$

in the domain G. Then there exists a unique solution $u \in W^{2,2}(G)$ of Problem (1.1), (1.4) $\forall f \in W^{1,2}(G)$.

Hereafter, $W^{n,2}(G)$ denotes the usual Sobolev space of functions with generalized derivatives up to the order $n = 1, 2, \ldots$, which are square-integrable over G.

In the three examples pointed out below, the solvability is true if $\alpha(x, y) > 1$ and $l = 2$, since $|k_x| = |2(x-1)| \leqslant 2$. Hence, $2\alpha \pm k_x > 0$, and the inequalities (1.5), (1.6) hold.

Example 1.2 (Figure 1.2a).

$$[(x-1)^2 + \tfrac{1}{4}y]u_{xx} + u_{yy} - \alpha(x, y)u_x = f(x, y).$$

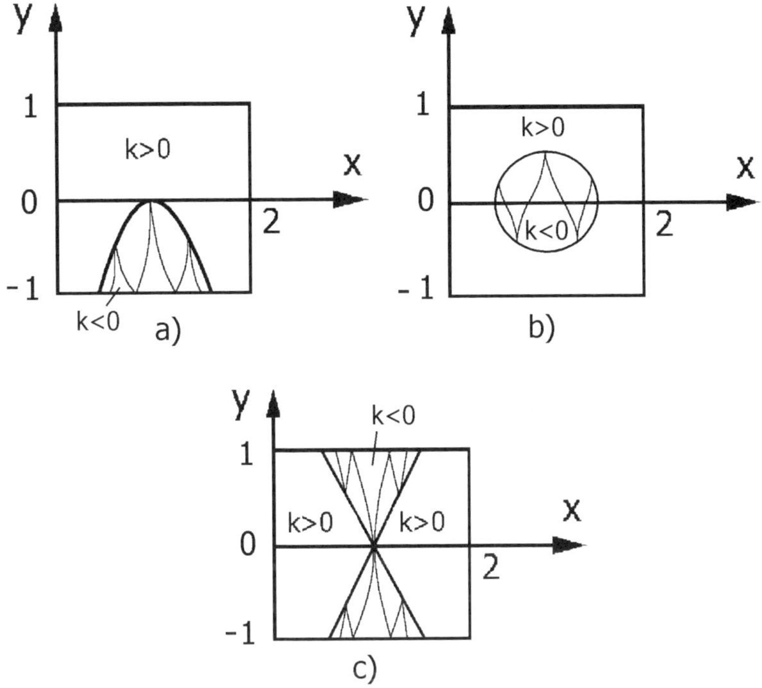

Figure 1.2 Location of the elliptic and hyperbolic subdomains in Examples 1.2–1.4.

Example 1.3 (Figure 1.2b).

$$[(x-1)^2 + y^2 - \tfrac{1}{4}]u_{xx} + u_{yy} - \alpha(x, y)u_x = f(x, y).$$

Example 1.4 (Figure 1.2c).

$$[(x-1)^2 - \tfrac{1}{4}y^2]u_{xx} + u_{yy} - \alpha(x, y)u_x = f(x, y).$$

The thin lines in Figure 1.2 outline the characteristics determined in the hyperbolic subdomains by solution of the ordinary differential equation

$$k(x, y)(\mathrm{d}y)^2 + (\mathrm{d}x)^2 = 0.$$

Theorem 1.1 was proved for a multidimensional equation with different techniques by Karatoprakliev (1977) and Vragov (1977). Below we use the method of singular perturbations combined with the Galerkin method suggested by Egorov (1987). However, we modify the proof by using simpler cut-off functions (Kuz'min 2000a).

Proof of Theorem 1.1.

Step 1 (Uniqueness). First, let us establish the estimate

$$\|u_x\|_0 \leqslant m\|Lu\|_0 \tag{1.7}$$

for any function $u \in W^{2,2}(G)$ satisfying the boundary conditions (1.4), where $\|\cdot\|_0$ is the norm of the space $L^2(G)$ and m is a positive constant independent of u. Integrating $Lu \cdot u_x$ over G, we obtain the identity

$$-2\int_G Lu \cdot u_x \,\mathrm{d}G = -2\int_G [ku_{xx} + (au_y)_y - \alpha u_x + cu]u_x \,\mathrm{d}G$$
$$= \int_G [-k(u_x^2)_x - 2u_x(au_y)_y + 2\alpha u_x^2 - c(u^2)_x]\,\mathrm{d}G,$$

where $\mathrm{d}G = \mathrm{d}x\,\mathrm{d}y$. Using Green's formula,

$$\int_\Gamma P(x, y)\,\mathrm{d}x + Q(x, y)\,\mathrm{d}y = \int_G (Q_x - P_y)\,\mathrm{d}G,$$

where the integration over the boundary $\Gamma = \partial G$ of G is performed in the direction that leaves the domain G on the left, we obtain

$$-2\int_G Lu \cdot u_x \,\mathrm{d}G = \int_G [(2\alpha + k_x)u_x^2 - a_x u_y^2 + c_x u^2]\,\mathrm{d}G$$
$$+ \int_\Gamma 2au_x u_y \,\mathrm{d}x - (ku_x^2 - au_y^2 + cu^2)\,\mathrm{d}y. \tag{1.8}$$

Owing to the boundary conditions (1.4), the identity (1.8) reduces to

$$-2\int_G Lu \cdot u_x \,\mathrm{d}G = \int_G [(2\alpha + k_x)u_x^2 - a_x u_y^2 + c_x u^2]\,\mathrm{d}G$$
$$+ \int_{-1}^{1} (au_y^2 - cu^2)|_{x=l}\,\mathrm{d}y + \int_{-1}^{1} (ku_x^2)|_{x=0}\,\mathrm{d}y. \tag{1.9}$$

Omitting the nonnegative terms on the right-hand side due to conditions (1.3), (1.6), we find

$$-2\int_G Lu \cdot u_x \,\mathrm{d}G \geqslant \int_G (2\alpha + k_x)u_x^2 \,\mathrm{d}G \geqslant \delta \int_G u_x^2 \,\mathrm{d}G.$$

The left-hand side can be estimated with Young's inequality

$$\frac{1}{\gamma}\|Lu\|_0^2 + \gamma\|u_x\|_0^2 \geqslant -2\int_G Lu \cdot u_x \,\mathrm{d}G \geqslant \delta \int_G -u_x^2 \,\mathrm{d}G, \tag{1.10}$$

where γ is a positive parameter. Then at $\gamma < \delta$, inequality (1.10) yields estimate (1.7).

Now, suppose that there exist two solutions $u_1(x, y)$ and $u_2(x, y)$ of Problem (1.1), (1.4). The difference $u = u_1 - u_2$ satisfies boundary conditions (1.4) and the equation $Lu = L(u_1 - u_2) = f - f = 0$. Therefore, estimate (1.7) yields $u_x = 0$ a.e. (almost everywhere) in G. Hence, $u(x, y)$ can depend only on y; due to the boundary condition $u(0, y) = 0$ we get $u = 0$ a.e. in G. This means that $u_1(x, y) = u_2(x, y)$, and the uniqueness is proved.

Step 2 (Existence of a weak solution). In order to prove the solvability of Problem (1.1), (1.4), we insert a singular perturbation term εu_{xxx} into Equation (1.1) and impose the extra condition $u_x = 0$ at $x = 0$:

$$L_\varepsilon u \equiv \varepsilon u_{xxx} + Lu = f, \qquad \varepsilon > 0, \tag{1.11}$$

$$u(0, y) = u_x(0, y) = 0 \qquad \text{at } x = 0, \tag{1.12}$$

$$u_x(l, y) = 0 \qquad \text{at } x = l, \tag{1.13}$$

$$u_y(x, \pm 1) = 0 \qquad \text{at } y = \pm 1. \tag{1.14}$$

An approximate solution $u^{N,\varepsilon}(x, y)$ of the resulting Problem (1.11)–(1.14) is sought in Galerkin's form (Ladyzhenskaya 1985, § 4.3),

$$u^{N,\varepsilon}(x, y) = \sum_{i=1}^{N} X_i^{N,\varepsilon}(x) Y_i(y), \tag{1.15}$$

where $Y_i(y)$, $i = 1, 2, \ldots, N$, is a complete system in $W^{2,2}(-1, 1)$, which is orthonormal in $L^2(-1, 1)$. The functions $X_i^{N,\varepsilon}(x)$ are to be found from the relations

$$\int_{-1}^{1} (L_\varepsilon u^{N,\varepsilon} - f) Y_j \, \mathrm{d}y = 0,$$

which reduce to

$$\int_{-1}^{1} [(\varepsilon u_{xxx}^{N,\varepsilon} + k u_{xx}^{N,\varepsilon} - \alpha u_x^{N,\varepsilon} + c u^{N,\varepsilon} - f) Y_j - a u_y^{N,\varepsilon} \cdot (Y_j)_y] \, \mathrm{d}y = 0,$$
$$j = 1, 2, \ldots, N, \tag{1.16}$$

if one integrates by parts the term $(a u_y^{N,\varepsilon})_y Y_j$ and omits $(a u_y^{N,\varepsilon} Y_j)|_{y=\pm 1}$ in view of (1.14).

Relations (1.16) are in fact a system of third-order ordinary differential equations with respect to $X_i^{N,\varepsilon}(x)$. Boundary conditions at $x = 0$ and $x = l$ are prescribed according to (1.12), (1.13):

$$X_i^{N,\varepsilon}(0) = \left.\frac{\mathrm{d}X_i^{N,\varepsilon}}{\mathrm{d}x}\right|_{x=0} = \left.\frac{\mathrm{d}X_i^{N,\varepsilon}}{\mathrm{d}x}\right|_{x=l} = 0. \tag{1.17}$$

The following *a priori* estimate is true,

$$\|u^{N,\varepsilon}\|_1^2 + \varepsilon \|u_{xx}^{N,\varepsilon}\|_0^2 \leqslant m \|f\|_0^2, \tag{1.18}$$

where $\|\cdot\|_p = \|\cdot\|_{W^{p,2}(G)}$ is the norm in the space $W^{p,2}(G)$ and the constant $m > 0$ is independent of $u^{N,\varepsilon}$. In order to prove (1.18), we multiply each of Equations (1.16) by $\mathrm{e}^{-\mu x}(X_j^{N,\varepsilon})_x$, $\mu > 0$, then sum up the results from 1 to N, and integrate with respect to x from 0 to l. This yields the relation

$$\int_G [(\varepsilon u_{xxx}^{N,\varepsilon} + k u_{xx}^{N,\varepsilon} - \alpha u_x^{N,\varepsilon} + c u^{N,\varepsilon}) u_x^{N,\varepsilon} - a u_y^{N,\varepsilon} u_{xy}^{N,\varepsilon}] e^{-\mu x} \, dG$$

$$= \int_G f u_x^{N,\varepsilon} e^{-\mu x} \, dG,$$

which makes it possible to derive (1.18) with the technique used for obtaining (1.7) from (1.8).

Estimate (1.18) proves the uniqueness of the solution to Problem (1.16), (1.17). The uniqueness ensures the existence of the solution to the system of N third-order equations endowed with $3N$ boundary conditions, as known from the theory of linear ordinary differential equations. Moreover, the solution $X_i^{N,\varepsilon}(x)$ belongs to the class of functions $C^4[0, l]$ due to the smoothness conditions imposed on the coefficients of Equation (1.1). Thus, the existence of the approximate solution (1.15) is established.

Consider the sequence of approximate solutions $u^{N,\varepsilon}(x, y)$ as $N \to \infty$. Owing to (1.18), the norm $\|u^{N,\varepsilon}\|_1$ is bounded uniformly in N. Therefore, due to the weak compactness of a bounded set in the Hilbert space, there exists a subsequence with a weak limit $u^\varepsilon(x, y) \in W^{1,2}(G)$.

In order to prove that $u^\varepsilon(x, y)$ is a solution of Equation (1.11), we multiply each of Equations (1.16) by a function $\chi_j(x) \in C^\infty(0, l)$ vanishing in the vicinities of $x = 0$, $x = l$, then sum up from $j = 1$ to $j = N$, and integrate with respect to x from 0 to l. This yields the identity

$$\int_G [(\varepsilon u_{xxx}^{N,\varepsilon} + k u_{xx}^{N,\varepsilon} - \alpha u_x^{N,\varepsilon} + c u^{N,\varepsilon} - f)\chi^N - a u_y^{N,\varepsilon} \chi_y^N] \, dG = 0, \qquad (1.19)$$

where $\chi^N(x, y) = \sum_{j=1}^N \chi_j(x) Y_j(y)$. Now we pass to the above-mentioned subsequence of $u^{N,\varepsilon}$, integrate by parts the first and second terms in (1.19), and let $N \to \infty$:

$$\int_G [\varepsilon u_x^\varepsilon \chi_{xx} - u_x^\varepsilon (k\chi)_x - a u_y^\varepsilon \chi_y + (-\alpha u_x^\varepsilon + c u^\varepsilon - f)\chi] \, dG = 0. \qquad (1.20)$$

Relation (1.20) is valid for any function $\chi \in W^{1,2}(G)$ vanishing at $x = 0$, $x = l$. Consequently, u^ε is a weak solution to Equation (1.11) (Ladyzhenskaya 1985). This solution satisfies the boundary conditions (1.12). Indeed, $u_{xx}^\varepsilon \in L^2(G)$ due to (1.18). Therefore, traces of u^ε, u_x^ε at $x =$ const. belong to $L^2(-1, 1)$ and satisfy the same conditions at $x = 0$ and $x = l$ as $u^{N,\varepsilon}$ and $u_x^{N,\varepsilon}$ do owing to (1.17). The boundary condition (1.14) is involved into the integral relation (1.20) as shown below (see (1.26)). Hence, $u^\varepsilon(x, y)$ is a weak solution to Problem (1.11)–(1.14) involving the singular perturbation εu_{xxx}.

Now, let us consider a sequence of 'viscous' solutions $u^\varepsilon(x, y)$ as $\varepsilon \to 0$. Due to estimate (1.18), the norm $\|u^\varepsilon\|_1$ is bounded uniformly in ε. Therefore, there exists a subsequence with a weak limit

$$u^0(x, y) \in W^{1,2}(G).$$

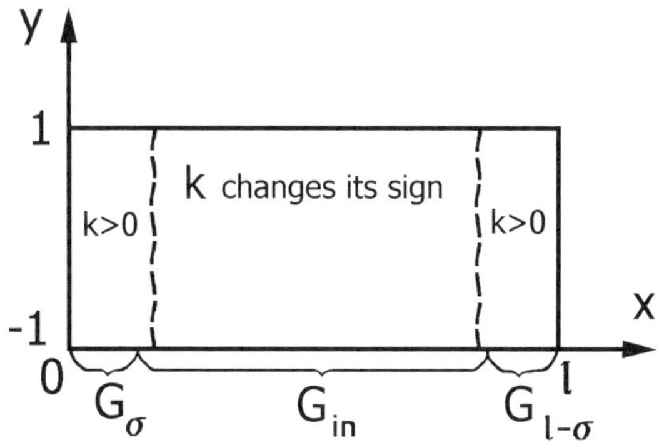

Figure 1.3 Subdomains $G_{l-\sigma}$, G_{in}, G_σ in which the estimates (1.22), (1.23) and the belonging (1.24) take place, respectively. (Reproduced from Kuz'min (2000a) by permission of Heldermann.)

This limit is a weak solution of the mixed-type Equation (1.1), as seen from (1.20) with $\varepsilon = 0$,

$$\int_G [-u_x^0(k\chi)_x - au_y^0\chi_y + (-\alpha u_x^0 + cu^0 - f)\chi]\,\mathrm{d}G = 0, \tag{1.21}$$

and it vanishes along with u^ε on the front side of $G : u^0(0, y) = 0$. In order to prove that the Neumann condition $u_x^0(l, y) = 0$ holds on the exit side of G, we need to use the estimate

$$\|u_x^\varepsilon\|_{W^{1,2}(G_{l-\sigma})} \leqslant m\|f\|_0 \quad \text{(in the vicinity of the exit side } x = l \text{ of } G\text{)}, \tag{1.22}$$

where $G_{l-\sigma} = G \cap (l - \sigma < x < l)$ (see Figure 1.3). Hereafter, we denote by m positive constants which are independent of u^ε and may be different in different formulae. The validity of (1.22) will be established below at Step 4 with a technique of cut-off functions in the same way as in the theory of elliptic equations. Estimate (1.22) shows that the weak solution u^0 satisfies the Neumann condition $u_x^0 = 0$ due to $u_x^\varepsilon = 0$ at $x = l$.

Step 3 (Regularity of the obtained solution). In order to prove the regularity, i.e. $u^0 \in W^{2,2}(G)$, we need one more estimate

$$\|u_{xx}^\varepsilon\|_{L^2(G_{\text{in}})} \leqslant m\|f\|_1, \tag{1.23}$$

where $G_{\text{in}} = G \cap (\sigma < x < l - \sigma)$. The validity of (1.23) under the additional condition $2\alpha - k_x \geqslant \delta > 0$ will be proved at Step 5. In the strip $G_\sigma = G \cap (0 < t < \sigma)$ adjacent to the front side of G, the type of Equation (1.1) is elliptic, so that $u^0 \in W^{1,2}(G_\sigma)$ yields

$$u^0 \in W^{2,2}(G_\sigma) \quad \text{(in the vicinity of the front side } x = 0 \text{ of } G\text{)}. \tag{1.24}$$

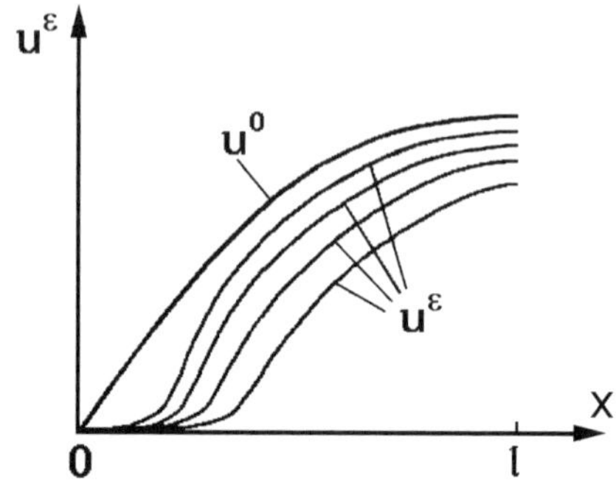

Figure 1.4 A sketch illustrating the manner of convergence $u^\varepsilon \to u^0$. (Reproduced from Kuz'min (2000a) by permission of Heldermann.)

By combining (1.22)–(1.24), we find $u^0_{xx} \in L^2(G)$. This enables one to represent (1.21) in the form

$$\int_G [u^0_{xx}\chi - au^0_y\chi_y + \underline{(ku^0_{xx} - u^0_{xx} - \alpha u^0_x + cu^0 - f)}\chi]\,\mathrm{d}G = 0. \qquad (1.25)$$

Having denoted by $-\tilde{f}$ the underlined terms, one can interpret $u^0(x, y)$ as a weak solution of the equation $u_{xx} + (au_y)_y = \tilde{f} \in L^2(G)$. At the same time, the latter equation endowed with the boundary conditions (1.4) has a solution $\tilde{u} \in W^{2,2}(G)$ as follows from the theory of equations of elliptic type (Ladyzhenskaya 1985). Because of the uniqueness in $W^{1,2}(G)$, we obtain $u^0 \equiv \tilde{u} \in W^{2,2}(G)$. As a consequence, the second term in (1.25) can be integrated by parts:

$$\int_G [ku^0_{xx} + (au^0_y)_y - \alpha u^0_x + cu^0 - f]\chi\,\mathrm{d}G$$
$$+ \int_0^l (au^0_y\chi)|_{y=-1}\,\mathrm{d}x - \int_0^l (au^0_y\chi)|_{y=1}\,\mathrm{d}x = 0. \qquad (1.26)$$

Due to the arbitrariness in the choice of $\chi \in W^{1,2}(G)$, we conclude that Equation (1.1) is satisfied a.e. in the domain G, and the boundary condition $u^0_y = 0$ is true a.e. at $y = \pm 1$.

We notice that estimate (1.23) is not valid in the whole domain G uniformly in ε. Hence, the boundary condition $u^\varepsilon_x(0, y) = 0$ does not yield the same condition with respect to u^0, and only $u^0(0, y) = 0$ is true at $x = 0$ (see Figure 1.4). In other words, the viscous solutions $u^\varepsilon(x, y)$ do not converge to $u^0(x, y)$ uniformly in the norm $\|\cdot\|_{W^{2,2}(G)}$.

Step 4 (A proof of estimate (1.22) for the viscous solution). The validity of the estimate (1.22) for the solution u^ε in the vicinity of the exit side of the rectangle G can be proved owing to the elliptic type of Equation (1.1) in that vicinity as follows.

Since $k(x, y) > 0$ at $x = 0$ and $x = l$, one can choose a small $\sigma > 0$ so that $k(x, y) > 0$ in the strips $0 \leqslant x \leqslant 2\sigma$ and $l - 2\sigma \leqslant x \leqslant l$, $|y| \leqslant 1$. Consider a nonnegative cut-off function $\eta(x) \in C^\infty[0, l]$ vanishing at $x \leqslant l - 2\sigma$ and equal to $e^{\mu x}$ at $l - \sigma \leqslant x \leqslant l$, where $\mu > 0$ is large enough so that $a_x\eta + a\eta_x \geqslant 0$ at $x = l$. Let us multiply (1.16) by $\eta(X_j^{N,\varepsilon})_{xx}$, then sum up over j from 1 to N, and integrate with respect to x from 0 to l. In this way, we arrive at

$$\int_G [(\varepsilon w_{xxx} + k w_{xx} - \alpha w_x + c w - f) w_{xx} - a w_y w_{xxy}]\eta \, \mathrm{d}G = 0, \tag{1.27}$$

where w denotes the approximate solution $u^{N,\varepsilon}$ for brevity. Relation (1.27) can be represented in the form

$$\int_G f w_{xx}\eta \, \mathrm{d}G = \int_G [\tfrac{1}{2}\varepsilon(w_{xx}^2)_x + k w_{xx}^2 - \alpha w_x w_{xx} + c w w_{xx} - a w_y w_{xxy}]\eta \, \mathrm{d}G. \tag{1.28}$$

Integrating by parts the first and the last terms in the right-hand side and taking into consideration that $w_{xy}(l, y) = 0$ due to (1.17), we obtain

$$\begin{aligned}
&\int_G f w_{xx}\eta \, \mathrm{d}G \\
&= \int_{-1}^{1} \tfrac{1}{2}\varepsilon w_{xx}^2(l, y)\eta(l) \, \mathrm{d}y \\
&\quad + \int_G [-\tfrac{1}{2}\varepsilon \eta_x w_{xx}^2 + k\eta w_{xx}^2 - \alpha\eta w_x w_{xx} + c\eta w w_{xx} + (a\eta w_y)_x w_{xy}] \, \mathrm{d}G \\
&\geqslant \int_G [-\tfrac{1}{2}\varepsilon \eta_x w_{xx}^2 + k\eta w_{xx}^2 - \alpha\eta w_x w_{xx} + c\eta w w_{xx} + (a\eta)_x w_y w_{xy} + a\eta w_{xy}^2] \, \mathrm{d}G.
\end{aligned} \tag{1.29}$$

The integral $\int[-\frac{1}{2}\varepsilon\eta_x w_{xx}^2] \, \mathrm{d}G$ can be estimated by $\|f\|_0^2$ in view of (1.18):

$$\begin{aligned}
&\int_G f w_{xx}\eta \, \mathrm{d}G + m\|f\|_0^2 \\
&\qquad \geqslant \int_G [k\eta w_{xx}^2 + a\eta w_{xy}^2 - \alpha\eta w_x w_{xx} + c\eta w w_{xx} + (a\eta)_x w_y w_{xy}] \, \mathrm{d}G.
\end{aligned} \tag{1.30}$$

Now integrating by parts the last three terms in (1.30), we find

$$\begin{aligned}
&\int_G f w_{xx}\eta \, \mathrm{d}G + m\|f\|_0^2 \\
&\qquad \geqslant \frac{1}{2}\int_{-1}^{1} (a\eta)_x w_y^2\big|_{x=l} \, \mathrm{d}y \\
&\qquad\qquad + \int_G [k\eta w_{xx}^2 + a\eta w_{xy}^2 + \tfrac{1}{2}(\alpha\eta)_x w_x^2 - (c\eta w)_x w_x - \tfrac{1}{2}(a\eta)_{xx} w_y^2] \, \mathrm{d}G.
\end{aligned} \tag{1.31}$$

The first integral in the right-hand side is nonnegative due to the choice of $\eta(x)$ and can be omitted. The terms involving the first-order derivatives of w can be estimated by $\|f\|_0^2$ owing to (1.18). That is why (1.31) reduces to

$$\int_G f w_{xx} \eta \, \mathrm{d}G + m\|f\|_0^2 \geqslant \int_G [k\eta w_{xx}^2 + a\eta w_{xy}^2] \, \mathrm{d}G \tag{1.32}$$

(we recall that $m > 0$ is independent of w and may be different in different formulae). Using Young's inequality in the left-hand side, we obtain

$$\frac{1}{2\gamma} \int_G f^2 \, \mathrm{d}G + \frac{\gamma}{2} \int_G w_{xx}^2 \eta^2 \, \mathrm{d}G + m\|f\|_0^2 \geqslant \int_G [k\eta w_{xx}^2 + a\eta w_{xy}^2] \, \mathrm{d}G. \tag{1.33}$$

Since $\eta \equiv 0$ at $x \leqslant l - 2\sigma$, the integration in the right-hand side is actually carried out over the strip $G_{l-2\sigma} = G \cap (x > l - 2\sigma)$. In this strip $k \geqslant k_{\min} > 0$, whence we arrive at the estimate

$$\frac{1}{2\gamma}\|f\|_0^2 + m\|f\|_0^2 \geqslant \int_{G_{l-2\sigma}} [(k_{\min}\eta - \gamma\eta^2/2) w_{xx}^2 + a\eta w_{xy}^2] \, \mathrm{d}G. \tag{1.34}$$

The parameter $\gamma > 0$ can be chosen small enough to provide $k_{\min}\eta - \gamma\eta^2/2 > 0$. Hence,

$$m\|f\|_0^2 \geqslant \int_{G_{l-\sigma}} [w_{xx}^2 + w_{xy}^2] \, \mathrm{d}G. \tag{1.35}$$

This means

$$\|w_x\|_{W^{1,2}(G_{l-\sigma})} \leqslant m\|f\|_0, \tag{1.36}$$

where $G_{l-\sigma} = G \cap (x > l - \sigma)$. Recalling that $w = u^{N,\varepsilon}$ and letting $N \to \infty$, we get (1.22) from (1.36).

Step 5 (A proof of estimate (1.23) for the viscous solution in the inner subdomain). Let σ be the same as in the previous section. We choose now a nonnegative cut-off function $\eta(x) \in C^\infty[0, l]$ equal to unity at $\sigma \leqslant x \leqslant l - \sigma$ and to zero at $0 \leqslant x \leqslant \sigma/2$ and $l - \sigma/2 \leqslant x \leqslant l$. Let us multiply (1.16) by $\eta(X_j^{N,\varepsilon})_{xxx}$, then sum up from 1 to N, and integrate with respect to x from 0 to l. In this way, we obtain

$$\int_G [(\varepsilon w_{xxx} + k w_{xx} - \alpha w_x + c w - f) w_{xxx} - a w_y w_{xxxy}] \eta \, \mathrm{d}G = 0, \tag{1.37}$$

i.e.

$$\begin{aligned} \int_G f w_{xxx} \eta \, \mathrm{d}G = \int_G [(\varepsilon w_{xxx}^2 &+ k w_{xx} w_{xxx} - \alpha w_x w_{xxx} \\ &+ c w w_{xxx} - a w_y w_{xxxy}] \eta \, \mathrm{d}G. \end{aligned} \tag{1.38}$$

By omitting the term εw_{xxx}^2 and integrating the others by parts, due to $\eta(0) = \eta(l) = 0$ we arrive at the estimate

$$-\int_G (f\eta)_x w_{xx}\,\mathrm{d}G \geqslant \int_G [-(k\eta)_x w_{xx}^2/2 + (\alpha w_x \eta)_x w_{xx} - (cw\eta)_x w_{xx} - (aw_y\eta)_{xx} w_{xy}]\,\mathrm{d}G. \tag{1.39}$$

Owing to the choice of $\eta(x)$, the derivatives η_x and η_{xx} vanish in G except for two strips $G \cap (\sigma/2 < x < \sigma)$ and $G \cap (l - \sigma < x < l - \sigma/2)$ in which $k \geqslant k_{\min} > 0$. The integrals over the second strip can be estimated by invoking inequality (1.36), while those over the first strip can be estimated using a similar inequality obtained in the same way as (1.36) with a cut-off function that vanishes at $x \leqslant \sigma/4$ and $x \geqslant 3\sigma/2$. Then (1.39) yields

$$-\int_G f_x w_{xx}\eta\,\mathrm{d}G + m\|f\|_0^2 \geqslant \int_G [\tfrac{1}{2}(2\alpha - k_x)w_{xx}^2 + \alpha_x w_x w_{xx} - (cw)_x w_{xx} - (aw_y)_{xx} w_{xy}]\eta\,\mathrm{d}G. \tag{1.40}$$

Now we use the inequality $2\alpha - k_x \geqslant \delta > 0$ valid due to the assumption (1.5):

$$-\int_G f_x w_{xx}\eta\,\mathrm{d}G + m\|f\|_0^2 \geqslant \int_G [\tfrac{1}{2}\delta w_{xx}^2 + \underline{\alpha_x w_x w_{xx}} - \underline{(c_x w + c w_x) w_{xx}} - (\underline{a_{xx} w_y} + 2a_x w_{xy} + \underline{a w_{xxy}}) w_{xy}]\eta\,\mathrm{d}G. \tag{1.41}$$

Integrating by parts the underlined terms in the right-hand side and estimating the terms with η_x again by $\|f\|_0^2$, we get

$$-\int_G f_x w_{xx}\eta\,\mathrm{d}G + m\|f\|_0^2 \geqslant \frac{1}{2}\int_G [\delta w_{xx}^2 - \alpha_{xx} w_x^2 + 2(c_x w)_x w_x + c_x w_x^2 + a_{xxx} w_y^2 - 3a_x w_{xy}^2]\eta\,\mathrm{d}G. \tag{1.42}$$

The integral of the terms involving the first-order derivatives of w can be estimated with (1.18). Also we notice that $a_x \leqslant 0$ due to conditions (1.5). That is why (1.42) reduces to

$$-\int_G f_x w_{xx}\eta\,\mathrm{d}G + m\|f\|_0^2 \geqslant \tfrac{1}{2}\delta \int_G w_{xx}^2 \eta\,\mathrm{d}G. \tag{1.43}$$

Using Young's inequality in the left-hand side, we find

$$\frac{1}{2\gamma}\|f_x\|_0^2 + \tfrac{1}{2}\gamma\|w_{xx}\eta\|_0^2 + m\|f\|_0^2 \geqslant \tfrac{1}{2}\delta \int_G w_{xx}^2 \eta\,\mathrm{d}G. \tag{1.44}$$

At $\gamma \leqslant \delta/2$, the second term on the left-hand side is obviously less than half the right-hand side:

$$m\|f\|_1^2 \geqslant \tfrac{1}{4}\delta \int_G w_{xx}^2 \eta\,\mathrm{d}G. \tag{1.45}$$

Since $\eta = 1$ in the subdomain $G_{\text{in}} = G \cap (\sigma < x < l - \sigma)$, we obtain

$$\|w_{xx}\|_{L^2(G_{\text{in}})} \leqslant m\|f\|_1, \tag{1.46}$$

where $w = u^{N,\varepsilon}$. At $N \to \infty$, (1.46) yields (1.23). □

Theorem 1.5 (Smoothness of the solution). *If the conditions of Theorem 1.1 are true and, moreover,*

$$2\alpha - 3k_x \geqslant \delta > 0, \qquad 2\alpha - 5k_x \geqslant \delta > 0, \tag{1.47}$$

in G *and* $f \in W^{3,2}(G)$, *then the solution* $u(x, y)$ *of Problem (1.1), (1.4) belongs to* $W^{4,2}(G_{\text{in}})$, *where* $G_{\text{in}} = G \cap (\sigma < x < l - \sigma)$, *for any small* $\sigma > 0$.

The assertion of the theorem follows from *a priori* estimates

$$\|u^{\varepsilon}_{xxx}\|_{L^2(G_{\text{in}})} \leqslant m\|f\|_2, \qquad \|u^{\varepsilon}_{xxxx}\|_{L^2(G_{\text{in}})} \leqslant m\|f\|_3, \tag{1.48}$$

which are valid under conditions (1.47) and can be obtained in the same way as (1.23). However, the third- and fourth-order derivatives of the solution $u(x, y)$ are not, in general, square integrable over the entire G because of singularities which can arise at the corner points (vertices) of the rectangle.

Under the conditions of Theorem 1.5, the solution $u(x, y)$ of Problem (1.1), (1.4) is classical, that is continuous in $\overline{G}$ along with the first- and second-order derivatives. This follows from the embedding theorems for a two-dimensional domain which yield $u \in C(\overline{G}) \cap C^2(G)$ if $u \in W^{2,2}(G) \cap W^{4,2}(G_{\text{in}})$.

Remark 1.6. In Problem (1.1), (1.4), we prescribed the Neumann condition $u_y = 0$ on the upper and lower sides $y = \pm 1$, $0 < x < l$, of the rectangle G because of applications in aerodynamics to be considered in the subsequent chapters. At the same time, Theorems 1.1 and 1.5 remain true if the Dirichlet condition $u = 0$ is prescribed on those sides of G. Then all *a priori* estimates obtained above remain valid, the only distinction is concerned with the choice of the complete system of functions $Y_i(y)$ involved in the sums (1.15), which must vanish at $y = \pm 1$ if Dirichlet's condition is imposed.

1.2 Equation of the Elliptic–Hyperbolic Type

In this section, we suppose that the coefficient $k(x, y)$ in Equation (1.1) satisfies the inequalities

$$k(0, y) > 0, \quad k(l, y) < 0 \quad \text{at } -1 \leqslant y \leqslant 1. \tag{1.49}$$

Therefore, the type of the equation is elliptic on the front side and hyperbolic on the exit side of the rectangle G.

Formulation of the problem is as follows. Find a solution $u(x, y)$ of Equation (1.1) in the rectangle $G = \{(x, y) \in \mathbb{R}^2 : 0 < x < l, -1 < y < 1\}$ subject to the conditions:

$$\left.\begin{array}{lll} u(0,y)=0 & \text{at } x=0, & -1<y<1, \\ u_y(x,\pm 1)=0 & \text{at } y=\pm 1, & 0<x<l. \end{array}\right\} \tag{1.50}$$

The exit side of G is free of boundary conditions, thus, in the case (1.49) the formulation of the problem is the same as the initial–boundary-value problem for an equation of parabolic type.

Theorem 1.7. *Let inequalities (1.2), (1.49) hold, and*

$$2\alpha \pm k_x \geqslant \delta > 0, \tag{1.51}$$

$$a_x \leqslant 0, \quad c \leqslant 0, \quad c_x \geqslant 0, \tag{1.52}$$

in the domain G. Then there exists a unique solution $u \in W^{2,2}(G)$ of Problem (1.1), (1.50) $\forall f \in W^{1,2}(G)$.

If, in addition,

$$2\alpha - 3k_x \geqslant \delta > 0, \qquad 2\alpha - 5k_x \geqslant \delta > 0, \tag{1.53}$$

in G and $f \in W^{3,2}(G)$, then the solution $u(x,y)$ belongs to $W^{4,2}(G_{\text{in}})$, where $G_{\text{in}} = G \cap (\sigma < x \leqslant l)$ for any small $\sigma > 0$.

The theorem can be illustrated by the equation

$$(\tfrac{3}{2} - x - y^2)u_{xx} + u_{yy} - \alpha(x,y)u_x = f(x,y), \tag{1.54}$$

considered in the rectangle G at $l > 3/2$ (see Figure 1.5). There exists a unique solution $u \in W^{2,2}(G)$ of Problem (1.54), (1.50) if $\alpha(x,y) > 1/2$ in G.

Proof. In order to prove the solvability, we choose $l' > l$ and smoothly continue the given functions $k(x,y)$, $a(x,y)$, $\alpha(x,y)$, $c(x,y)$, $f(x,y)$ into the domain $\{(x,y) : l < x < l',\ -1 < y < 1\}$ in such a way that $k(x,y)$ becomes positive at $x = l'$, $-1 \leqslant y \leqslant 1$, and the inequalities (1.2), (1.51), (1.53) remain true in the extended domain $G' = \{(x,y) : 0 < x < l',\ -1 < y < 1\}$. Since $k(l',y) > 0$, the type of Equation (1.1) is elliptic on the right border of G', that is why we prescribe the extra boundary condition $u_x(l',y) = 0$.

Now all the inequalities required in Theorems 1.1 and 1.5 with respect to the coefficients of the equation hold in the extended domain G'. Therefore, there exists a solution $u \in W^{2,2}(G') \cap W^{4,2}(G'_{in})$ to the problem in question in the domain G'. This solution also solves the original Problem (1.1), (1.50) in the domain G. Moreover, the third- and fourth-order derivatives of the solution are square integrable over G with the exception of the vicinities of two corner points $x = 0$, $y = \pm 1$.

In order to prove the uniqueness, we consider the difference $u = u_1 - u_2$ of two solutions which satisfies the homogeneous equation $Lu = 0$. Owing to the boundary

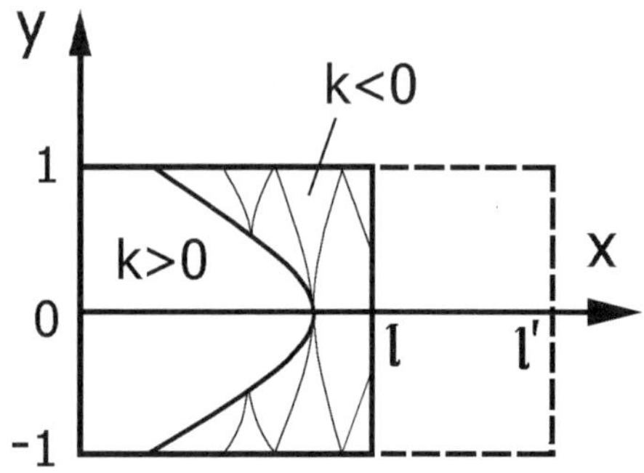

Figure 1.5 Location of elliptic and hyperbolic subdomains in case of Equation (1.54).

conditions (1.50), the identity (1.8) reduces to

$$-2\int_G Lu \cdot u_x \, \mathrm{d}G = \int_G [(2\alpha + k_x)u_x^2 - a_x u_y^2 + c_x u^2] \, \mathrm{d}G$$
$$+ \int_{-1}^{1} (-ku_x^2 + au_y^2 - cu^2)\Big|_{x=l} \mathrm{d}y + \int_{-1}^{1} (ku_x^2)\Big|_{x=0} \mathrm{d}y.$$

All the terms in the right-hand side are nonnegative due to the conditions with respect to the coefficients (1.51), (1.52) and $k(l, y) < 0$ in the case under consideration. Since Lu vanishes, we find that so does each of the integrals in the right-hand side. Hence, $\int_G (2\alpha + k_x)u_x^2 \, \mathrm{d}G = 0$ and $(2\alpha + k_x)u_x^2 = 0$ a.e. in G. Recalling that $2\alpha + k_x > 0$, we obtain $u_x \equiv 0$. Due to $u(0, y) = 0$, one arrives at $u(x, y) = 0$ in G, that is $u_1(x, y) = u_2(x, y)$. □

The chosen way of continuation of the coefficients into the extended domain G' in the proof of the solvability does not influence the solution in the original domain G. This can be accounted for by propagation of perturbations in the hyperbolic subdomains of Equation (1.1) in the positive direction of the x-axis as discussed in Kuz'min (1987). Due to (1.49), the equation is hyperbolic at $x = l$. Consequently, perturbations cannot propagate across the section $x = l$ in the negative direction of the x-axis, and values of the coefficients k, a, α, c, f at $x > l$ cannot influence the solution at $x < l$, i.e. in the domain G.

The equation

$$[1 - x + 0.5\sin(2\pi y)]u_{xx} + u_{yy} - \alpha(x, y)u_x = f(x, y) \qquad (1.55)$$

provides another example illustrating Theorem 1.7 (see Figure 1.6). Again, the solvability is true at $\alpha(x, y) > 1/2$ in G.

We note that the term 'elliptic–hyperbolic type' in the title of the section should not be associated with the unique change of the sign of $k(x, y)$ with x increasing from 0 to l. This term only implies the validity of the conditions (1.49), i.e. that the type is

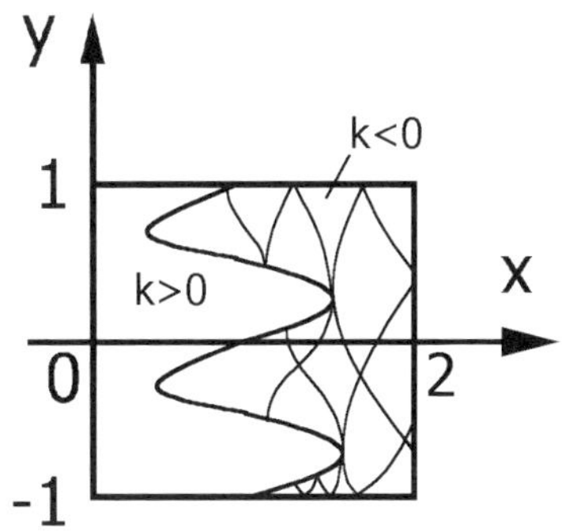

Figure 1.6 Location of elliptic and hyperbolic subdomains in case of Equation (1.55).

elliptic at $x = 0$ and hyperbolic at $x = l$. Inside the domain G, the type may change a finite or even infinite number of times ($k(x, y)$ may vanish on an arbitrary set). The following example demonstrates a triple change of type:

$$(3 - 4x + y)(4 - 4x + y)(5 - 4x + y)u_{xx} + u_{yy} - \alpha(x, y)u_x = f(x, y). \quad (1.56)$$

Simple estimates show that $|k_x| < 384$ if $l = 2$. Therefore, the solvability of Problem (1.50), (1.56) is true at least for $\alpha > 192$.

In this chapter, we do not analyse the influence of characteristic lines of Equation (1.1) on the solvability of boundary-value problems, though the behaviour of the characteristics in the hyperbolic subdomains may be different for very similar equations. For instance, in the case of the equation

$$(\tfrac{1}{2} - x + py^2)u_{xx} + u_{yy} - \alpha(x, y)u_x = f(x, y), \quad (1.57)$$

the behaviour of the characteristics essentially depends on the parameter p. For $p < 0$, the pattern of the characteristics of Equation (1.57) is the same as that in Figure 1.5. For $0 < p \leqslant 1/16$ and $p > 1/16$ the qualitative patterns are sketched in Figure 1.7. A detailed analysis is presented in Kuz'min (1992a).

1.3 Equation of the Hyperbolic–Elliptic Type

In this section, we suppose that the type of the equation is elliptic on the exit side and hyperbolic on the front side of the rectangle G:

$$k(0, y) < 0, \quad k(l, y) > 0 \quad \text{at} \ -1 \leqslant y \leqslant 1. \quad (1.58)$$

Formulation of the problem is as follows. Find a solution $u(x, y)$ of Equation (1.1) in the rectangle $G = \{(x, y) \in \mathbb{R}^2 : \ 0 < x < l, \ -1 < y < 1\}$ endowed with the

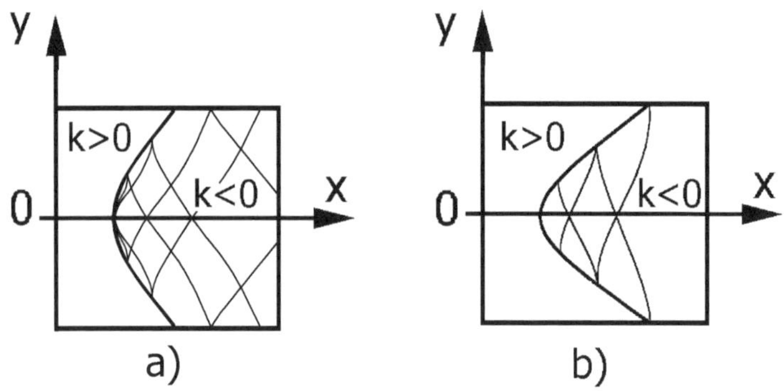

Figure 1.7 Qualitative patterns of the behaviour of characteristic lines in the case of Equation (1.57) with the parameter p: (a) $0 < p \leqslant 1/16$, (b) $p > 1/16$.

initial–boundary conditions:

$$\left.\begin{aligned} u(0,y) = u_x(0,y) &= 0 \quad \text{at } x = 0, \quad -1 < y < 1,\\ u_x(0,y) &= 0 \quad \text{at } x = l, \quad -1 < y < 1,\\ u_y(x,\pm 1) &= 0 \quad \text{at } y = \pm 1, \quad 0 < x < l. \end{aligned}\right\} \tag{1.59}$$

Theorem 1.8. *Let inequalities (1.2), (1.58) hold, and*

$$2\alpha \pm k_x \geqslant \delta > 0, \tag{1.60}$$

$$a_x \leqslant 0, \quad c \leqslant 0, \quad c_x \geqslant 0, \tag{1.61}$$

in the domain G. Then there exists a unique solution $u \in W^{2,2}(G)$ of Problem (1.1), (1.59) $\forall f \in W^{1,2}(G)$.

If, in addition,

$$2\alpha - 3k_x \geqslant \delta > 0, \qquad 2\alpha - 5k_x \geqslant \delta > 0, \tag{1.62}$$

in G and $f \in W^{3,2}(G)$, moreover, the compatibility conditions for initial and boundary data hold at the points $x = 0$, $y = \pm 1$, then the solution $u(x, y)$ belongs to $W^{4,2}(G_{\text{in}})$, where $G_{\text{in}} = G \cap (0 \leqslant x \leqslant l - \sigma)$ for any small $\sigma > 0$.

The theorem can be illustrated by the equation (Figure 1.8)

$$(-\tfrac{3}{2} + x + y^2)u_{xx} + u_{yy} - \alpha(x, y)u_x = f(x, y). \tag{1.63}$$

The solvability of Problem (1.59), (1.63) in $W^{2,2}(G)$ is true if $\alpha(x, y) > 1/2$ in $\overline{G}$, $l > 3/2$.

Proof. Without loss of generality, we suppose that $f(x, y)$ vanishes in a small vicinity of the segment $x = 0, -1 \leqslant y \leqslant 1$ (the general case reduces to this one by subtracting

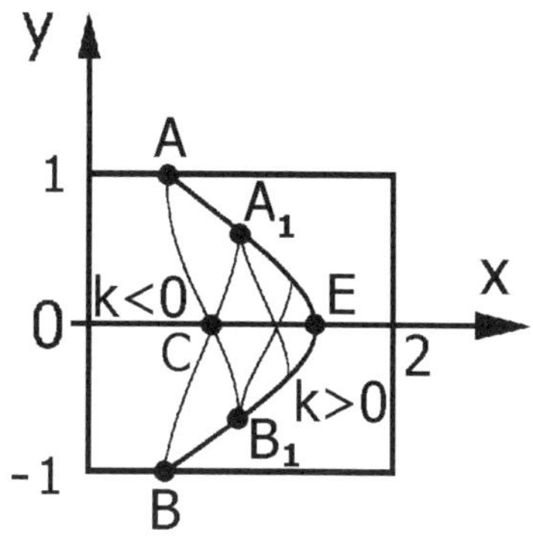

Figure 1.8 Location of elliptic and hyperbolic subdomains in case of Equation (1.63).

from $u(x, y)$ the solution of the initial–boundary-value Problem (1.59) for Equation (1.1) of hyperbolic type in the vicinity of the front side).

In order to prove the solvability, we choose $l' < 0$ and continue the given functions $k(x, y)$, $a(x, y)$, $\alpha(x, y)$, $c(x, y)$, $f(x, y)$ into the domain $\{(x, y) : l' < x < 0, -1 < y < 1\}$ in such a way that $k(x, y) > 0$ at $x = l'$, $-1 \leqslant y \leqslant 1$, $f(x, y) \equiv 0$ at $l' < x < 0$, $-1 \leqslant y \leqslant 1$, and inequalities (1.2), (1.60), (1.62) as well as the conditions for the smoothness of the coefficients remain true in the extended domain $G' = \{(x, y) : l' < x < l, -1 < y < 1\}$. Owing to $k(l', y) > 0$, the type of Equation (1.1) is elliptic on the left border of G', that is why we prescribe the boundary condition $u(l', y) = 0$.

In the extended domain G', all the inequalities required in Theorems 1.1 and 1.5 with respect to the coefficients of the equation hold. Therefore, there exists a solution $u \in W^{2,2}(G') \cap W^{4,2}(G'_{in})$ to the problem in the extended domain G'. This solution also solves the original Problem (1.1), (1.59) in the domain G. Moreover, the third- and fourth-order derivatives of the solution are square integrable over G, except possibly in the vicinities of two corner points $x = l$, $y = \pm 1$.

The chosen way of continuation of the coefficients into the region $x < 0$ does not influence the solution in the original domain G. This is accounted for by the fact that, for any continuation, one actually obtains $u \equiv 0$ at $x < 0$ due to Theorem 1.7.

In order to prove the uniqueness, we consider the difference of two solutions $u = u_1 - u_2$ which satisfies the equation $Lu = 0$ and the initial–boundary conditions (1.59). Owing to the initial and boundary data, the line integral in the identity (1.8) vanishes, and it reduces to

$$-2\int_G Lu \cdot u_x \, \mathrm{d}G = \int_G [(2\alpha + k_x)u_x^2 - a_x u_y^2 + c_x u^2] \, \mathrm{d}G.$$

The terms in the right-hand side are nonnegative due to the conditions (1.60), (1.61). Since Lu vanishes, we obtain $\int_G (2\alpha + k_x)u_x^2 \, \mathrm{d}G = 0$ and $u_x = 0$ a.e. in G. Consequently, $u_1(x, y) = u_2(x, y)$. □

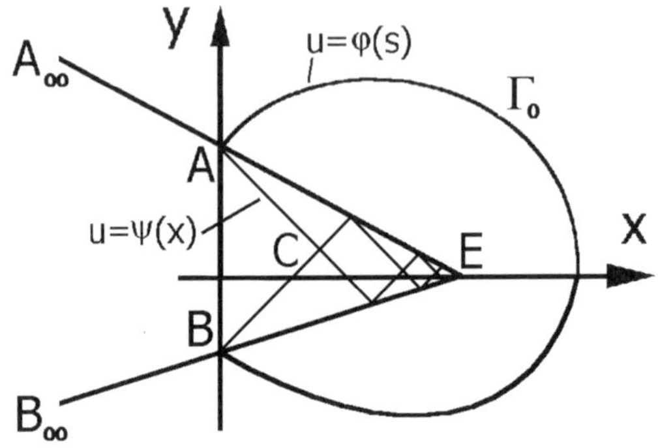

Figure 1.9 A model problem concerned with the supersonic jet noise.

Again, the term 'hyperbolic–elliptic type' implies merely the validity of conditions (1.58), i.e. that the type is hyperbolic at $x = 0$ and elliptic at $x = l$. Inside the domain G the type may change a finite or even infinite number of times as x increases from 0 to l.

One can treat Problem (1.59), (1.63) in a different way as follows. Denote by $\tau(y)$ the values of the solution $u(x, y)$ on the line of type degeneracy $x = \frac{3}{2} - y^2$. If $\tau(y)$ were given as well as the boundary conditions (1.59), then one would obtain the initial–boundary-value problem for Equation (1.63) in the hyperbolic subdomain $G^- = G \cap (x < \frac{3}{2} - y^2)$, and the boundary-value problem with the mixed Dirichlet–Neumann conditions for Equation (1.63) in the elliptic subdomain $G^+ = G \cap (x > \frac{3}{2} - y^2)$. At a given arbitrary $\tau(y)$, however, the limit values of the normal derivative of solutions $u^-(x, y)$ and $u^+(x, y)$ of the above problems on the type degeneracy line are in general different. Therefore, we obtain the problem of finding a function $\tau(y)$ which would provide the matching of the normal derivatives (or the derivatives with respect to the x-coordinate) of $u^-(x, y)$ and $u^+(x, y)$ along the type degeneracy line.

For this method, an important point is the existence of characteristic lines intersecting the type degeneracy line twice. If such characteristics exist, the above method leads to a singular integral equation with a shift. Indeed, in the case of Equation (1.63), the initial–boundary conditions (1.59) enable one to find $u(x, y)$ in the subdomain bounded by the front section $x = 0$ and the characteristic lines AC and BC (see Figure 1.8). Denote by $u(x, y) = \psi(y)$ the obtained solution on ACB. In the triangular domain ACA_1, following the classical method of integral equations, one can represent the solution $u(x, y)$ of the Cauchy problem with initial data $\tau(y)$, $\nu(y)$ on the arc AA_1 as a sum of integrals of $\tau(y)$ and $\nu(y)$. After that, setting $u(x, y)$ on AC equal to the known $\psi(y)$, we obtain an integral relation involving $\tau(y)$ and $\nu(y)$ on AA_1. In the same way, we obtain an integral relation between τ and ν on BB_1.

Two more integral relation arise due to the existence of characteristics intersecting the line of type degeneracy twice. The solution of the Cauchy problem in G^- with data $u = \tau(y)$ and $\partial u/\partial n = \nu(y)$ on the type degeneracy line is single-valued if and only if $\tau(y)$ and $\nu(y)$ satisfy certain consistency conditions. In order to get these

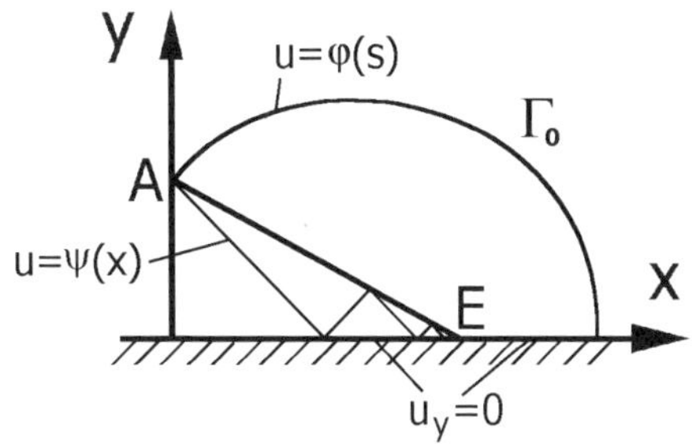

Figure 1.10 A simplified problem for the model equation.

conditions, consider, for example, the segment CE connecting points $C(x_C, 0)$ and $E(3/2, 0)$. Solving separately two Cauchy problems with initial data on AA_1E and BB_1E, we can obtain the solution $u^-(x, y)$ in the subdomains ACE and BCE located above and below the segment CE, respectively. After that one can match the solution $u^-(x, y)$ and its normal derivative u_y^- along the segment CE and, thus, obtain the third and fourth relations between $\tau(y)$ and $\nu(y)$, which arise on the type degeneracy line AEB due to the hyperbolic subdomain G^-. As in the classical Tricomi problem, one more integral relation between $u = \tau(y)$ and $\partial u/\partial n = \nu(y)$ on the line AEB can be obtained from the condition that $u(x, y)$ satisfies Equation (1.1) in the elliptic subdomain G^+ and the given boundary conditions on the upper, lower and right borders of this subdomain. Then eliminating $\tau(y)$ from the obtained relations, one can arrive at a singular integral equation with respect to $\nu(y)$.

The above way of reducing the boundary-value problem for the equation of mixed type to a singular integral equation was realized by Kuz'min (1975) for the model equation

$$k(x, y)u_{xx} + u_{yy} = 0, \tag{1.64}$$

where

- $k(x, y) = 1$ in the sector S^+ bounded by two rays EA_∞, EB_∞ with vertices at a point E of the x-axis and located on the right of the rays, and
- $k(x, y) = -1$ in the sector S^- located on the left of the rays (see Figure 1.9).

We assume that the rays make angles $\theta_1 < \pi/4$ and $\theta_2 < \pi/4$ with the negative direction of the x-axis. Let Γ_0 be a simple curve located in S^+ with endpoints A and B on EA_∞ and EB_∞, respectively. Consider the domain G bounded by the curve Γ_0 and the characteristic segments AC, BC. A model boundary-value problem concerned with acoustical radiation of supersonic jets consists of finding a solution of Equation (1.64) in G which satisfies the Dirichlet condition on AC and BC and either the Dirichlet or Neumann condition on Γ_0. The equivalent singular integral equation on the line of change of type takes the most simple form

$$U(x) + \frac{R(x)}{\pi}\int_{-p}^{q}\frac{U(t)}{t - g(x)}\,dt = f(x), \qquad -p < x < q, \tag{1.65}$$

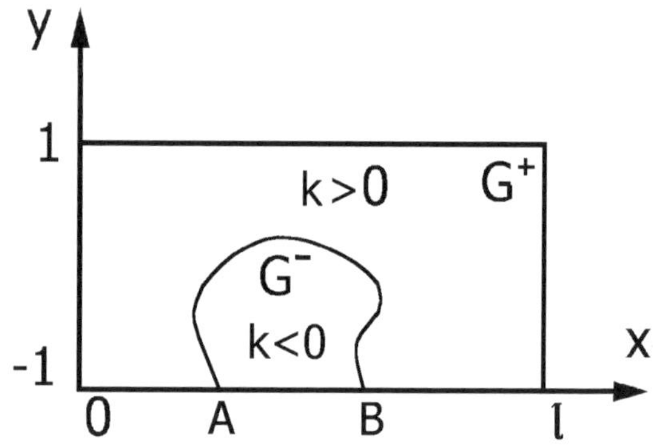

Figure 1.11 A local hyperbolic subdomain.

if the elliptic subdomain G^+ coincides with the full sector S^+, so that Γ_0 is constituted by the two rays AA_∞ and BB_∞. Solvability of the integral Equation (1.65) and the pointed out boundary-value problem for the mixed-type Equation (1.64) was proven under the condition that the parameters θ_1 and θ_2 are close enough to $\pi/4$. If both the geometry of the domain G and the boundary conditions are symmetric with respect to the x-axis, then the problem simplifies, because $u(x, y) = u(x, -y)$, and one can concentrate on solving the problem in the half domain $G \cap (y > 0)$ under the prescribed Neumann condition $u_y = 0$ on the x-axis (Figure 1.10).

1.4 Equation with a Local Hyperbolic Subdomain

As in the basic problem, we suppose that

$$k(0, y) > 0, \quad k(l, y) > 0 \quad \text{at } -1 \leqslant y \leqslant 1,$$

therefore, the type of Equation (1.1) is elliptic on both the front and the exit sides of the rectangle $G = \{(x, y) \in \mathbb{R}^2 : \ 0 < x < l, \ -1 < y < 1\}$. We now suppose that the type degeneracy set determined by the equation $k(x, y) = 0$ is a simple curve with endpoints A and B on the lower side $y = -1, 0 \leqslant x \leqslant l$ of G (Figure 1.11).

The boundary conditions are the same as in Section 1.1:

$$\left.\begin{aligned} u(0, y) &= 0 && \text{at } x = 0, && -1 < y < 1, \\ u_x(0, y) &= 0 && \text{at } x = l, && -1 < y < 1, \\ u_y(x, \pm 1) &= 0 && \text{at } y = \pm 1, && 0 < x < l. \end{aligned}\right\} \tag{1.66}$$

Owing to Theorem 1.1, there exists a unique solution of Problem (1.1), (1.66) in the space $W^{2,2}(G)$ if the conditions (1.5), (1.6) hold. However, in the problem of transonic flow with a local supersonic region, the condition (1.5) cannot not be true in the whole domain because of the relation $\alpha = -k_x$ between the coefficients of the linearized equation governing small perturbation in the flow,

$$(1-\varphi_x)u_{xx} + \frac{1}{1+\gamma}u_{yy} - \varphi_{xx}u_x = 0 \tag{1.67}$$

(see (3)). In this case, the sign of the expression $2\alpha \pm k_x = -2k_x \mp k_x$ is determined by the derivative $-k_x$, which is obviously positive in the vicinity of the point A but inevitably negative in the vicinity of the point B.

Therefore, of importance is an analysis of the solvability of Problem (1.1), (1.66) under weaker conditions as compared with (1.5). Below, in Theorem 1.9, we prove the solvability of Problem (1.1), (1.66) under the condition $2\alpha \pm k_x > 0$ valid only on the type degeneracy line AB, not in the whole domain G. In Theorem 1.10, we require that this condition hold only at the endpoints A and B of the type degeneracy line.

For simplicity, we consider Equation (1.1) in the special case $a(x,y) \equiv 1, c(x,y) \equiv 0$, and analyse the equation with a parameter λ:

$$Lu - \lambda u := k(x,y)u_{xx} + u_{yy} - \alpha(x,y)u_x - \lambda u = f(x,y). \tag{1.68}$$

Theorem 1.9. *Let the line of type degeneracy of Equation (1.68) determined by the equation $k(x,y) = 0$ be a simple smooth curve with endpoints A and B on the lower side $y = -1, 0 \leqslant x \leqslant l$ of G. If*

$$2\alpha \pm k_x > 0 \quad \textit{on } \mathrm{AB}, \tag{1.69}$$

and λ is large enough, then there exists a unique solution $u \in W^{2,2}(G)$ of Problem (1.66), (1.68) $\forall f \in W^{1,2}(G)$.

At an arbitrary λ, the Fredholm alternative is true: either there exists a unique solution $u \in W^{2,2}(G)$ of Problem (1.66), (1.68) for any $f \in W^{1,2}(G)$, and the homogeneous problem with $f(x,y) \equiv 0$ has no solutions distinct from the identical zero, or the latter problem has a finite number of linearly independent nonzero solutions. In the second case, the nonhomogeneous Problem (1.66), (1.68) is solvable provided the right-hand side $f(x,y)$ satisfies the same number of integral relations.

Proof. Following Kuz'min (1985, 1992a), we shall analyse the identity obtained with Green's formula,

$$\begin{aligned}
&-\int_G (Lu - \lambda u)\cdot(2u_x + rku)\mathrm{e}^{-\mu x}\,\mathrm{d}G \\
&= \left(\int_{G^-} + \int_{G^+}\right)[(2\alpha + k_x - \mu k + rk^2)u_x^2 + (\mu + rk)u_y^2 \\
&\qquad + rk_x u_x u + rk(\alpha - \mu k + 2k_x)u_x u + \lambda(\mu + rk)u^2]\mathrm{e}^{-\mu x}\,\mathrm{d}G \\
&\quad + \int_\Gamma [(2u_x u_y + rku_y u)\mathrm{e}^{-\mu x}\,\mathrm{d}x - (ku_x^2 - u_y^2 - \lambda u^2 + rk^2 u_x u)\mathrm{e}^{-\mu x}\,\mathrm{d}y],
\end{aligned} \tag{1.70}$$

where G^- and G^+ are the hyperbolic and elliptic parts of G in which $k < 0$ and $k > 0$, respectively. We choose the coefficient $r(x, y)$ to vanish in $\overline{G}^-$, while $r(x, y) = \text{const.} > 0$ in G^+.

The positive parameters μ and λ will be chosen so as to provide the positive definiteness of the form with respect to u_x and u_y in the integral over G. Then identity (1.70) will yield the estimate

$$\|u\|_1 \leqslant m\|Lu - \lambda u\|_0, \qquad m > 0, \tag{1.71}$$

for each function $u \in W^{2,2}(G)$ satisfying boundary conditions (1.66). In (1.71), as usual, $\|\cdot\|_p$ denotes the norm $\|\cdot\|_{W^{p,2}(G)}$. A proof of the estimate (1.71) is split into four steps.

Step 1. Denote by I_{G^-}, I_{G^+}, I_Γ the three integrals on the right-hand side of (1.70). Since r vanishes in $\overline{G}^-$, from (1.70) we obtain

$$I_{G^-} = \int_{G^-} [(2\alpha + k_x - \mu k)u_x^2 + \mu u_y^2 + \lambda\mu u^2]\mathrm{e}^{-\mu x}\,\mathrm{d}G.$$

The inequality $2\alpha \pm k_x \geqslant \delta > 0$ is true on the type degeneracy line AB by the assumption (1.69) of Theorem 1.9. Therefore, the coefficient in front of u_x^2 in identity (1.70) is positive on AB. In the hyperbolic subdomain G^-, we have $k < 0$. Hence, by choosing a sufficiently large μ, due to the term $-\mu k$ one can provide the positivity of $(2\alpha + k_x - \mu k)$ in G^- and obtain the estimate for the integral I_{G^-}:

$$\begin{aligned} I_{G^-} &\geqslant \int_{G^-} [m(u_x^2 + u_y^2) + \lambda\mu u^2]\mathrm{e}^{-\mu x}\,\mathrm{d}G \\ &\geqslant m\int_{G^-} (u_x^2 + u_y^2 + u^2)\,\mathrm{d}G. \end{aligned} \tag{1.72}$$

Hereafter m denotes positive constants which are independent of u and, in general, different in different formulae.

Step 2. The integration over the boundary Γ in (1.70) is carried out in the direction which leaves the domain G on the left. Due to the condition $u_y = 0$ at $y = \pm 1$, the integral I_Γ reduces to

$$\begin{aligned} I_\Gamma = &\int_{-1}^{1} -(ku_x^2 - u_y^2 - \lambda u^2 + rk^2 u_x u)\big|_{x=l}\mathrm{e}^{-\mu l}\,\mathrm{d}y \\ &+ \int_{1}^{-1} -(ku_x^2 - u_y^2 - \lambda u^2 + rk^2 u_x u)\big|_{x=0}\,\mathrm{d}y. \end{aligned}$$

Using the boundary conditions $u = 0$ at $x = 0$ and $u_x = 0$ at $x = l$, we find

$$I_\Gamma = \int_{-1}^{1} (u_y^2 + \lambda u^2)\big|_{x=0}\mathrm{e}^{-\mu l}\,\mathrm{d}y + \int_{-1}^{1} (ku_x^2)\big|_{x=0}\,\mathrm{d}y \geqslant 0. \tag{1.73}$$

Step 3. In the elliptic subdomain, one can provide the positivity of $(2\alpha + k_x - \mu k + rk^2)$ by choosing a sufficiently large r_0 under the fixed μ. Then the integral I_{G^+} in (1.70)

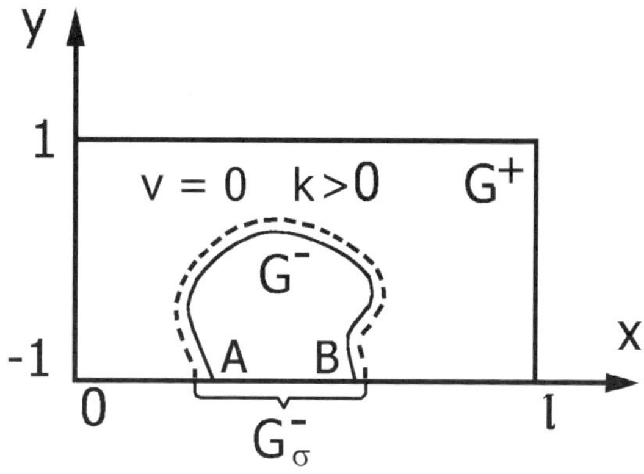

Figure 1.12 Extension of the local hyperbolic subdomain.

can be estimated as

$$I_{G^+} \geqslant \int_{G^+} [m(u_x^2 + u_y^2) + rak_y u_y u + rk(\alpha - \mu k + 2k_x)u_x u + \lambda(\mu + rk)u^2]\mathrm{e}^{-\mu x}\,\mathrm{d}G.$$

The coefficients in front of $u_x u$ and $u_y u$ do not involve λ, while the coefficient in front of u^2 increases with increasing λ. Therefore, at sufficiently large λ, we obtain

$$I_{G^+} \geqslant m \int_{G^+} (u_x^2 + u_y^2 + u^2)\,\mathrm{d}G. \tag{1.74}$$

By summing up inequalities (1.72)–(1.74), we arrive at the expression

$$-\int_G (Lu - \lambda u)\cdot(2u_x + rku)\mathrm{e}^{-\mu x}\,\mathrm{d}G \geqslant m \int_G (u_x^2 + u_y^2 + u^2)\,\mathrm{d}G = m\|u\|_1^2. \tag{1.75}$$

Now, using Young's inequality, we obtain estimate (1.71).

Step 4. The second *a priori* estimate

$$\|u\|_2 \leqslant m\|Lu - \lambda u\|_1 \tag{1.76}$$

is true for sufficiently large λ and each function $u \in W^{2,2}(G)$ satisfying boundary conditions (1.66), and such that $Lu \in W^{1,2}(G)$. To prove this, we denote by G_σ^- the extension of G^- obtained by attaching a narrow strip located in G^+ and adjacent to the line $k(x, y) = 0$: $G_\sigma^- = \{(x, y) \in G : k(x, y) < \sigma\}$.

Consider the product $v = u\zeta$, where the cut-off function $\zeta(x, y)$ equals unity on G_σ^- and vanishes on $G\backslash G_{2\sigma}^-$ (Figure 1.12). In addition, we let ζ depend only on y in the small vicinities of points A and B. Using Green's formula, we obtain

$$
\begin{aligned}
&-2\int_G (Lv-\lambda v)_x \cdot v_{xx} e^{-\mu x}\,dG \\
&\quad = \int_G [(2\alpha - k_x - \mu k)v_{xx}^2 + \mu v_{xy}^2 + (\lambda + \alpha_x)\mu v_x^2] e^{-\mu x}\,dG \\
&\qquad + \int_\Gamma [2v_{xx}v_{xy} e^{-\mu x}\,dx - [k v_{xx}^2 - v_{xy}^2 - (\lambda+\alpha_x)v_{xx}^2] e^{-\mu x}\,dy]. \quad (1.77)
\end{aligned}
$$

Due to vanishing v outside $G_{2\sigma}^-$, the boundary integral reduces to the expression

$$
I_\Gamma = \int_0^l 2v_{xx}v_{xy}|_{y=-1} e^{-\mu x}\,dx
$$

similar to (1.73). Since $v_y(x,-1)=0$, which holds owing to the boundary condition $u_y(x,-1)=0$ and $\zeta_y(x,1)=0$, we obtain $v_{xy}=0$ at $y=-1$. Consequently, $I_\Gamma = 0$.

Thus, (1.77) yields

$$
\begin{aligned}
&-2\int_G (Lv-\lambda v)_x \cdot v_{xx} e^{-\mu x}\,dG \\
&\quad = \int_G [(2\alpha - k_x - \mu k)v_{xx}^2 + \mu v_{xy}^2 + (\lambda + \alpha_x)\mu v_x^2] e^{-\mu x}\,dG \\
&\quad \geqslant \int_G [(2\alpha - k_x - \mu k)v_{xx}^2 + \mu v_{xy}^2] e^{-\mu x}\,dG - m\|v\|_1^2.
\end{aligned}
$$

Recalling that $\zeta = v = 0$ outside $G_{2\sigma}^-$, we obtain

$$
\begin{aligned}
&-2\int_{G_{2\sigma}^-} (Lv-\lambda v)_x \cdot v_{xx} e^{-\mu x}\,dG \\
&\quad \geqslant \int_{G_{2\sigma}^-} [(2\alpha - k_x - \mu k)v_{xx}^2 + \mu v_{xy}^2] e^{-\mu x}\,dG - m\|v\|_1^2. \quad (1.78)
\end{aligned}
$$

Let us choose μ sufficiently large so that the coefficient $(2\alpha - k_x - \mu k)$ be positive in the hyperbolic subdomain up to the type degeneracy line AB. By the continuity, $(2\alpha - k_x - \mu k)$ is positive in the small vicinity of G^-. Under the fixed μ, we choose σ small enough, so that $2\alpha - k_x - \mu k > 0$ in $G_{2\sigma}^-$. Then (1.78) yields

$$
-2\int_{G_{2\sigma}^-} (Lv-\lambda v)_x \cdot v_{xx} e^{-\mu x}\,dG \geqslant m_1 \int_{G_{2\sigma}^-} (v_{xx}^2 + v_{xy}^2) e^{-\mu x}\,dG - m\|v\|_1^2. \tag{1.79}
$$

Using Young's inequality in the left-hand side, we arrive at the estimate

$$
m_1 \int_{G_{2\sigma}^-} (v_{xx}^2 + v_{xy}^2)\,dG - m\|v\|_1^2 \leqslant \|(Lv-\lambda v)_x\|_0^2.
$$

It can be seen from (1.68) that

$$\begin{aligned}(Lv - \lambda v)_x &= [k(u\zeta)_{xx} + (u\zeta)_{yy} - \alpha(u\zeta)_x - \lambda u\zeta]_x \\ &= [(Lu - \lambda u)\zeta]_x + [k(-2u_x\zeta_y + u\zeta_{xx}) + 2u_y\zeta_y + u\zeta_{yy} - \alpha u\zeta_x]_x.\end{aligned}$$

Therefore,

$$\begin{aligned}&m_1 \int_{G_{2\sigma}^-} (v_{xx}^2 + v_{xy}^2)\, \mathrm{d}G \\ &\quad \leqslant \|(Lu - \lambda u)_x\|_0^2 + \|[k(2u_x\zeta_x + u\zeta_{xx}) + 2u_y\zeta_y + u\zeta_{yy} - \alpha u\zeta_x]_x\|_0^2 + m\|v\|_1^2.\end{aligned}$$

The derivatives of ζ vanish outside the strip $G_{2\sigma}^- \backslash G_\sigma^-$. Inside the strip, the operator Lu is strictly elliptic. Hence, due to the well-known local properties of elliptic equations, the second term on the right-hand side can be estimated by $\|u\|_1$ and $\|Lu\|_0$:

$$m_1 \int_{G_{2\sigma}^-} (v_{xx}^2 + v_{xy}^2)\, \mathrm{d}G \leqslant \|(Lu - \lambda u)_x\|_0^2 + \|Lu\|_0^2 + m\|u\|_1^2.$$

The norm $\|u\|_1$ can be estimated by (1.71):

$$m_1 \int_{G_{2\sigma}^-} (v_{xx}^2 + v_{xy}^2)\, \mathrm{d}G \leqslant \|(Lu - \lambda u)_x\|_0^2 + \|Lu - \lambda u\|_0^2.$$

Recalling that $\zeta \equiv 1$ in G_σ^-, we find

$$m_1 \int_{G_\sigma^-} (u_{xx}^2 + u_{xy}^2)\, \mathrm{d}G \leqslant \|(Lu - \lambda u)_x\|_0^2 + \|Lu - \lambda u\|_0^2.$$

Since the operator Lu is elliptic in the subdomain $G \backslash G_\sigma^-$, the norm $\|u_x\|_{W^{1,2}(G\backslash G_\sigma^-)}$ is bounded from above by $\|u\|_1 + \|Lu\|_0$. Therefore, we get

$$m\|u_x\|_1 \leqslant \|Lu - \lambda u\|_1.$$

Using the original Equation (1.68), one can estimate the norm $\|u_{yy}\|_0$ by $\|u_{xx}\|_0$ and $\|Lu - \lambda u\|_0$. Therefore, we finally obtain (1.76).

Step 5. The estimate (1.76) enables one to establish the solvability of Problem (1.68), (1.66) with the method of prolongation by a parameter. In order to do that, we use the solvability of the equation

$$L^0 u := ku_{xx} + u_{yy} - \alpha_0 u_x - \lambda u = f \tag{1.80}$$

endowed with the boundary conditions (1.66). We suppose that α_0 and λ are large enough, so that the inequalities (1.5), (1.6) and the estimate (1.76) are true.

Consider the family of equations

$$L^\gamma u := L^0 u + \gamma(\alpha_0 - \alpha)u_x = f \in W^{1,2}(G), \qquad 0 \leqslant \gamma \leqslant 1, \tag{1.81}$$

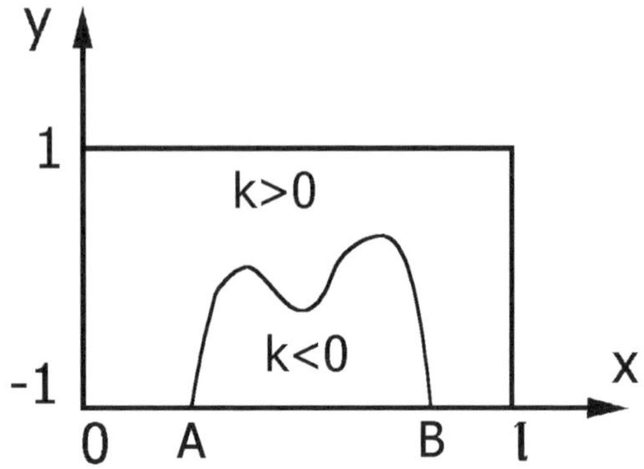

Figure 1.13 Local hyperbolic subdomain with $k_y > 0$ on the type degeneracy line.

under the conditions (1.66). The inverse operator $(L^0)^{-1}$ from the whole $W^{1,2}(G)$ into $W^{2,2}(G)$ is bounded due to (1.76). Hence, Equation (1.81) can be represented in the equivalent form

$$u + \gamma(L^0)^{-1}[(\alpha - \alpha_0)u_x] = (L^0)^{-1} f.$$

The above equation is solvable in $W^{2,2}(G)$ for $0 \leqslant \gamma \leqslant \gamma_1$ if γ_1 is small enough, since

$$\|(L^0)^{-1}[(\alpha - \alpha_0)u_x]\|_2 \leqslant m\|(\alpha - \alpha_0)u_x\|_1 \leqslant m\|u\|_2.$$

In the same way, using the estimate $\|(L^\gamma u)^{-1}\|_2 \leqslant m\|u\|_1$, which is uniform with respect to γ, one can prove the solvability of the equation $L^\gamma u = f$ for $\gamma_1 \leqslant \gamma \leqslant 2\gamma_1$. Thus, in a finite number of steps, one gets the solvability of the equation $L^\gamma u = f$ at $\gamma = 1$, i.e. the solvability of Problem (1.66), (1.68).

In order to prove the Fredholm alternative, we represent Equation (1.68) at any λ in the form $L_{\lambda_0} u := Lu - \lambda_0 u = f + (\lambda - \lambda_0)u$. This equation is equivalent to $u = (\lambda - \lambda_0)L_{\lambda_0}^{-1} u + L_{\lambda_0}^{-1} f$ at sufficiently large λ_0 as shown above. Taking into consideration the compactness of the embedding operator from $W^{2,2}(G)$ into $W^{1,2}(G)$, we obtain that $L_{\lambda_0}^{-1}$ is a compact operator in $W^{2,2}(G)$, which yields the Fredholm solvability of Problem (1.66), (1.68). □

In the following theorem, we assume that the condition $2\alpha \pm k_x > 0$ with respect to the coefficients of Equation (1.68) holds only at the endpoints A and B of the type degeneracy line. At the same time, we use the additional assumption $k_y \geqslant \delta > 0$ on the line AB (Figure 1.13).

Theorem 1.10. *Let the line of type degeneracy determined by the equation $k(x, y) = 0$ be a simple smooth curve with endpoints* A *and* B *on the lower side $y = -1$, $0 \leqslant x \leqslant l$, of G, and $k_y \geqslant \delta > 0$ on this line. If*

$$2\alpha \pm k_x > 0 \quad \textit{at the endpoints A, B,} \tag{1.82}$$

and λ is large enough, then there exists a unique solution $u \in W^{2,2}(G)$ of Problem (1.66), (1.68) $\forall f \in W^{1,2}(G)$. At an arbitrary λ, the Fredholm alternative formulated in Theorem 1.9 is true.

Proof. A crucial point is the derivation of a necessary *a priori* estimate by the analysis of the expression

$$-\int_G (Lu - \lambda u) \cdot (2u_x - 2qu_y + rku)\mathrm{e}^{-\mu x}\, \mathrm{d}G,$$

where the special coefficient,

$$q(x, y) = q_0 \cdot (1 - y^2) \exp[-\varepsilon\mu k^2(x, y)], \tag{1.83}$$

is introduced in order to benefit from the positivity of k_y instead of $2\alpha \pm k_x$ on the type degeneracy line (Kuz'min 1985). Using Green's formula, we obtain the identity

$$\begin{aligned}
&-\int_G (Lu - \lambda u) \cdot (2u_x - 2qu_y + rku)\mathrm{e}^{-\mu x}\, \mathrm{d}G \\
&\quad= \left(\int_{G^-} + \int_{G^+}\right)[(2\alpha + k_x + qk_y + kq_y - \mu k + rk^2)u_x^2 \\
&\qquad\qquad - 2(\alpha q - \mu kq + qk_x + kq_x)u_x u_y + (\mu + q_x + rk)u_y^2 \\
&\qquad\qquad + rk_x u_x u + rk(\alpha - \mu k + 2k_x)u_x u \\
&\qquad\qquad + \lambda(\mu + q_y + rk)u^2]\mathrm{e}^{-\mu x}\, \mathrm{d}G \\
&\qquad + \int_\Gamma [(qku_x^2 + 2u_x u_y - qu_y^2 + q\lambda u^2 + rku_y u)\mathrm{e}^{-\mu x}\, \mathrm{d}x \\
&\qquad\qquad - (ku_x^2 - 2qku_x u_y - u_y^2 - \lambda u^2 + rk^2 u_x u)\mathrm{e}^{-\mu x}\, \mathrm{d}y],
\end{aligned} \tag{1.84}$$

where G^- and G^+ are the hyperbolic and elliptic parts of G in which $k < 0$ and $k > 0$, respectively. We require that the coefficient $r(x, y)$ vanish in $\overline{G}^-$, while $r(x, y) = \text{const.} > 0$ in G^+.

The parameters q_0, ε, μ, λ will be chosen so as to provide sufficiently large values of $q(x, y)$ on the type degeneracy line AB and sufficiently small $|q(x, y)|$ outside a vicinity of AB. Then identity (1.84) will yield the estimate

$$\|u\|_1 \leqslant m\|Lu - \lambda u\|_0, \tag{1.85}$$

for each function $u \in W^{2,2}(G)$ satisfying boundary conditions (1.66).

Step 1. It can be seen from (1.83) that the coefficient $q(x, y)$ reduces to $q_0(1 - y^2)$ on the type degeneracy line AB. Therefore, one can fix q_0 large enough so that

$$2\alpha \pm k_x + q_0(1 - y^2)k_y \geqslant \delta > 0$$

on AB. In the vicinities of points A and B, where the term $q_0(1 - y^2)k_y$ is small, this inequality holds by the assumption (1.82) of the theorem. Hence, the coefficient in front of u_x^2 in the second line of identity (1.84) is positive on the type degeneracy line.

Denote by I_{G^-}, I_{G^+}, I_Γ the three integrals on the right-hand side of (1.84). Since r vanishes in $\overline{G^-}$, from (1.84) we obtain

$$I_{G^-} = \int_{G^-} [(2\alpha + k_x + qk_y + kq_y - \mu k + rk^2)u_x^2 \\ - 2(\alpha q - \mu kq + qk_x + kq_x)u_x u_y \\ + (\mu + q_x)u_y^2 + \lambda(\mu + q_y)u^2]\mathrm{e}^{-\mu x}\,\mathrm{d}G.$$

By choosing a sufficiently large μ and small $\varepsilon > 0$, due to the term μ in front of u_y^2 and the term $-\mu k$ in front of u_x^2, where $-k > 0$, we can obtain a necessary estimate for the integral over the hyperbolic subdomain I_{G^-}. Indeed, let us represent the coefficient in front of u_x^2 in the form

$$(2\alpha + k_x + qk_y + \mu|k|/2) + |k|(\mu/2 - q_y), \tag{1.86}$$

and choose ε so small that the sum $-qk_x + \mu|k|/2$ becomes an increasing function of $\mu, \forall(x, y) \in G^-$. At sufficiently large μ, we obtain $(2\alpha + k_x + qk_y + \mu|k|/2) \geqslant \delta > 0$ in G^-. After that, we choose ε so small that the inequality

$$\mu/2 \pm q_x = \mu(\tfrac{1}{2} \mp 2\varepsilon kk_x q) + O(1) \geqslant \mu/3 + O(1)$$

is true in G in addition to the previous one. Thus, (1.86) becomes positive in $\overline{G^-}$. The coefficient in front of $u_x u_y$ in (1.84) is bounded from above by a constant independent of μ, because the functions $-\mu kq$ and kq_x are bounded at $0 \leqslant \mu < \infty$. At the same time, the coefficient in front of u_y^2 increases with increasing μ. Therefore, under sufficiently large μ, the functions u_x and u_y constitute a positive definite bilinear form:

$$I_{G^-} \geqslant \int_{G^-} [m(u_x^2 + u_y^2) + \lambda(\mu q_y)u^2]\mathrm{e}^{-\mu x}\,\mathrm{d}G \geqslant m\|u\|_1^2. \tag{1.87}$$

Step 2. The integration in I_Γ is carried out in the counterclockwise direction. Owing to the Dirichlet condition $u = 0$ at $x = 0$, the Neumann one at $x = l$, and the vanishing $q(x, y)$ at $y = \pm 1$, we find

$$I_\Gamma = \int_{-1}^{1} ku_x^2|_{x=0}\,\mathrm{d}y + \int_{-1}^{1} (u_y^2 + \lambda u^2)|_{x=l}\mathrm{e}^{-\mu l}\,\mathrm{d}y \geqslant 0 \tag{1.88}$$

at $\lambda \geqslant 0$.

Step 3. For the integral over the elliptic subdomain I_{G^+} in (1.84), choosing sufficiently large r_0 under fixed μ, we obtain

$$I_{G^+} \geqslant \int_{G^+} [m(u_x^2 + u_y^2) + rk_y u_y u \\ + rk(\alpha - \mu k + 2k_x)u_x u + \lambda(\mu + q_y + rk)u^2]\mathrm{e}^{-\mu x}\,\mathrm{d}G. \tag{1.89}$$

The coefficients in front of $u_x u$ and $u_y u$ do not involve λ, while the coefficient in front of u^2 increases with increasing λ because $\mu + q_y \geqslant \mu/2 + O(1) > 0$ (see Step 1). Therefore, if λ is sufficiently large, we obtain $I_{G^+} \geqslant m\|u\|_1^2$. Also, $I_\Gamma \geqslant 0$ owing to

(1.88). That is why all the integrals on the right-hand side of (1.84) are nonnegative at sufficiently large μ, r_0, λ. Moreover, by summing up (1.87), (1.88), (1.89), we obtain

$$-\int_D (Lu - \lambda u) \cdot (2u_x - 2qu_y + rku)\mathrm{e}^{-\mu x}\,\mathrm{d}G \geqslant m\|u\|_1^2. \tag{1.90}$$

Now, using Young's inequality, we arrive at the inequality (1.85).

Step 4. The second *a priori* estimate

$$\|u\|_2 \leqslant m\|Lu - \lambda u\|_1, \tag{1.91}$$

is true for sufficiently large λ and each function $u \in W^{2,2}(G)$ satisfying boundary conditions (1.66), and such that $Lu \in W^{1,2}(G)$. To prove this, we again denote by G_σ^- the extension of the hyperbolic subdomain G^- obtained by adding a narrow strip located in G^+ and adjacent to the line $k(x, y) = 0$: $G_\sigma^- = \{(x, y) \in G : k(x, y) < \sigma\}$.

Consider the product $v = u\zeta$, where the cut-off function $\zeta(x, y)$ equals unity on G_σ^- and vanishes on $G\backslash G_{2\sigma}^-$. Green's formula yields

$$\begin{aligned}
&-2\int_G (Lv - \lambda v)_x \cdot (v_{xx} - qv_{xy})\mathrm{e}^{-\mu x}\,\mathrm{d}G \\
&\quad = \int_G [(2\alpha - k_x + qk_y + kq_y - \mu k)v_{xx}^2 - 2(\alpha q - \mu kq + kq_x)v_{xx}v_{xy} \\
&\qquad + (\mu - q_y)v_{xy}^2 + (\lambda + \alpha_x)(\mu + q_y)v_x^2]\mathrm{e}^{-\mu x}\,\mathrm{d}G \\
&\qquad + \int_\Gamma [qkv_{xx}^2 + 2v_{xx}v_{xy} - qv_{xy}^2 + q(\lambda + \alpha_x)v_x^2)\mathrm{e}^{-\mu x}\,\mathrm{d}x \\
&\qquad\quad + (kv_{xx}^2 - 2qkv_{xx}v_{xy} - v_{xy}^2 - (\lambda + \alpha_x)v_{xx}^2)\mathrm{e}^{-\mu x}\,\mathrm{d}x],
\end{aligned} \tag{1.92}$$

where $q(x, y)$ is defined by (1.83). Due to the vanishing v outside $G_{2\sigma}^-$, the boundary integral reduces to the expression

$$I_\Gamma = \int_0^l [qkv_{xx}^2 + 2v_{xx}v_{xy} - qv_{xy}^2 + q(\lambda + \alpha_x)v_x^2]\big|_{y=-1}\mathrm{e}^{-\mu x}\,\mathrm{d}x,$$

similar to (1.88). Since $q = 0$ and $v_{xy} = 0$ at $y = -1$, we obtain $I_\Gamma = 0$.

Therefore, (1.92) yields

$$\begin{aligned}
&-2\int_G (Lv - \lambda v)_x \cdot (v_{xx} - qv_{xy})\mathrm{e}^{-\mu x}\,\mathrm{d}G \\
&\quad \geqslant \int_G [(2\alpha - k_x + qk_y + kq_y - \mu k)v_{xx}^2 \\
&\qquad - 2(\alpha q - \mu kq + kq_x)v_{xx}v_{xy} + (\mu - q_y)v_{xy}^2]\mathrm{e}^{-\mu x}\,\mathrm{d}G - m\|v\|_1^2.
\end{aligned}$$

Let us choose q_0 and μ to be sufficiently large, and ε to be sufficiently small. Then, as at Step 1, we obtain that the integrand on the right-hand side is a positive definite bilinear form in G^-:

$$(2\alpha - k_x + qk_y + kq_y - \mu k)v_{xx}^2 - 2(\alpha q - \mu kq + kq_x)v_{xx}v_{xy} + (\mu - q_y)v_{xy}^2 \geqslant m(v_{xx}^2 + v_{xy}^2).$$

Due to the continuity of all the coefficients, the above is true in the extension $G_{2\sigma}^-$ of the subdomain G^- if σ is sufficiently small. Recalling that $\zeta = v = 0$ outside $G_{2\sigma}^-$, we get the estimate

$$-2\int_{G_{2\sigma}^-} (Lv - \lambda v)_x \cdot (v_{xx} + qv_{xy})\mathrm{e}^{-\mu x}\,\mathrm{d}G \geqslant m_1 \int_{G_{2\sigma}^-} (v_{xx}^2 + v_{xy}^2)\,\mathrm{d}G - m\|v\|_1^2.$$

Using Young's inequality, we arrive at

$$m_1 \int_{G_{2\sigma}^-} (v_{xx}^2 - v_{xy}^2)\,\mathrm{d}G \leqslant \|(Lv - \lambda v)_x\|_0^2 + m\|v\|_1^2.$$

The rest of the derivation of (1.91) and Step 5 follow the proof of Theorem 1.9. □

As mentioned in the beginning of this section, for Equation (1.67) the condition $2\alpha \pm k_x > 0$ cannot be true at the endpoint B of the type degeneracy line. The violation of the condition $2\alpha \pm k_x$ at point B accounts for the appearance of a singularity of the solution to Equation (1.67) at that point. This corresponds to formation of a shock wave in the vicinity of the downstream endpoint B of the supersonic region in the case of Equation (3) governing the nonlinear perturbation.

We shall continue the study of problems with a local hyperbolic subdomain for Equation (1.67) in Section 3.4. In order to avoid working in $W^{1,2}(G)$ or in weighted Sobolev spaces, which are necessary to capture the singularity at the point B, we shall modify the boundary condition on the lower side of the domain G so as to prevent the arising of the singularity and to provide the existence of a regular solution.

1.5 Other Developments

If the type of Equation (1.1) is hyperbolic on both the front and exit sides of the rectangle G

$$k(0, y) < 0, \quad k(l, y) < 0, \quad -1 \leqslant y \leqslant 1, \tag{1.93}$$

then the initial–boundary conditions read

$$\left.\begin{aligned} u(0, y) = u_x(0, y) = 0 \qquad & \text{at } x = 0, \quad -1 < y < 1, \\ u_y(x, \pm 1) = 0 \qquad & \text{at } y = \pm 1, \quad 0 < x < l. \end{aligned}\right\} \tag{1.94}$$

There exists a unique solution of Problem (1.1), (1.94) in $W^{2,2}(G)$ under the conditions (1.5), (1.6). This can be proved by continuation of the coefficients of the equation across the front and exit sides into an extended domain in the same way as in Sections 1.2 and 1.3.

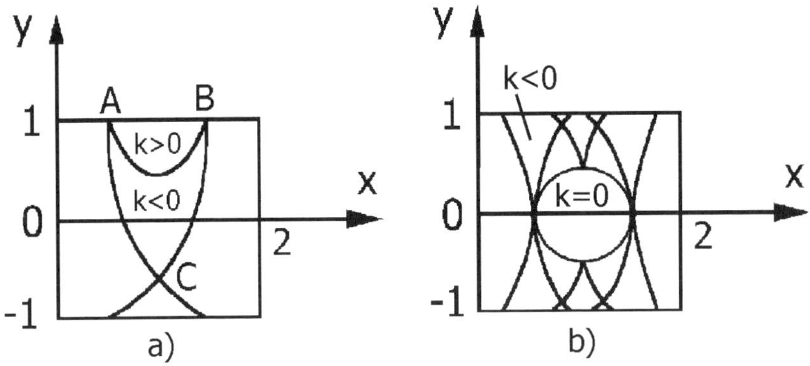

Figure 1.14 Examples of equations which are of hyperbolic type at both the front and exit sides of the rectangle G.

An example of the equation of mixed type in the case (1.93) is as follows:

$$[y - \tfrac{3}{4} - (x-1)^2]u_{xx} + u_{yy} - \alpha(x, y)u_x = f(x, y). \tag{1.95}$$

The type degeneracy line $k(x, y) = 0$ meets the upper side of the rectangle G of length $l = 2$ at the points A(1/2, 1) and B(3/2, 1). Thus, we get a local elliptic subdomain adjacent to the upper side (see Figure 1.14a). The unique solvability of Problem (1.94), (1.95) in the space $W^{2,2}(G)$ is true if $2\alpha \pm k_x \equiv 2\alpha \pm 2(x-1) > 0$ in $\overline{G}$. This inequality holds, e.g. if $\alpha(x, y) > 1$.

An example of an equation of the *parabolic–hyperbolic* type can be given by (1.1) with $a \equiv 1$, $c \equiv 0$ and

$$k(x, y) = [\tfrac{1}{4} - (x-1)^2 - y^2]^4 \qquad \text{if } (x-1)^2 + y^2 > \tfrac{1}{4},$$
$$k(x, y) = 0 \qquad \text{if } (x-1)^2 + y^2 \leqslant \tfrac{1}{4}$$

(Figure 1.14b). As simple estimates show, $|k_x(x, y)| \leqslant 7^3/8$ if $l = 2$, therefore, the condition $2\alpha \pm k_x > 0$ in G holds if $\alpha(x, y) > 7^3/16$. Then Problem (1.1), (1.94) has a unique solution $u \in W^{2,2}(G)$.

We note that in the subdomain where $k = 0$ and the type of the equation is parabolic, the condition $2\alpha \pm k_x > 0$ reduces to the inequality $\alpha > 0$, which is evidently of principal importance for equations of the parabolic type.

More general sufficient conditions of the solvability as compared with those given by Theorem 1.10 were formulated by Kuz'min (1985, 1992a). The derivative k_y was allowed to change the sign on the type degeneracy line, which was constituted by a finite number of smooth closed curves or curves with endpoints on the upper or lower sides of the rectangle. Let the condition

$$2\alpha \pm k_x > 0 \quad \text{on the set } K \tag{1.96}$$

hold, where K is the set of all characteristic points and the endpoints of the type degeneracy line (Figure 1.15). Then the equation $Lu - \lambda u = f$ endowed with appro-

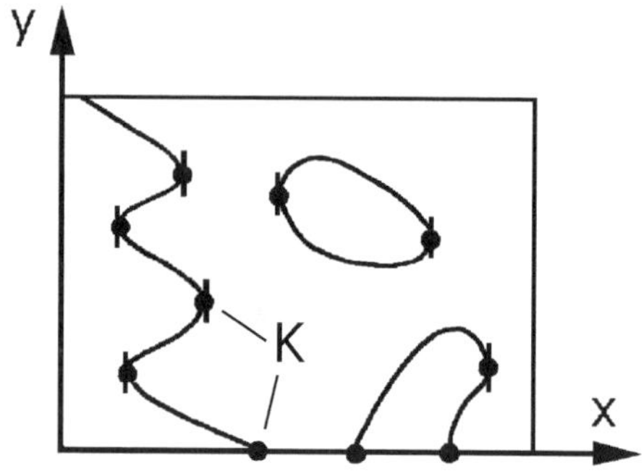

Figure 1.15 A set K on which $2\alpha \pm k_x > 0$ is supposed to hold.

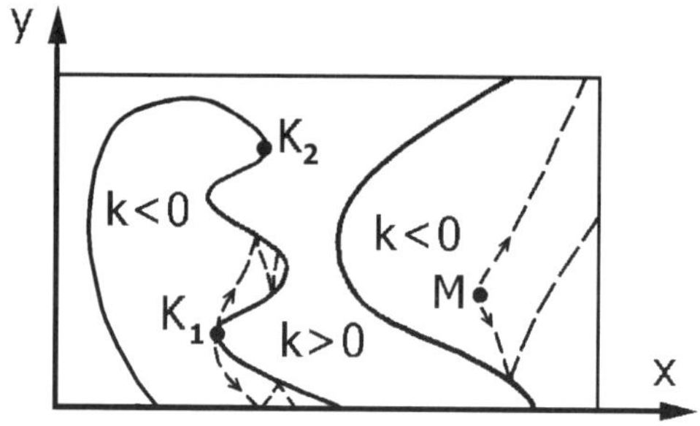

Figure 1.16 Propagation of a nonsmoothness in the hyperbolic subregion.

priate initial–boundary conditions (which depend on the signs of k at $x = 0$ and $x = l$, see (1.4), (1.50), (1.59), and (1.94)) is uniquely solvable in $W^{2,2}(G)$ if λ is large enough. At an arbitrary λ, the Fredholm alternative is true.

The validity of condition (1.96) is essential for the existence and uniqueness of regular solutions in the space $W^{2,2}(G)$. The points of the set K where $k_x > 0$ (hyperbolic–elliptic transition) are responsible for the appearance of singularities which can arise if (1.96) is violated. The points of the set K where $k_x < 0$ (elliptic–hyperbolic transition) are responsible for the nonuniqueness which can occur if the condition (1.96) is not true.

It can be shown that weak singularities or nonsmoothness of the solutions to the above problems propagate in the positive direction of the x-axis (Kuz'min 1987). For example, let the conditions of Theorem 1.5 be true with the exception of $f \in W^{3,2}(G)$, which is replaced by $f \in W^{3,2}(G \backslash G_{\mathrm{M}})$, where M is a point located in a hyperbolic subdomain, G_{M} is a small circle with the centre at M (see Figure 1.16). Then a weak singularity propagates within the subdomain bounded from the left by the characteristic lines tangent to G_{M}. Also, one can examine the propagation of weak singularity of the solution along characteristic lines emanating from a point K_1 of the type degeneracy line at which the conditions (1.47) are violated and $k_x > 0$ (Figure 1.16). At the same time, if there are no characteristic lines emanating from a point K_2 of the type degeneracy line at which the conditions (1.47) are violated and $k_x > 0$, this produces an isolated singularity at that point.

The developed theory was extended to embrace equations with the term containing the mixed derivative u_{xy}:

$$k(x, y)u_{xx} + 2b(x, y)u_{xy} + [a(x, y)u_y]_y - \alpha(x, y)u_x + \beta(x, y)u_y + c(x, y)u = f(x, y). \tag{1.97}$$

In this case, the characteristic direction $dx/dy = b/a$ on the type degeneracy line is not parallel to the y-axis if $b \neq 0$. Consequently, the set K of characteristic points of the type degeneracy line is not simply the set where the tangent to the line is parallel to the y-axis. That is why the left-hand side of the inequality (1.96) changes and involves the coefficients $\beta(x, y)$ and $b(x, y)$ (Kuz'min 1986b).

As mentioned in Remark 1.6, the replacement of the Neumann condition by the Dirichlet one on the upper and lower sides $y = \pm 1$ of the rectangle does not influence *a priori* estimates and the obtained results except for, possibly, details of the behaviour of the solution in the vicinities of the corner points of the domain G.

A different situation takes place if the Dirichlet condition is replaced by the Neumann one on the front side of G (if $k > 0$ there) or vice versa on the exit side. This violates Theorems 1.1 and 1.5, so that conditions (1.5), (1.6) do not guarantee the *unique* solvability of the modified Problem (1.1), (1.4). At the same time, the Fredholm solvability of the equation $Lu - \lambda u = f$ at any λ as well as the unique solvability at sufficiently large λ remain true if (1.96) holds.

A generalized solution from the space $W^{1,2}(G)$ can be introduced with integral relations as follows. Consider, for example, Problem (1.1), (1.50) for the equation of elliptic–hyperbolic type. The function $u \in W^{1,2}(G)$ is called a generalized solution of this problem if

$$\int_G [-u_x(k\chi)_x - au_y\chi_y + (-\alpha u_x + cu - f)\chi]\, dG = 0 \tag{1.98}$$

is true for any $\chi \in C^\infty(G)$ vanishing at $x = 0$ and in the arbitrary small vicinity of $x = l$. The relation is obtained by application of Green's formula to the expression $\int_G (Lu - f)\chi\, dG$ according to the general approach of the theory of partial differential equations. The condition of the Fredholm solvability of boundary-value problems in the space $W^{1,2}(G)$ is the inequality $2\alpha + k_x \geqslant \delta > 0$ which is to hold in the domain G or, if the set of type degeneracy consists of a finite number of smooth curves, at the set K of the characteristic points and endpoints of those curves (Kuz'min 1992a). As a matter of fact, solutions from $W^{1,2}(G)$ belong to an *energy class* $V^1(G)$ such that $W^{2,2}(G) \subseteq V^1(G) \subset W^{1,2}(G)$ and that there exists the energy integral $\int_{-1}^{1}(u^2 + |k|u_x^2 + au_y^2)\, dy$ over any section $x = \text{const.}$ of the rectangle G.

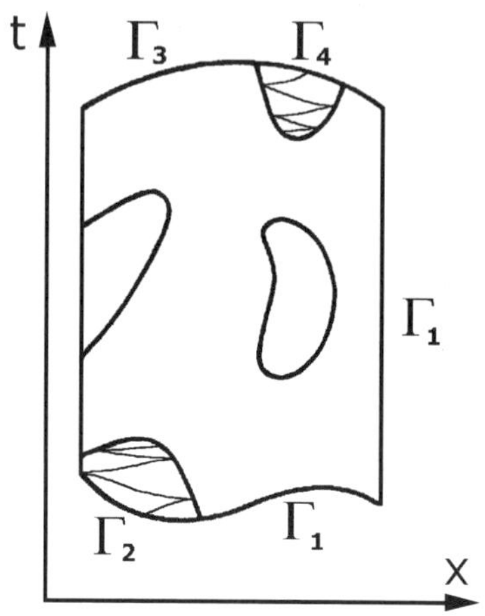

Figure 1.17 A domain in which a general problem for the multidimensional Equation (1.99) is formulated.

A multidimensional equation of mixed type

$$k(x,t)u_{tt} + \sum_{i,j=1}^{n} [a_{ij}(x,t)u_{x_i}]_{x_j} - \alpha(x,t)u_t + \sum_{i=1}^{n} \beta_i(x,t)u_{x_i} + c(x,t)u = f(x,t), \qquad (1.99)$$

was tackled by Karatoprakliev (1976, 1977), Vragov (1977), Dachev (1982) and Larkin *et al.* (1983). In Equation (1.99), the differential operator $\sum a_{ij}(x,t)u_{x_ix_j}$ with respect to the spatial variables $x = (x_1, x_2, \dots, x_n)$ is supposed to be *uniformly elliptic*, i.e. the form $\sum a_{ij}\xi_i\xi_j$ is positive definite in a given domain, while the coefficient $k(x,t)$ may change its sign.

Formulation of a general boundary-value problem proposed by Karatoprakliev (1976) is as follows. Consider Equation (1.99) in a domain D of the $(n+1)$-dimensional Euclidean space $\mathbb{R}^{n+1}$ bounded by a piecewise-smooth surface Γ. Let $H(x) = k\nu_{n+1}^2 + \sum_{i,j=1}^{n} a_{ij}\nu_i\nu_j$, where $\nu = (\nu_1, \dots, \nu_{n+1})$ is the inner unit normal to Γ, and the component ν_{n+1} corresponds to the t-axis. Denote

$$\begin{aligned}
&\Gamma_1 : (x,t) \in \Gamma, \quad \text{where } \nu_{n+1} \geqslant 0,\ H \geqslant 0,\\
&\Gamma_2 : (x,t) \in \Gamma, \quad \text{where } \nu_{n+1} > 0,\ H < 0,\\
&\Gamma_3 : (x,t) \in \Gamma, \quad \text{where } \nu_{n+1} < 0,\ H > 0,\\
&\Gamma_4 : (x,t) \in \Gamma, \quad \text{where } \nu_{n+1} < 0,\ H \leqslant 0.
\end{aligned}$$

The problem consists of finding a solution of Equation (1.99) that satisfies the conditions

$$u = 0 \quad \text{for } (x,t) \in \Gamma_1, \qquad u_t = 0 \quad \text{for } (x,t) \in \Gamma_3,$$
$$u = 0 \quad \text{and} \quad u_t = 0 \quad \text{for } (x,t) \in \Gamma_2$$

(see Figure 1.17). Karatoprakliev introduced a strong solution in $W^{1,2}(D)$ as a function which belongs to the domain of definition of a closed differential operator. Under certain restrictions on the coefficients of the equation, he proved the uniqueness of the strong solution and the existence of a weak solution to the boundary-value problem. The fact that the weak solution coincides with the strong one was proved in the case when the equation is of elliptic–hyperbolic type, so that $\Gamma = \Gamma_1 \cup \Gamma_4$ and $\Gamma_2 = \Gamma_3 = \emptyset$. In this case the boundary conditions reduce to $u = 0$ for $(x,t) \in \Gamma_1$. The proof was performed with a technique of smoothing operators.

Vragov (1977) employed the method of singular perturbations in order to study boundary-value problems for Equation (1.99) in a domain $D = \Omega \times (0,T)$ bounded by the cylindrical surface $\partial\Omega \times (0,T)$ and the two planes $\{(x,t) \in \mathbb{R}^{n+1} : x \in \Omega,\ t = 0\}$, $\{(x,t) \in \mathbb{R}^{n+1} : x \in \Omega,\ t = T\}$.

Dachev (1991) formulated conditions of the existence, uniqueness, and smoothness of a strong solution to boundary-value problems for the equation of elliptic–hyperbolic type with the mixed derivative u_{tx_i}:

$$Lu - \lambda u := ku_{tt} + 2\sum_{i=1}^{n} b_i u_{tx_i} + \sum_{i,j=1}^{n} (a_{ij}u_{x_i})_{x_j} - \alpha u_t + \sum_{i=1}^{n} \beta_i u_{x_i} + (c-\lambda)u$$
$$= f \tag{1.100}$$

in a cylindrical domain $D = \{(x,t) \in \mathbb{R}^{n+1} : \varphi_1(x) < t < \varphi_2(x)\}$, where $\varphi_i(x) \in C^\infty(\overline{\Omega})$ $(i = 1,2)$ and Ω is a bounded domain in $\mathbb{R}^n$. Define $\Gamma_i = \{(x,t) \in \mathbb{R}^{n+1} : x \in \overline{\Omega},\ t = \varphi_i(x)\}$ and $\Gamma_3 = \Gamma \backslash (\Gamma_1 \cup \Gamma_2)$, where Γ is the full boundary of the domain D, and Γ_3 is the lateral boundary. The main result is the following theorem. Let L be elliptic in the vicinity of Γ_1 and

$$H = k\nu_0^2 + 2\sum_{i=1}^{n} b_i \nu_i \nu_0 + \sum_{i,j=1}^{n} a_{ij}\nu_i\nu_j \leqslant 0$$

on Γ_2, where ν is the inner unit normal to Γ. Then there exists a constant λ_0 such that for each $\lambda \geqslant \lambda_0$ and $f \in L^2(D)$ Equation (1.100) endowed with the boundary condition

$$u = 0 \quad \text{on } \Gamma_1 \cup \Gamma_3$$

admits a unique strong solution in $W^{1,2}(D)$ provided that

$$2\left[\alpha - \sum_{i=1}^{n} (b_i)_{x_i}\right] + k_t > 0 \tag{1.101}$$

on the type degeneracy set $D_0 = D \cap (k = 0)$. Moreover, if $f \in W^{s,2}(D)$ and $2[\alpha - \sum_{i=1}^{n}(b_i)_{x_i}] - (2p-1)k_t > 0$ on D_0 $(p = 0, 1, \ldots, s)$, then the solution belongs to $W^{s+1,2}(D)$. In order to prove the theorem, Dachev used a technique of smoothing operators developed by Karatoprakliev (1977). In addition, he discussed some existence results for degenerate hyperbolic equations.

Finite differences with respect to the t-variable were employed for the study of Equation (1.99) by Terekhov (1979, 1980), who considered a problem with Dirichlet's condition on both foundations of the cylindrical domain in the case $k(x, 0) > 0$, $k(x, T) > 0$. Romanova (1994, 1995) studied the basic problem with the condition $u_t = 0$ at $t = T$ for the equation

$$k(x,t)u_{tt} + \sum_{i=1}^{n} u_{x_i x_i} - \alpha(x,t)u_t - c(x,t)u = f(x,t).$$

She introduced the same singular perturbation term εu_{ttt} as in Equation (1.11) and considered the finite-difference scheme

$$\varepsilon u_{\tau\bar{\tau}\tau} + k(x,t)u_{\tau\bar{\tau}} + \sum_{i=1}^{n} u_{h_i \bar{h}_i} - \alpha(x,t)u_\tau - c(x,t)u = f(x,t),$$

where

$$u_\tau = \frac{u(x, t+\tau) - u(x,t)}{\tau}, \qquad u_{\bar{\tau}} = \frac{u(x,t) - u(x, t-\tau)}{\tau},$$

and similar expressions are implied for u_{h_i} and $u_{\bar{h}_i}$. The parameters τ and h_i are the mesh cell sizes along the axes t and x_i, respectively. Under the conditions $2\alpha + k_t \geqslant \delta > 0$, $c_t \leqslant 0$, $c \geqslant c_0$ in D, $f \in L^2(D)$, where c_0 is sufficiently large positive constant, Romanova proved the existence of a solution $u \in W^{1,2}(D)$ of the basic problem and the convergence of the approximate solutions to the solution $u(x, y)$ as $\tau, h_i, \varepsilon \to 0$. For the hyperbolic–elliptic equation $(k(x, 0) < 0,\ k(x, T) > 0)$, the solvability in $W^{1,2}(D)$ and the convergence of finite-difference approximations of the initial–boundary-value problem were proved by introducing the singular perturbation term of fourth order $-\varepsilon u_{tttt}$. Furthermore, the convergence of the approximate solutions to $u(x, t)$ in the norm of $W^{2,2}(D)$ was established under the additional condition $2\alpha - k_t \geqslant \delta > 0$.

Remark 1.11. Throughout this book, we use the frame of reference (x, y) relevant for applications in aerodynamics, but different from the frame (x, t), in which Equations (1.99), (1.100) have been written. As easily seen, in the case of two independent variables $(n = 1)$ and $\beta_1(x_1, t) \equiv 0$, Equation (1.99) reduces to (1.1) by the substitution x for t, then $-y$ for x_1 and by rotation of the frame of reference in the clockwise direction (Figure 1.18).

Remark 1.12. In the majority of the above-mentioned papers concerned with Equation (1.99), the conditions with respect to the coefficients were applied in the whole domain under consideration. The developed theory of characteristic points K of the

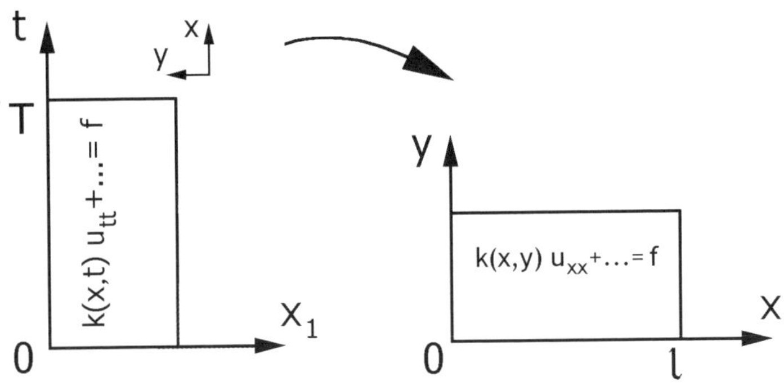

Figure 1.18 The transition from the frame of reference (x_1, t) to (x, y).

type degeneracy line (Kuz'min 1985, 1987) has considerably weakened restrictions for the coefficients of the equation. Actually, the theory exempted the coefficients from any conditions except (1.96) on the set K, though at the expense of additional assumptions on the geometry of the set where $k(x, y)$ vanishes.

The condition (1.96) remains invariant with respect to the multiplication of Equation (1.1) by a smooth positive function (which obviously cannot influence the solvability issue) in contrast to the inequalities (1.5), (1.6), which in this case change in a certain way and become more complex. Also, the condition (1.96) shows the points of the domain at which singularities of the solution may arise. Finally, if the hyperbolic subdomains contract to zero so that the type of the equation becomes elliptic over G, then the condition (1.96) disappears, while (1.5), (1.6) remain the same and irrelevant for the equation of elliptic type.

The developed theory of characteristic points K of the type degeneracy line provides conditions of the Fredholm solvability of the boundary-value problems, which are quite similar to the results of the classical theory for purely elliptic equations. Moreover, the obtained results and the role of $2\alpha + k_x$ on the set K are in accordance with the theory worked out by Fichera (1960) and Kohn and Nirenberg (1967) for equations of elliptic–parabolic type. On the other hand, in a special case when $k(x, y) < 0$ over G, i.e. Equation (1.1) is purely hyperbolic, the set K is empty and we obtain the unique solvability of the initial–boundary-value problem for the equation of hyperbolic type.

In addition, the approach based on the condition (1.96) turns out to be useful for applications, since the aerodynamicist usually deals with one or several sonic lines and/or shock waves, not with an arbitrary set of type degeneracy. Moreover, some hints on the location and shape of the sonic line are often available.

Mixed-type equations of orders higher than two have attracted much attention from mathematicians since the late 1980s. Egorov (1987) considered the equation

$$\sum_{i=1}^{2s} k_i(x, t)\frac{\partial^i u}{\partial t^i} + M_{2m}u - c(x)u = f(x, t), \qquad x = (x_1, \ldots, x_n) \qquad (1.102)$$

in $D = \Omega \times (0, T)$, where $\Omega \in \mathbb{R}^n$ is a bounded domain with a smooth boundary, M_{2m} is a uniformly elliptic operator in x of order $2m$, while $k_{2s}(x, t)$ may change its sign in the cylindrical domain D. Having defined subsets of the foundations of D as follows,

$$P_0^+ = \{x \in \Omega : (-1)^{s-1} k_{2s}(x, 0) > 0\},$$
$$P_{\mathrm{T}}^- = \{x \in \Omega : (-1)^{s-1} k_{2s}(x, T) < 0\},$$

Egorov prescribed the boundary conditions

$$\left.\begin{aligned} \partial^i u/\partial t^i &= 0 \quad \text{for } t = 0 \text{ and } t = T \quad (i = 0, \dots, s-1), \\ \partial^s u/\partial t^s &= 0 \quad \text{for } t = 0, x \in P_0^+ \text{ and } t = T, x = P_{\mathrm{T}}^-, \\ \partial^i u/\partial \nu^i &= 0 \quad \text{on } \partial\Omega \times [0, T] \quad (i = 0, \dots, m-1), \end{aligned}\right\} \tag{1.103}$$

and proved the following theorem. If $c(x)$ is sufficiently large, $f \in L^2(D)$, and $(-1)^{s-1}[2k_{2s-1} + (1-2s)k_{2st} + \mu(2s-1)k_{2s}] > \delta > 0$ in $\overline{D}$ for some constants μ and δ, then Problem (1.102), (1.103) admits a generalized solution u in the anisotropic Sobolev space $W_2^{m,s}(D)$ which involves the L^2-norms of all derivatives of u with respect to x and t of the orders $\leqslant m$ and $\leqslant s$, respectively.

Egorov and Fedorov (1995) studied the regularity of the generalized solutions to (1.102) and also considered boundary-value problems for the equation of an odd order in t,

$$\sum_{i=1}^{2s+1} k_i(x, t) \frac{\partial^i u}{\partial t^i} + M_{2m} u = f(x, t),$$

where the coefficient k_{2s+1} is of variable sign in D. Further developments concerned with the mixed-type equations of higher order are presented in a book of Egorov *et al.* (2000).

1.6 The Tricomi Equation

In the theory described above, the unique solvability of boundary-value problems in $W^{2,2}(G)$ or $W^{1,2}(G)$ has been established provided the coefficient $k(x, y)$ of Equation (1.1) *does not* change its sign on the sides $x = 0$ and $x = l$ of the rectangle G, or the coefficient $k(x, t)$ of Equation (1.99) does not change its sign on the foundations $t = 0$ and $t = T$ of the cylindrical domain D. Therefore, the presented theorems do not enable one to deduce the unique solvability of boundary-value problems, e.g. for the equation

$$y u_{xx} + u_{yy} + \alpha(x, y) u_x = f(x, y), \tag{1.104}$$

which differs from the classical Tricomi equation only by the extra term αu_x and the right-hand side f.

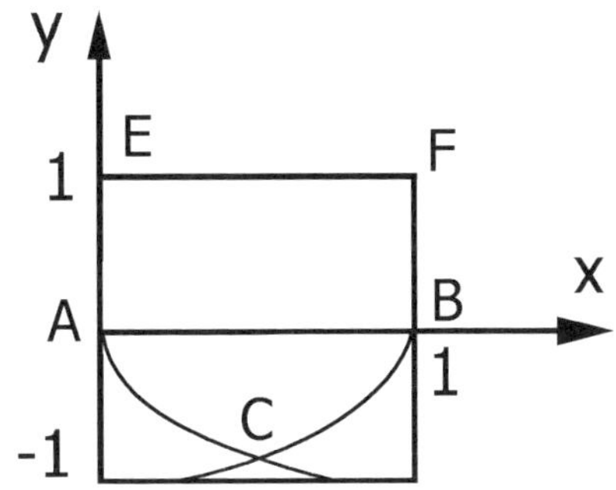

Figure 1.19 The rectangle in which Problem (1.104)–(1.106) is considered.

Consider Equation (1.104) in the rectangle $G = \{(x, y) : -1 < y < 1,\ 0 < x < 1\}$ (Figure 1.19) endowed with the boundary conditions on the elliptic part of the boundary

$$u = 0 \quad \text{on AE, EF}, \qquad u_x = 0 \quad \text{on BF}, \tag{1.105}$$

and the initial–boundary conditions on the hyperbolic part of the boundary:

$$\left.\begin{aligned} u = u_x &= 0 \qquad \text{at } x = 0, \qquad -1 < y < 0, \\ u &= 0 \qquad \text{at } y = -1, \qquad 0 < x < 1. \end{aligned}\right\} \tag{1.106}$$

If $f(x, y) \equiv 0$ in the subdomain below the characteristic line AC, then the initial–boundary conditions (1.106) for the hyperbolic subdomain yield $u \equiv 0$ below AC up to AC. In this case, Problem (1.104)–(1.106) reduces to finding a solution of Equation (1.104) in the domain bounded by the characteristics AC, CB, and the segments AE, EF, BF, under the conditions (1.105) on the elliptic part of the boundary and the prescribed

$$u = 0 \quad \text{on the characteristic line AC.} \tag{1.107}$$

If Problem (1.104)–(1.106) were solvable, this would result in the solvability of the Tricomi-like Problem (1.104), (1.105), (1.107) with the mixed Dirichlet–Neumann condition on the elliptic part of the boundary. The uniqueness of the solution $u \in W^{2,2}(G)$ to Problem (1.104)–(1.106) immediately follows from the identity (1.8) if $\alpha(x, y) > 0$ in G. However, the coefficient y in front of u_{xx} changes its sign on the left- and right-hand sides of the domain. As a consequence, the solvability results formulated in Theorems 1.1, 1.9 are not applicable in this case. The method used in Section 1.1 for establishing the existence of the solution to the basic problem fails as well.

Solvability of a problem which resembles the Tricomi one can be demonstrated if one considers Equation (1.95) in the domain G shown in Figure 1.14a and imposes $f(x, y) \equiv 0$ below the characteristic line AC. The boundary conditions (1.94), the given initial data on the front side of G and the condition $u_y = 0$ at $y = \pm 1$ make it possible to solve the problem in a part of the hyperbolic subdomain and yield $u = 0$ on the characteristic line AC emanating from the point A of the type degeneracy line. Thus, in the domain bounded by the characteristics AC, BC and the segment AB we obtain a problem similar to the Tricomi one (with some discrepancies in the shapes of the type degeneracy line and the elliptic part of the boundary). Due to Theorem 1.1, there exists a unique solution $u \in W^{2,2}(G)$ of this problem if $f \in W^{1,2}(G)$ and $\alpha > 1$ in G.

The existence of a *weak* solution $u \in W^{1,2}(G)$ to Problem (1.104)–(1.106) can be established with *a priori* estimates for solutions of the adjoint problem (Vragov 1977, Theorem 3). A similar problem with the Dirichlet condition over the full elliptic part of the boundary was studied by Aziz *et al.* (1982). The authors considered the equation

$$k(y)u_{xx} + u_{yy} + c(x, y)u = f(x, y), \tag{1.108}$$

with $k(0) = 0$, $\operatorname{sgn} k(y) = \operatorname{sgn} y$, in the rectangle shown in Figure 1.19 under the conditions (1.106) and the Dirichlet condition

$$u = 0 \quad \text{on AE, EF, BF}, \tag{1.109}$$

replacing (1.105). The 'a,b,c' method of Friedrichs was used in the proof of the uniqueness of a quasi-regular solution and in obtaining *a priori* estimate for the adjoint problem. This ensures the existence of a generalized solution $u \in W^{-1,2}(G)$ of Problem (1.106), (1.109). Here $W^{-1,2}(G)$ is the space dual to $W^{1,2}(G)$ with respect to the $L^2(G)$ inner product.

The *classical Tricomi problem* is formulated as follows. Let Γ_0 be a piecewise-smooth curve in the half-plane $y > 0$ with endpoints A(0, 0) and B(1, 0) on the x-axis. Let Γ_1, Γ_2 be the characteristics of (1.108) emanating from A and B, as defined by the equations $\mathrm{d}x + \sqrt{-k}\,\mathrm{d}y = 0$ and $\mathrm{d}x - \sqrt{-k}\,\mathrm{d}y = 0$, respectively. These characteristics intersect at a point C. Denote by G a simply connected domain in $\mathbb{R}^2$ bounded by the curves Γ_0, Γ_1 and Γ_2. The Tricomi problem consists of finding a solution of Equation (1.108) in the domain G which satisfies the boundary condition

$$u = 0 \quad \text{on } \Gamma_0 \cup \Gamma_1. \tag{1.110}$$

Studies of the Tricomi problem with the methods of functional analysis have demonstrated significant progress since the beginning of the 1980s. Kal'menov (1983) considered the Tricomi problem for the equation

$$Lu := \operatorname{sgn} y \cdot |y|^m u_{xx} + u_{yy} = f(x, y),$$

in the case when Γ_0 is a smooth Lyapunov curve terminating at $y = 0$ by arcs of the 'normal contour' given by the algebraic equation

$$(x - \tfrac{1}{2})^2 + \left(\frac{2}{m+2}\right)^2 y^{m+2} = \tfrac{1}{4}.$$

He introduced a strong solution $u \in L^2(G)$ as a function which belongs to the domain of definition of a closure of the differential operator Lu on the subset of functions

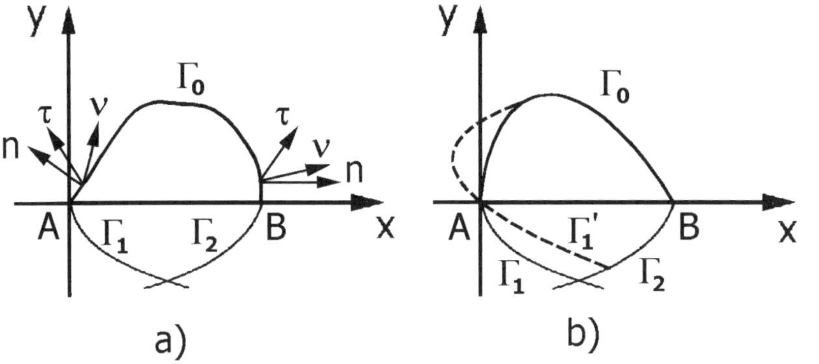

Figure 1.20 Configurations of the elliptic part Γ_0 of the boundary.

$u \in C^2(\overline{G})$ satisfying the boundary condition (1.110). Kal'menov proved the strong solvability of the Tricomi problem $\forall f \in L^2(G)$ at $m < 2$.

Another approach to the Tricomi problem was developed by Aziz *et al.* (1990), who considered the nonlinear equation

$$yu_{xx} + u_{yy} + c(x, y)u = f(x, y, u), \tag{1.111}$$

and introduced a generalized solution $u \in W^{1,2}(G)$ through the integral relation $\int_G (yu_x v_x + u_y v_y + fv)\,\mathrm{d}G = 0$ valid for all $v \in W^{1,2}(G)$, $v|_{\Gamma_0 \cup \Gamma_2} = 0$. The curve Γ_0 was assumed to satisfy the following inequalities

$$\nu_1 n_1 + \nu_2 n_2 \geqslant 0 \quad \text{and} \quad \tau_1 n_1 + \tau_2 n_2 \geqslant 0 \quad \text{along } \Gamma_0, \tag{1.112}$$

where (n_1, n_2) is the outward normal to Γ_0, while the vectors (ν_1, ν_2) and (τ_1, τ_2) are determined by their components (Figure 1.20a):

$$\nu_1 = 2(x - 1) - (\tfrac{3}{4})^{1/3}, \qquad \nu_2 = 1 + y,$$
$$\tau_1 = 2x - (\tfrac{3}{4})^{1/3}, \qquad \tau_2 = 1 + y.$$

Under the conditions (1.112), Aziz *et al.* established the existence and uniqueness of the generalized solution to the Tricomi problem. Also, Galerkin-type numerical procedures were considered and convergence rates of finite-element solutions were obtained. A regularity statement for the generalized solution of Problem (1.110), (1.111) was proved by Schneider (1992) under the conditions (1.112) and additional smoothness assumptions with respect to the curve Γ_0 and the right-hand side $f(x, y)$.

The *generalized Tricomi problem* consists of solving Equation (1.114) under the boundary condition

$$u = 0 \quad \text{on } \Gamma_0 \cup \Gamma_1', \tag{1.113}$$

where Γ_1' is a piecewise-smooth curve emanating from the origin A and lying inside the characteristic triangle ABC, so that $\mathrm{d}x + \xi(x, y)\,\mathrm{d}y = 0$, $\xi \geqslant \sqrt{-k}$ along Γ_1' (see the dashed curve in Figure 1.20b). The domain under consideration is bounded by Γ_0, Γ_1' and the characteristic line Γ_2 emanating from the point B. If $\xi \equiv \sqrt{-k}$, then the curve Γ_1' coincides with the characteristic line AC, and the above problem reduces to the classical Tricomi Problem (1.110).

Dachev (1987) studied the generalized Tricomi problem for the equation

$$\begin{aligned} Lu - \lambda u &:= k(y)u_{xx} + u_{yy} + c_1(x, y)u_x + c_2(x, y)u_y + c(x, y)u - \lambda u \\ &= f(x, y), \end{aligned} \tag{1.114}$$

where $k(0) = 0$, $k'(y) > 0$, $f \in L^2(G)$. He pointed out conditions for the coefficients of the equation and the curve Γ_0 under which the strong solution of Problem (1.113), (1.114) is unique, and there exists a weak solution $u \in W^{1,2}(G)$, which turns out to coincide with the strong one. Also, the regularity of the solutions to the Tricomi problem with Neumann boundary conditions was investigated. The results were obtained under minimal restrictions concerning the coefficients. At the endpoints A and B of the type degeneracy line, the inequalities $2c_1(0, 0) + k'(0)\tan\theta_{\mathrm{A}} > 0$, $2c_1(1, 0) + k'(0)\tan\theta_{\mathrm{B}} > 0$ have to be fulfilled, where θ_{A} and θ_{B} are the angles made by the curve Γ_0 with the positive direction of the x-axis. In the case of the Dirichlet boundary condition on Γ_0, only the first inequality is assumed to hold. After performing a coordinate transformation and formulating auxiliary problems, Dachev derived the necessary *a priori* estimates, from which the existence of strong solutions to the auxiliary problems and their regularity were obtained. From these results the conclusions concerning solvability of the boundary-value problems in question for Equation (1.114) were deduced.

Sun (1985) considered a similar equation

$$\begin{aligned} Lu - \lambda u &:= k(x, y)u_{xx} + u_{yy} - \alpha(x, y)u_x + \beta(x, y)u_y + c(x, y)u - \lambda u \\ &= f(x, y), \end{aligned} \tag{1.115}$$

where $k(x, 0) = 0$, $\operatorname{sgn} k(x, y) = \operatorname{sgn} y$, and $f \in L^2(G)$. He applied the 'a,b,c' method to prove the estimate $\|u\|_1 \leqslant m\|(L - \lambda)u\|_0$ for any function $u \in W^{2,2}(G)$ satisfying the boundary condition (1.113). This estimate holds under a sufficiently large λ and the condition on the type degeneracy line

$$2\alpha(x, 0) + k_x(x, 0) + \frac{1}{\sigma}k_y(x, 0) \geqslant \delta > 0, \qquad x_{\mathrm{A}} \leqslant x \leqslant x_{\mathrm{B}}, \tag{1.116}$$

where δ is an arbitrary small positive constant, while σ is the positive constant involved in the inequality

$$\max\left(\max_{\overline{\Gamma_1'}} \xi(x, y), \max_{\overline{G^-}} \sqrt{-k(x, y)}\right) < \sigma.$$

The function $\xi(x, y)$ determines the configuration of the curve Γ_1' as pointed out above; $G^- = G \cap (y < 0)$ is the hyperbolic part of the domain G. The term $k_x(x, 0)$

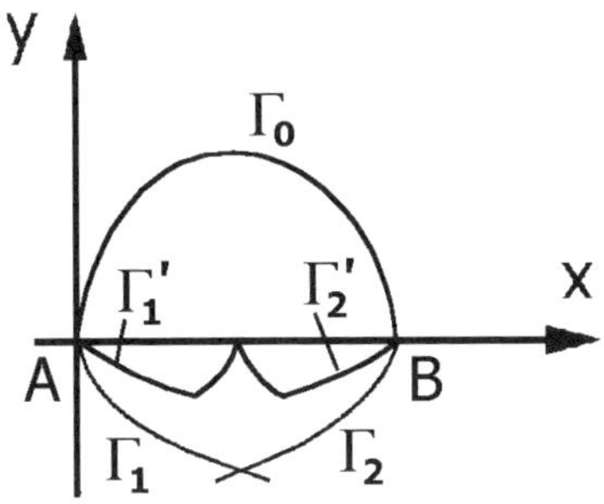

Figure 1.21 A domain in which the Frankl problem is formulated.

in (1.116) may be omitted since $k(x, 0) = 0$. The restrictions on the elliptic part of the boundary Γ_0 are set by the inequality

$$n_2 - \sigma n_1 > 0 \quad \text{on } \Gamma_0,$$

where (n_1, n_2) is the outward normal to Γ_0. The obtained *a priori* estimates enable one to establish the existence and uniqueness of a strong solution $u \in W^{1,2}(G)$ to the generalized Tricomi Problem (1.115), (1.113) at a sufficiently large λ. Also, the Fredholm property at any λ was proved. Under an additional assumption on the smoothness of $\Gamma_0 \cup \Gamma_1'$, Sun stated the regularity of the obtained solution referring to the technique of smoothing operators worked out by Karatoprakliev (1977).

The admissible curves Γ_0 in the above-mentioned papers of Dachev (1987), Sun (1985), Aziz *et al.* (1990) cannot approach the x-axis in the normal direction at both points A and B (Figure 1.20) in distinction to the normal contours considered by Kal'menov (1983), and the ones presented in classical works on equations of mixed type (Ferrari and Tricomi 1968; Smirnov 1978) where the method of integral equations was employed. This is accounted for by singularities of the first-order derivatives that arise at the endpoint B of Γ_0 and by the necessity of using weighted Sobolev spaces if Γ_0 approaches the x-axis in the normal direction. Sabitov (1992) used the alternating Schwartz method in order to weaken restrictions on the behaviour of the curve Γ_0 in the vicinity of the endpoints A and B.

The equation of mixed type with a weak nonlinearity,

$$\begin{aligned} Lu &\equiv k(x, y)u_{xx} + u_{yy} + \alpha(x, y)u_x + \beta(x, y)u_y + c(x, y)u - |u|^\rho u \\ &= f(x, y), \end{aligned}$$

where $\operatorname{sgn} k(x, y) = \operatorname{sgn} y$, $k(x, 0) = 0$, $f \in L^2(D)$, $\rho = \text{const.} > 0$, was considered by Sun and Xiao (1989). The authors obtained conditions of the existence and uniqueness of a strong solution to the generalized Tricomi problem by applying the fixed-point principle.

Larkin and Schneider (1990) proved the existence of a generalized solution to the Tricomi problem for the equation

$$Tu + \lambda l(u) := yu_{xx} + u_{yy} + \lambda l(u) = f(x, y),$$

where $l = \alpha^1 \partial/\partial x + \alpha^2 \partial/\partial y$ is a special differential operator and $\lambda \geqslant 0$ is a constant. Furthermore, they established the solvability of an initial–boundary-value problem for the evolution equation $Tu + \partial l(u)/\partial t = F(x, y, t)$ by an approximation method. The generalized solution of the evolution problem was shown to converge to the solution of the Tricomi problem $Tu = f$ as $t \to \infty$. The rate of convergence was estimated. Using the Faedo–Galerkin method, Larkin and Schneider (1994) proved the existence of a generalized solution to the initial–boundary-value problem for the nonlinear equation

$$Tu + R(x, y, t)u|u|^{\rho} + \partial l(u)/\partial t = F(x, y, t, u),$$

$0 \leqslant \rho \leqslant 2$, in a cylinder $D = \Omega \times (0, t_1)$. The authors also proved the existence of a generalized solution to the corresponding stationary boundary-value problem $Tu + r(x, y)u|u|^{\rho} = f(x, y, u)$ in Ω. The convergence of the approximate generalized solution of the evolution problem to the approximate generalized solution of the corresponding stationary problem as $t \to \infty$ was established. In a subsequent paper, Larkin and Schneider (1995) gave sufficient conditions for the uniqueness of these solutions.

The *Frankl problem* consists of finding a solution of the equation $k(y)u_{xx} + u_{yy} = f(x, y)$ or Equation (1.115) in the domain which is bounded by the curve Γ_0 located in the upper half-plane with endpoints A and B on the y-axis, two noncharacteristic curves Γ_1' and Γ_2' emanating from points A and B, and by two characteristic lines emanating from a point of the segment AB (Figure 1.21). The boundary conditions are

$$u = 0 \quad \text{on } \Gamma_0 \cup \Gamma_1' \cup \Gamma_2'. \tag{1.117}$$

We adopt this formulation of the Frankl problem following Morawetz (1954), Mises (1958), Ferrari and Tricomi (1968), and keeping in mind the original paper of Frankl (1947) devoted to transonic flow with a local supersonic region. We notice that some authors associate the term 'Frankl's problem' with the generalized Tricomi problem, at which he drew attention in the preface to the Russian translation of Tricomi's book (1947) and which he tackled later in a mathematical paper (Frankl 1951).

The uniqueness of the Frankl problem was established by Morawetz (1954, 1956) under certain restrictions on the coefficient $k(y)$ and the domain G. This result made it possible to prove that initially smooth transonic flow of inviscid gas over an airfoil cannot remain smooth, but must contain shock waves, if a supersonic portion of the airfoil is varied while the speed at infinity is kept fixed. For analysis of the problem, the hodograph variables were employed in which the governing equations become linear. The proof was quite complicated and later characterized by Morawetz (1964) as 'tediously long'. A simpler proof was presented by Morawetz (1964, 1966) under more restrictive assumptions which, however, covered a physically interesting situation.

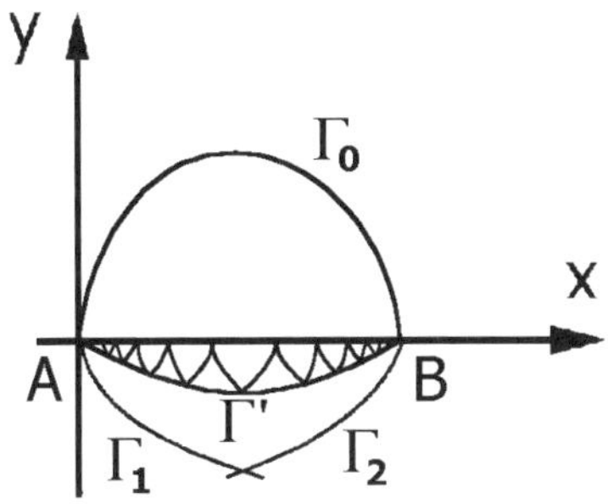

Figure 1.22 A domain in which the Dirichlet problem is considered.

Cook (1978) studied the Frankl problem for the linearized von Kármán Equation (1.67) directly in the plane (x, y) using the technique developed by Morawetz and asymptotic expansions for transonic flow at large distances from the airfoil.

The *Dirichlet problem* for the mixed-type Equation (1.115) can be considered as a limit of the Frankl problem when the endpoint of the curve Γ_1' approaches the point B along a path in the hyperbolic region (Figure 1.22). Morawetz (1970a,b) proved the existence of a weak solution to the Dirichlet problem for the Tricomi equation admitting a sufficiently *strong singularity* at one of the two parabolic points A and B of the boundary. She considered a special case when a smooth boundary is defined in the elliptic domain by an algebraic function that contains the 3/2 power of its ordinate. The obtained result is consonant to the discussion in the end of Section 1.4 regarding the singularity of a solution at the endpoint B of the type degeneracy line.

Soldatov (1994a–c) studied the Dirichlet problem for the model equation $\operatorname{sgn} y \cdot u_{xx} + u_{yy} = 0$ in the domain D bounded by smooth curves Γ and γ located in $y > 0$ and $y < 0$, respectively, with common endpoints $(0, 0)$ and $(1, 0)$, where γ is assumed not to be tangent to the characteristics $x \pm y = \text{const.}$ and to lie inside the characteristic triangle with the base interval $[0, 1]$. Soldatov investigated the uniqueness and existence of solutions with certain power behaviours near $A(0, 0)$ and $B(1, 0)$. He employed the requirement that $u(x, y)$ be in the weighted Hölder spaces $H_{\mu,\nu}(\overline{G}) = H_{\mu,\nu}(\overline{G}, 0, 1)$, $\nu = (\nu_0, \nu_1)$. By definition, the spaces consist of functions

$$\varphi(x, y) = (x^2 + y^2)^{(\nu_0-\mu)/2}[(x-1)^2 + y^2]^{(\nu_1-\mu)/2}\varphi_*(x, y),$$

where $\varphi_* \in H_\mu(\overline{G})$ and $\varphi_*(0, 0) = \varphi_*(1, 0) = 0$. The question of the regularity of a solution $u \in H_{\mu,\nu}$ was discussed, and the derivatives u_x and u_y were shown to admit weak singularities at the point A or B depending on the formulation of the problem.

We remark that in the Tricomi problem and the Dirichlet one, the strength of the singularity of the solution at point B depends on the coefficients in front of first-order derivatives if one considers Equations (1.104), (1.115). However, in the latter problem, the singularity appears to be stronger because of the accumulative phenomenon, whose significance was emphasized by Busemann (1949), Guderley

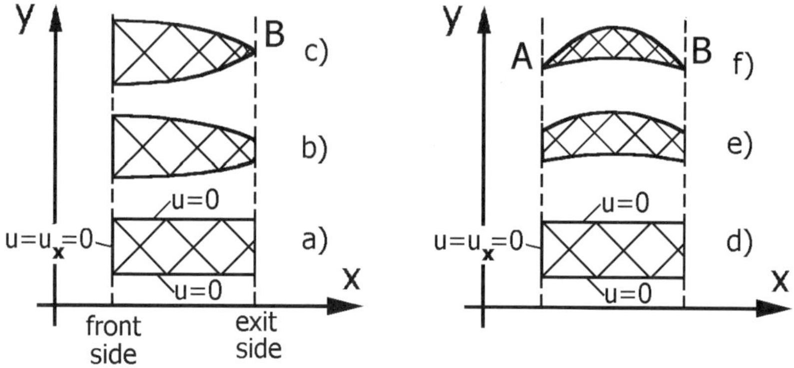

Figure 1.23 Initial–boundary-value problems for an equation of hyperbolic type in domains with cross-sections contracting to zero as $x \to l$.

(1953) and Mises (1958), and which will be discussed in Section 3.3. The nature of the accumulative phenomenon is the same as that in the initial–boundary-value problem for the most simple equation $u_{xx} - u_{yy} = 0$ considered in a domain which, being initially a rectangle, changes in such a way that the exit section contracts to a point B which closes the domain as the left-hand plots in Figure 1.23 show. The right-hand plots illustrate the case when the front section of the rectangle contracts to a point as well, so that we obtain a boundary-value problem without initial data. Such problems for equations of hyperbolic type were treated by Goursat (1933), Peyser (1961), Holten (1962) and Melnik (1981).

A large class of linear mixed-type equations of Tricomi type was studied by Payne (1997). He established theorems on the local and semi-global solvability using a theory of Hörmander and own results about the nontrapping of null-bicharacteristics over compact sets. Also, Payne (1998) studied propagation of singularities for mixed-type equations with the Dirichlet boundary condition. The obtained results included the exposure of the boundary effects of reflection, diffraction, and trapped gliding rays.

Lupo and Payne (2000a) investigated the Tricomi problem in domains which are normal in the sense that Γ_0 is orthogonal to the x-axis at the points A and B. For the pure Tricomi equation, i.e. (1.104) with $\alpha \equiv 0$, they proved the existence and uniqueness of generalized solutions lying in a weighted counterpart $\tilde{W}^{1,2}(G)$ of the space $W^{1,2}(G)$ equipped with the norm

$$\|u\|^2_{\tilde{W}^{1,2}(G)} = \int_G (|y|u_x^2 + u_y^2 + u^2)\,\mathrm{d}x\,\mathrm{d}y.$$

Also, an associated maximum/minimum principle coupled with known results of Krein–Rutman type was proven. Exploiting these results on the generalized solvability, Lupo and Payne (2000b) proved the existence of a principal eigenvalue for the Tricomi problem in normal domains, that is, the existence of a positive eigenvalue of minimum magnitude with an associated positive eigenvector for the equation $yu_{xx} + u_{yy} + \lambda u = 0$ under the boundary condition (1.110).

For the model equation $\operatorname{sgn} y \cdot u_{xx} + u_{yy} + \lambda u = 0$, the existence of a nontrivial eigenfunction of the Tricomi problem was proved by Kal'menov (1977) with the theory of positive operators on the cone.

2

Flow That Accelerates from Subsonic to Supersonic Velocity in a Channel

In this chapter we mainly examine inviscid shock-free flow with positive acceleration in a channel. We consider boundary-value problems with respect to linear or nonlinear perturbations of a given transonic flow. The condition of positive acceleration provides the inequality $2\alpha \pm k_x > 0$ which is of importance for the unique solvability of problems as established in the previous chapter. A few cases of flow with local regions of negative acceleration will be discussed in Section 2.9.

2.1 A Uniqueness Theorem

2.1.1 Governing equations

Consider steady, plane, isentropic, compressible flow which is governed by the continuity equation and the condition of irrotationality,

$$(\rho u)_x + (\rho v)_y = 0, \qquad u_y - v_x = 0, \tag{2.1}$$

where (x, y) are the Cartesian coordinates, $u(x, y)$ and $v(x, y)$ are the nondimensional components of the flow velocity in the directions x and y, while $\rho(x, y)$ is the nondimensional density. The velocity components and the density are normalized by the critical speed of sound a_* and the critical density ρ_*, respectively, which are supposed to be constant over the flow field. In the isentropic flow, the density $\rho(x, y)$ is expressed in terms of the velocity magnitude λ according to the Bernoulli law,

$$\rho = [1 + \tfrac{1}{2}(\gamma - 1)(1 - \lambda^2)]^{1/(\gamma-1)}, \qquad \lambda^2 = u^2 + v^2, \tag{2.2}$$

where $\gamma > 1$ is the Poisson adiabatic constant. As known, the Mach number is related to the velocity λ by the expression

$$M^2 = 2\lambda^2[\gamma + 1 - (\gamma - 1)\lambda^2]^{-1}. \tag{2.3}$$

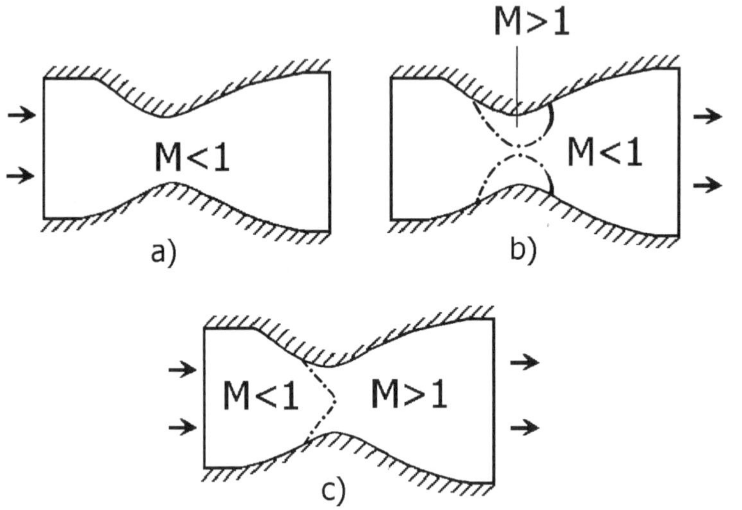

Figure 2.1 Different regimes of steady flow in a nozzle: (a) $M_{\text{out}} < M'$, (b) $M' < M_{\text{out}} < M'' < 1$, (c) $M_{\text{out}} > 1$.

The occurrence of subsonic or transonic flow regime in a channel and, consequently, the type of the system of nonlinear Equations (2.1) depend on both the geometry of the channel and the prescribed boundary conditions. In case of a Laval nozzle, i.e. a channel of variable cross-sectional area which has a minimum at some value of the longitudinal coordinate x, at least three regimes of steady transonic flow through the nozzle can occur.

(i) If the inclination angle θ is given at the inlet, the slip condition is used on the solid walls, and the Mach number M_{out} is prescribed at the outlet, M_{out} being small enough, then there exists a unique solution of the boundary-value problem which describes purely subsonic flow in the channel (Figure 2.1a) as rigorous mathematical studies showed (Bers 1958; Feistauer 1984).

(ii) If the prescribed Mach number M_{out} is not small and belongs to a range of subsonic values: $M' < M_{\text{out}} < M'' < 1$, then we obtain transonic flow with local supersonic subdomains adjacent to the walls near the nozzle throat (Figure 2.1b) as documented by many experimental and numerical studies; M' is a so-called *critical* Mach number.

(iii) If the condition of supersonic flow $M_{\text{out}} > 1$ is imposed at the outlet, then a flow with the subsonic–supersonic transition takes place, as numerous experimental and computational studies demonstrated. In this case, the sonic line extends from one wall of the nozzle to the other one (Figure 2.1c). Values of the Mach number at the outlet, where the governing equations are of hyperbolic type, are not specified, they depend on the geometry of the channel and can only be determined *a posteriori*. In practice, instead of the condition $M_{\text{out}} > 1$ one provides a condition of sufficiently small pressure in the outer media.

For a discussion of flow regimes with strong recompression shock waves caused by the large backpressure in the divergent part of the nozzle, the reader is referred, for instance, to Romine (1998).

Thus, the problem of flow through the nozzle of a given geometry is expected to have several solutions which depend on the assigned conditions at the inlet and outlet. If one treats *transonic* flow in a channel, then the formulation of the problem pointed out in item (iii) is typically implied, i.e. the condition $M_{\text{out}} > 1$ is assumed at the outlet.

If boundary conditions are imposed in terms of λ, not M, then the above flow regimes occur under similar conditions with respect to λ: (i) $\lambda_{\text{out}} \ll 1$, (ii) $\lambda' < \lambda_{\text{out}} < \lambda'' < 1$, (iii) $\lambda_{\text{out}} > 1$. This immediately follows from (2.3). In cases (i) and (ii), prescription of the Mach number (or λ) at the outlet and θ at the inlet is associated with conditions of transonic flow through turbomachinery cascades.

In this chapter, we concentrate on regime (iii), i.e. the one with a supersonic velocity at the outlet. Flow with local supersonic regions (Figure 2.1b) will be treated in Chapter 3.

Equations (2.1) make it possible to introduce the velocity potential $\varphi(x, y)$ and the stream function $\psi(x, y)$ by the relations

$$\varphi_x = u, \quad \varphi_y = v, \quad \psi_x = -\rho v, \quad \psi_y = \rho u, \tag{2.4}$$

i.e.

$$\varphi_x = \lambda \cos\theta, \quad \varphi_y = \lambda \sin\theta, \quad \psi_x = -\rho\lambda \sin\theta, \quad \psi_y = \rho\lambda\cos\theta, \tag{2.5}$$

where $\theta(x, y)$ is the angle of the velocity vector inclination to the x-axis. We now reduce system (2.4) to Chaplygin's equations with φ and ψ as independent variables. This will enable us to simplify the geometry of domains where boundary-value problems are to be considered. Definition (2.4) yields

$$\mathrm{d}\varphi = u\,\mathrm{d}x + v\,\mathrm{d}y, \qquad \mathrm{d}\psi = -\rho v\,\mathrm{d}x + \rho u\,\mathrm{d}y.$$

Having resolved the latter system with respect to $\mathrm{d}x$ and $\mathrm{d}y$, we obtain

$$\mathrm{d}x = \frac{\rho u\,\mathrm{d}\varphi - v\,\mathrm{d}\psi}{\rho(u^2 + v^2)}, \qquad \mathrm{d}y = \frac{\rho v \mathrm{d}\varphi + u\,\mathrm{d}\psi}{\rho(u^2 + v^2)}.$$

Replacing u by $\lambda\cos\theta$ and v by $\lambda\sin\theta$, we find

$$\mathrm{d}x = \frac{\cos\theta}{\lambda}\,\mathrm{d}\varphi - \frac{\sin\theta}{\lambda\rho}\,\mathrm{d}\psi, \qquad \mathrm{d}y = \frac{\sin\theta}{\lambda}\,\mathrm{d}\varphi + \frac{\cos\theta}{\lambda\rho}\,\mathrm{d}\psi. \tag{2.6}$$

The necessary condition for $\mathrm{d}x$ to be the full differential is the equality

$$\left(\frac{\cos\theta}{\lambda}\right)_\psi = \left(-\frac{\sin\theta}{\lambda\rho}\right)_\varphi, \qquad \forall(\varphi, \psi),$$

which can be easily transformed to the form

$$\sin\theta\left[\frac{\theta_\psi}{\lambda}+\lambda_\varphi\frac{(\lambda\rho)_\lambda}{\lambda^2\rho^2}\right]+\cos\theta\left[\frac{\lambda_\psi}{\lambda^2}-\frac{\theta_\varphi}{\lambda\rho}\right]=0.$$

Similarly, the necessary condition for $\mathrm{d}y$ in (2.6) to be the full differential yields

$$\cos\theta\left[\frac{\theta_\psi}{\lambda}+\lambda_\varphi\frac{(\lambda\rho)_\lambda}{\lambda^2\rho^2}\right]-\sin\theta\left[\frac{\lambda_\psi}{\lambda^2}-\frac{\theta_\varphi}{\lambda\rho}\right]=0.$$

The joint consideration of these two identities shows that the expressions in the square brackets must vanish:

$$\theta_\psi+\lambda_\varphi\frac{(\lambda\rho)_\lambda}{\lambda\rho^2}=0,\qquad \frac{\lambda_\psi}{\lambda}-\frac{\theta_\varphi}{\rho}=0. \tag{2.7}$$

Thus, we have arrived at *Chaplygin's equations* with respect to λ and θ in the plane (φ,ψ). The dependence $\rho(\lambda)$ is given by the Bernoulli law (2.2). Hence,

$$(\lambda\rho)_\lambda=\tfrac{1}{2}(\gamma+1)(1-\lambda^2)\rho^{2-\gamma}=(1-M^2)\rho.$$

Due to (2.5), the Jacobian

$$\frac{\partial(\varphi,\psi)}{\partial(x,y)}=\varphi_x\psi_y-\varphi_y\psi_x=\rho\lambda^2$$

does not vanish if $\rho\lambda\neq 0$. Therefore, the mapping $(x,y)\leftrightarrow(\varphi,\psi)$ establishes a one-to-one correspondence at least in the absence of stagnation points.

In the *nearsonic approximation*, using the expansions

$$\rho=1+o(|\lambda-1|),\qquad (\lambda\rho)_\lambda=(\gamma+1)(\lambda-1)+o(|\lambda-1|)$$

and neglecting the terms of the order $o(|\lambda-1|)$, from (2.7) we obtain the equations

$$(\gamma+1)(\lambda-1)\lambda_\varphi=\theta_\psi,\qquad \lambda_\psi=\theta_\varphi. \tag{2.8}$$

Since $\varphi_x=\lambda\cos\theta$ and $\psi_y=\rho\lambda\cos\theta$, one gets $\varphi\simeq x$, $\psi\simeq y$ in the approximation under consideration, whence in the (x,y)-plane the system assumes the same form as (2.8)

$$(\gamma+1)(\lambda-1)\lambda_x=\theta_y,\qquad \lambda_y=\theta_x, \tag{2.9}$$

Due to $\varphi_x\simeq\lambda$, $\varphi_y\simeq\theta$, one arrives at the second-order von Kármán equation with respect to the potential $\varphi(x,y)$:

$$(\gamma+1)(1-\varphi_x)\varphi_{xx}+\varphi_{yy}=0. \tag{2.10}$$

This equation can also be obtained by employment of asymptotic expansions in a small parameter δ directly in the plane (x,y), where δ represents the thickness of an airfoil or the maximum deviation of the nozzle walls from straight segments. As a matter of fact, Equation (2.10) follows from the continuity equation $(\rho u)_x+(\rho v)_y=0$ that takes the form

$$\rho_x u+\rho u_x+\rho v_y+o(|\lambda-1|)=0 \tag{2.11}$$

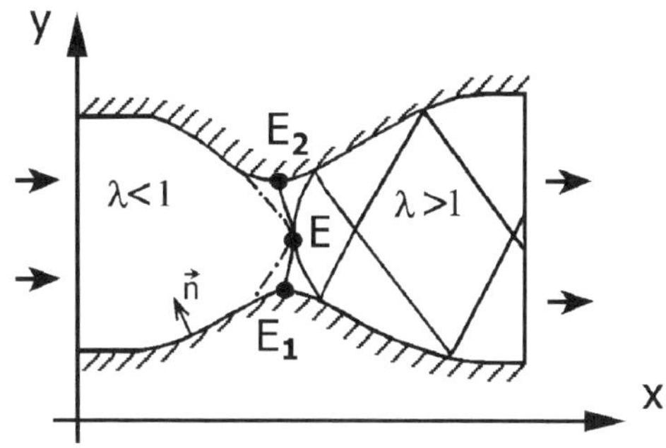

Figure 2.2 A sketch of flow through a Laval nozzle in the physical plane.

in the nearsonic approximation in which $u = 1 + o(|\lambda - 1|)$, $\rho = 1 + o(|\lambda - 1|)$, $v = o(|\lambda - 1|)$. Indeed, the Bernoulli law yields $\rho_x = -\rho^{2-\gamma} u u_x + \cdots$, where the dots denote terms of higher order in $|\lambda - 1|$. Therefore, by substituting this expression into (2.11), we obtain

$$(1 - \rho^{1-\gamma} u^2) u_x + v_y + \cdots = 0.$$

Now, the approximation

$$\rho^{\gamma-1} \simeq 1 + \tfrac{1}{2}(\gamma - 1)(1 - u^2) \simeq 1 + (\gamma - 1)(1 - u)$$

yields

$$\left[1 - \frac{u^2}{1 + (\gamma - 1)(1 - u)}\right] u_x + v_y + \cdots = 0,$$
$$(\gamma + 1)(1 - u) u_x + v_y + \cdots = 0.$$

Omitting the dots, one arrives at the first equation of system (2.9). We refer the reader to Cole and Cook (1986) for a methodology of asymptotic expansions in the (x, y)-plane.

2.1.2 Formulation of the problem

If one considers the flow in a nozzle with variable cross-section in the physical plane (Figure 2.2), then in the plane (φ, ψ) the domain under consideration is a rectangle whose upper and lower sides correspond to the nozzle walls (Figure 2.3). A typical pattern of the flow field includes a point E at which two Mach lines (i.e. characteristic lines of the system (2.8)) are tangent to the sonic line and continue in both directions from that point. The characteristic lines are defined in the supersonic region by solutions of the ordinary differential equation $\mathrm{d}\varphi/\mathrm{d}\psi = \pm[(\gamma + 1)(1 - \lambda)]^{1/2}$, which yields $\mathrm{d}\varphi/\mathrm{d}\psi = 0$ on the sonic line $\lambda - 1 = 0$. Hence, at the contact point E, the sonic line is tangent to the vertical line $\varphi = \varphi_{\mathrm{E}}$ and orthogonal to the streamline $\psi = \psi_{\mathrm{E}}$.

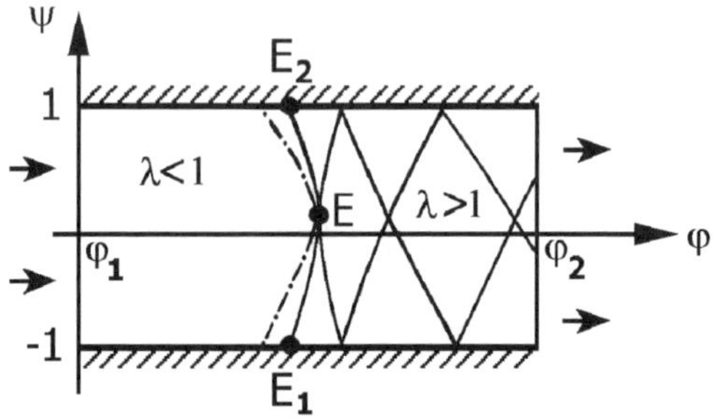

Figure 2.3 A sketch of flow through a Laval nozzle in the streamline coordinates (φ, ψ).

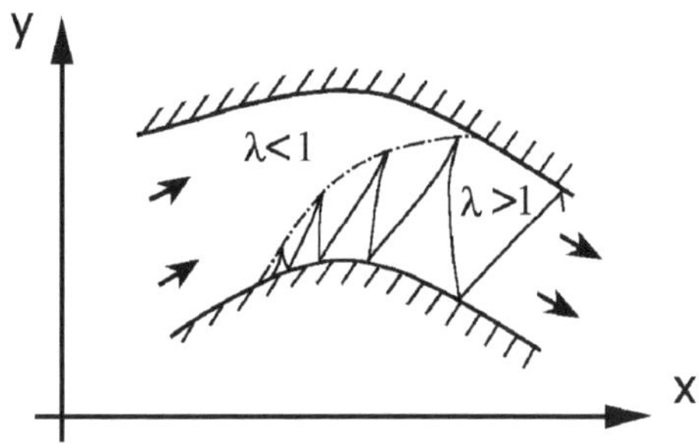

Figure 2.4 Accelerating flow without points of contact of the sonic line and the characteristic lines.

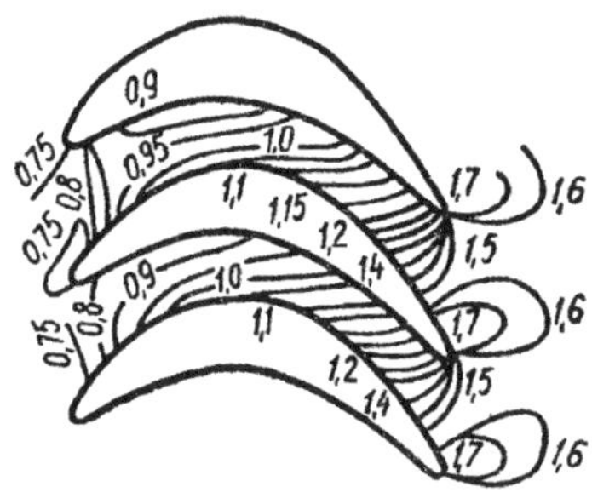

Figure 2.5 Isotachs λ = const. of transonic flow in a turbomachinery cascade. Computations of Sokolovsky and Gnesin (1986).

Also, a pattern of accelerating flow without contact points can occur (Figure 2.4). Such a pattern can be realized in transonic flow through a curved channel or through a turbomachinery blade row, as particular solutions and numerical simulations demonstrate (Figure 2.5).

Formulation of a boundary-value problem for the accelerating flow in the streamline coordinates (φ, ψ) is as follows. Find a solution $\lambda(\varphi, \psi)$, $\theta(\varphi, \psi)$ of system (2.7) or

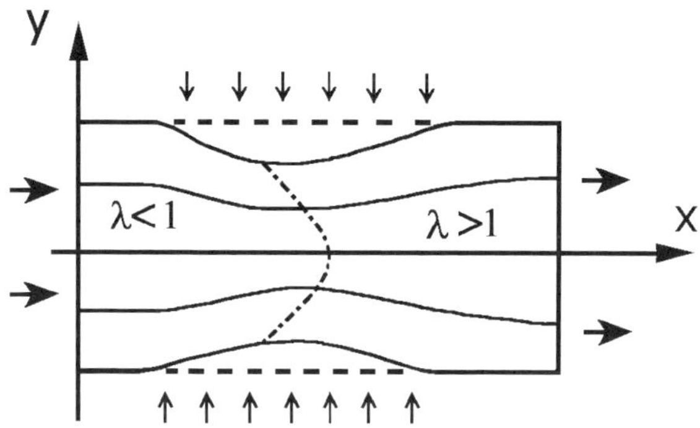

Figure 2.6 Modelling of flow in a channel of variable cross-section by a blowing which is distributed along the upper and lower sides of a rectangle in the physical plane.

(2.8) in the rectangle $G = \{(\varphi, \psi) \in \mathbb{R}^2;\ \varphi_1 < \varphi < \varphi_2,\ -1 < \psi < 1\}$ under the boundary conditions

$$\left.\begin{array}{lllll} \theta = h(\psi), & \lambda < 1 & \text{at the inlet} & \varphi = \varphi_1, & -1 < \psi < 1, \\ & \lambda > 1 & \text{at the outlet} & \varphi = \varphi_2, & -1 < \psi < 1, \\ \theta = \theta^{\pm}(\varphi) & & \text{at} & \psi = \pm 1, & \varphi_1 < \varphi < \varphi_2, \end{array}\right\} \tag{2.12}$$

where h, $\theta^{\pm}$ are given functions. According to (2.12), the flow velocity is to be subsonic at the inlet and supersonic at the outlet of the channel. However, its values are not prescribed. The given functions $\theta^{\pm}(\varphi)$ define *a posteriori* the shape of the walls in the plane (x, y) as discussed below in Section 2.3.

In the *nearsonic approximation*, the problem for flow in a nozzle can be formulated in a rectangle of the physical plane (x, y) instead of (φ, ψ) if appropriate boundary conditions on the upper and lower sides of the rectangle are imposed. The standard slip condition on the rigid walls reads $\varphi_n = (\nabla\varphi, \boldsymbol{n}) = 0$, where $\boldsymbol{n}$ is the normal vector (Figure 2.2). This condition can be rewritten in the form $u\cos(\boldsymbol{n}, x) + v\cos(\boldsymbol{n}, y) = 0$ or $v = -u\cos(\boldsymbol{n}, x)/\cos(\boldsymbol{n}, y) = 0$. Setting $u \cong 1$, we obtain $\varphi_y = v = \theta^{\pm}(x)$ on the walls where the inclination angles $\theta^{\pm}(x)$ are defined by the tangent to the profile of the given walls. As asymptotic expansions show, the condition $\varphi_y = \theta^{\pm}(x)$ can be prescribed at the segments $y = \pm 1, 0 < x < l$ instead of the walls if the deviation of the walls configuration from the straight lines $y = \pm 1$ is small. The obtained boundary condition $\varphi_y(x, \pm 1) = \theta^{\pm}(x)$ physically means the existence of a blowing which is distributed along the segments $y = \pm 1, 0 < x < l$. Interaction of the blowing with the flow entering the nozzle through the inlet determines the streamlines which approximate the configuration of the given walls (Figure 2.6).

Therefore, the problem can be formulated in the plane (x, y) as follows. Find a solution $\varphi(x, y)$ of the von Kármán Equation (2.10) in the rectangle $G = \{(x, y) \in$

$\mathbb{R}^2; 0 < x < l,\ -1 < y < 1\}$ under the boundary conditions

$$\left.\begin{array}{lllll} \varphi = h(y), & \varphi_x < 1 & \text{at the inlet} & x = 0, & -1 < y < 1, \\ & \varphi_x > 1 & \text{at the outlet} & x = l, & -1 < y < 1, \\ \varphi_y = \theta^{\pm}(x) & & \text{at} & y = \pm 1, & 0 < x < l, \end{array}\right\} \tag{2.13}$$

where $h, \theta^{\pm}$ are given functions.

2.1.3 Uniqueness theorem in case of the von Kármán equation

In this section, we prove the uniqueness of the solution to Problem (2.10), (2.13) posed in the plane (x, y), though a similar result is valid for Problem (2.8), (2.12) formulated in the plane of the streamline coordinates (φ, ψ) (Shifrin 1978).

Theorem 2.1. *Let there exist a classical solution $\varphi(x, y) \in C^1(\overline{G}) \cap C^2(G)$ or a solution $\varphi(x, y)$ from the Sobolev space $W^{2,2}(G)$ of the nonlinear Problem (2.10), (2.13), and the second-order derivative of the velocity potential be positive in G: $\varphi_{xx}(x, y) > 0$, i.e. the flow acceleration be positive over the channel. Then this solution is unique in the above-mentioned classes of functions.*

Proof. Suppose there are two solutions $\varphi(x, y)$ and $\hat{\varphi}(x, y)$ of Problem (2.10), (2.13). By subtracting Equation (2.10) written in terms of $\hat{\varphi}$ from that written for φ, we obtain

$$(\gamma + 1)[(1 - \varphi_x)\varphi_{xx} - (1 - \hat{\varphi}_x)\hat{\varphi}_{xx}] + \varphi_{yy} - \hat{\varphi}_{yy} = 0. \tag{2.14}$$

The expression in the square brackets can be represented as follows,

$$\begin{aligned} (1 - \varphi_x)\varphi_{xx} - (1 - \hat{\varphi}_x)\hat{\varphi}_{xx} &= (1 - \varphi_x)\varphi_{xx} - (1 - \hat{\varphi}_x)(\hat{\varphi}_{xx} - \varphi_{xx} + \varphi_{xx}) \\ &= (\hat{\varphi}_x - \varphi_x)\varphi_{xx} - (1 - \hat{\varphi}_x)(\hat{\varphi}_{xx} - \varphi_{xx}) \\ &= -u_x\varphi_{xx} + (1 - \hat{\varphi}_x)u_{xx}, \end{aligned}$$

where $u = \varphi - \hat{\varphi}$. Therefore, Equation (2.14) becomes

$$(1 - \hat{\varphi}_x)u_{xx} - \varphi_{xx}u_x + \frac{1}{\gamma + 1}u_{yy} = 0. \tag{2.15}$$

By adding and subtracting the term pu_xu_{xx} with a parameter p, we arrive at the equation

$$[1 - p\varphi_x - (1 - p)\hat{\varphi}_x]u_{xx} - [(1 - p)\varphi_{xx} + p\hat{\varphi}_{xx}]u_x + \frac{1}{\gamma + 1}u_{yy} = 0. \tag{2.16}$$

Both φ and $\hat{\varphi}$ satisfy boundary conditions (2.13), hence the difference u satisfies the homogeneous conditions

$$\left.\begin{array}{llll} u = 0 & \text{at} & x = 0, & -1 < y < 1, \\ u_y = 0 & \text{at} & y = \pm 1, & 0 < x < l. \end{array}\right\} \tag{2.17}$$

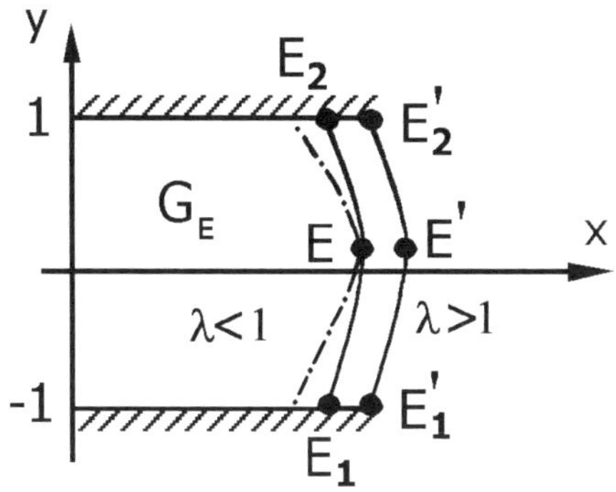

Figure 2.7 A reduced domain in which the condition of positive acceleration is required.

Using the notation

$$k(x, y) = 1 - p\varphi_x - (1-p)\hat{\varphi}_x, \qquad \alpha(x, y) = (1-p)\varphi_{xx} + p\hat{\varphi}_{xx},$$

we cast Equation (2.16) into the form (1.1). Due to the results obtained in Section 1.2, the solution of Problem (2.16), (2.17) is identically zero $u \equiv 0$ if the condition $2\alpha + k_x \equiv (\gamma+1)[(2-3p)\varphi_{xx} + (3p-1)\hat{\varphi}_{xx}] > 0$ is true in the domain G. At $p = 1/3$, the latter condition reduces to the inequality $\varphi_{xx} > 0$, which holds by the assumptions of Theorem 2.1. Thus, we have $u \equiv 0$. Consequently, $\varphi(x, y) \equiv \hat{\varphi}(x, y)$ in G. □

Remark 2.2. The assertion of Theorem 2.1 remains true if the condition of positive acceleration of the given solution $\varphi_{xx} \geqslant \delta > 0$ holds only in the domain G_E bounded by the inlet $x = 0$, $-1 \leqslant y \leqslant 1$, the segments $y = +1$, $0 \leqslant x \leqslant l$, and the characteristic lines EE_1 and EE_2 emanating from the contact point E in the upstream direction (Figure 2.7). However, in this case an extra assumption is to be imposed on the solution $\hat{\varphi}$.

In order to treat the case, one can multiply Equation (2.16) by u_x and integrate the result over G_E. In this way we obtain at $p = 1/3$

$$\int_{G_E} \left[(1 - \tfrac{1}{3}\varphi_x - \tfrac{2}{3}\hat{\varphi}_x)\tfrac{1}{2}(u_x^2)_x - (\tfrac{2}{3}\varphi_{xx} + \tfrac{1}{3}\hat{\varphi}_{xx})u_x^2 + \frac{1}{\gamma+1}u_{yy}u_x \right] dG = 0.$$

Using Green's formula, we find

$$\int_{G_E} \left[(\tfrac{1}{3}\varphi_{xx} + \tfrac{2}{3}\hat{\varphi}_{xx})\tfrac{1}{2}u_x^2 - (\tfrac{2}{3}\varphi_{xx} + \tfrac{1}{3}\hat{\varphi}_{xx})u_x^2 - \frac{1}{\gamma+1}u_y u_{xy} \right] dG$$
$$+ \int_{\partial G_E} (1 - \tfrac{1}{3}\varphi_x - \tfrac{2}{3}\hat{\varphi}_x)\tfrac{1}{2}u_x^2\, dy - \frac{1}{\gamma+1}u_y u_x\, dx = 0.$$

After that, we use Green's formula again to transform the integral of $u_y u_{xy}$ and note that, due to the boundary conditions (2.17), the line integral vanishes over the boundary with the exception of E_1EE_2:

$$-\int_{G_E} \tfrac{1}{2}\varphi_{xx}u_x^2\,dG + \int_{E_1EE_2} (1 - \tfrac{1}{3}\varphi_x - \tfrac{2}{3}\hat{\varphi}_x)\tfrac{1}{2}u_x^2\,dy - \frac{1}{\gamma+1}u_y u_x\,dx - \frac{1}{2(\gamma+1)}u_y^2\,dy = 0.$$

By the definition of the characteristic lines, we have $dx = [(\gamma+1)(\varphi_x - 1)]^{1/2}\,dy$ along E_1E and $dx = -[(\gamma+1)(\varphi_x - 1)]^{1/2}\,dy$ along EE_2. Hence, the last identity can be written in the form

$$-\int_{G_E} \tfrac{1}{2}\varphi_{xx}u_x^2\,dG + \int_{E_1EE_2} \left[(1 - \tfrac{1}{3}\varphi_x - \tfrac{2}{3}\hat{\varphi}_x)\tfrac{1}{2}u_x^2 \pm \frac{1}{\gamma+1}u_y u_x\sqrt{(\gamma+1)(\varphi_x - 1)} - \frac{1}{2(\gamma+1)}u_y^2\right] dy = 0,$$

i.e.

$$-\int_{G_E} \tfrac{1}{2}\varphi_{xx}u_x^2\,dG + \int_{E_1EE_2} \left[\tfrac{1}{3}(\varphi_x - \hat{\varphi}_x)u_x^2 - \frac{1}{2}\left(u_x\sqrt{\varphi_x - 1} \mp \frac{u_y}{\sqrt{\gamma+1}}\right)^2\right] dy = 0. \quad (2.18)$$

If $\varphi_x \leqslant \hat{\varphi}_x$ along E_1EE_2, then all terms on the left-hand side of (2.18) are nonpositive, whence $u \equiv 0$ in G_E. Now, solving a sequence of Goursat problems downstream of E_1EE_2, one can easily establish that $u \equiv 0$ in the whole G. If the inequality $\varphi_x \leqslant \hat{\varphi}_x$ is not true along E_1EE_2, then one can replace the characteristics E_1EE_2 by a line $E_1'E'E_2'$ determined by the equation $dx/dy = (-1)^i\sqrt{(\gamma+1)(\varphi_x/3 + 2\hat{\varphi}_x/3 - 1)}$, where $i = 0$ at $y < y_{E'}$, $i = 1$ at $y > y_{E'}$, and assume that $\varphi_{xx} > 0$ upstream of that line. Then instead of (2.18) we obtain the relation

$$-\int_{G_{E'}} \tfrac{1}{2}\varphi_{xx}u_x^2\,dG - \frac{1}{2}\int_{E_1'E'E_2'} \left(u_x\sqrt{\tfrac{1}{3}\varphi_x + \tfrac{2}{3}\hat{\varphi}_x - 1} \mp \frac{u_y}{\sqrt{\gamma+1}}\right)^2 dy = 0,$$

which again yields $u \equiv 0$.

It is noteworthy that the necessity of prescribing a sole boundary condition at the subsonic inlet in the case of two-dimensional inviscid flow can be motivated by general arguments associated with the concept of characteristics for equations of hyperbolic type. These arguments were often used in numerical studies of boundary-value problems for equations governing inviscid flow (Fletcher 1991, Section 11.6). At the same time, there were doubts whether the uniqueness is true in all feasible cases. Pirumov and Roslyakov (1990), for example, wrote: 'The idea of the nonuniqueness of the critical flux, especially in the case of flow with nonequilibrium physical–chemical reactions appears to be quite plausible.'

Mathematical studies of the direct problem for transonic flow in a nozzle, i.e. the problem with a given geometry of the nozzle, were first undertaken by Nikolsky (1948). He treated small perturbation of a given inviscid transonic flow and derived linearized equations governing the perturbation in the hodograph plane (λ, θ) and also in the plane of streamline coordinates (φ, ψ). Frankl (1959) proved the existence of a weak solution to the linearized problem in the hodograph plane under an arbitrarily prescribed value of the flux variation involved in the boundary conditions. However, any attempts to construct the corresponding flow in the physical plane faced enormous difficulties. As a consequence, the question of the uniqueness and existence of the solution to the direct problem of the Laval nozzle theory was open for a long time. Lifshitz and Ryzhov (1966) proved the uniqueness of the flow flux through a nozzle of a given geometry under the assumption of a fixed asymptotic type of flow in the vicinity of the point E at which the characteristic direction is tangent to the sonic line. From this result it follows immediately that the solution of the linearized direct problem (under a prescribed variation of the shape of walls) is unique. A similar result was obtained by Oryol (1971) with another technique in the nearsonic approximation. Sevost'yanov (1973) studied a direct problem for the third-order equation $\varepsilon(\varphi_{xx} + \varphi_{yy})_x + (\gamma + 1)(1 - \varphi_x)\varphi_{xx} + \varphi_{yy} = 0$ involving the streamline viscosity term. He proved the uniqueness of the solution in the class of flows with positive acceleration using a technique similar to that employed for the proof of Theorem 2.1.

2.2 Solvability of a Nonlinear Perturbation Problem

2.2.1 Particular solutions that describe accelerating flow

A particular exact solution to the von Kármán Equation (2.10) is given by the expression

$$\varphi(x, y) = x + \tfrac{1}{2}Ax^2 + \tfrac{1}{2}A^2(\gamma + 1)xy^2 + \tfrac{1}{24}A^3(\gamma + 1)^2y^4, \tag{2.19}$$

which can be verified by substitution. The parameter $A > 0$ is the flow acceleration at the origin, the sonic line $\varphi_x - 1 = 0$ is the parabola $x + A(\gamma + 1)y^2/2 = 0$. Two characteristic lines $x - A(\gamma + 1)y^2/2 = 0$ and $x + A(\gamma + 1)y^2/4 = 0$ of Equation (2.10) contact the parabola at its vertex E(0, 0). The velocity components are given by $u = 1 + Ax + A^2(\gamma + 1)y^2/2$ and $v = A^2(\gamma + 1)y[x + A(\gamma + 1)y^2/6]$. Solution (2.19) represents an example of flow in a nozzle in the nearsonic approximation if two streamlines $\psi(x, y) = \pm\text{const.}$ are chosen for the walls of the nozzle. Also, solution (2.19) can represent flow in a curved channel, as sketched in Figure 2.4, if two streamlines located below the x-axis are chosen for walls.

A solution to the full system of continuity and irrotationality equations (2.1) governing inviscid two-dimensional flow can be constructed by solving a Cauchy problem with initial data on the x-axis. Meyer (1908) represented a given velocity distribution along the x-axis by a power series in x, and sought the solution as a power series in x and y with indeterminate coefficients. Substitution of the series into the full-potential Equation (2.38) that follows from (2.1) enabled Meyer to obtain recurrence relations for the coefficients. The existence of the solution to the above problem in a small vicinity of the origin follows from the classical Cauchy–Kowalewsky Theorem (Courant and Hilbert 1962).

Frankl (1945), Lighthill (1947) and Cherry (1959, 1960) developed a technique of constructing particular solutions using the hodograph plane in which the governing equations become linear. The equations of motion in the hodograph plane follow from (2.7) if one uses the velocity magnitude λ and the inclination angle θ for independent variables, while φ and ψ for dependent ones. By separating variables, the solution can be obtained as a trigonometrical series in θ with hypergeometric functions as coefficients. If the given data are values of the velocity along the zero streamline, then in the hodograph plane these are equivalent to imposing the coordinate along the boundary $\theta = 0$ as a function of λ, thus prescribing sufficient Cauchy data for the dependent variable. Frankl obtained a two-parametric formula describing flow in the subsonic region

$$\psi(\lambda, \theta) = -\theta - A \sum_{n=1}^{\infty} \lambda^{2nm} \frac{F_{2nm}(\lambda^2(\gamma - 1)/(\gamma + 1))}{F_{2nm}((\gamma - 1)/(\gamma + 1))} \frac{\sin 2nm\theta}{m^{4/3}},$$

where A and m are parameters, $F_\nu(\tau) = F(a_\nu, b_\nu, \nu + 1, \tau)$ is the hypergeometric function,

$$a_\nu + b_\nu = \nu - \frac{1}{\gamma - 1}, \qquad a_\nu b_\nu = -\frac{\nu(\nu + 1)}{2(\gamma - 1)}.$$

Also, Frankl proved that this flow may be continued across the sonic line as a smooth supersonic flow even if the given velocity distribution along the x-axis in the supersonic region is nonanalytic.

In a review of contributions to the nozzle design problem made in 1950s, Halt (1962) wrote:

> ... in a region just downstream from the sonic line, at the throat of the nozzle, the velocity vector takes the same supersonic value at three distinct points, so that this region is mapped three times on the same corresponding region in the hodograph plane. A line of branch points separates this region from the single valued region. Lighthill handles this difficulty by expanding ψ in powers of θ in the region, in which there is a one-to-one correspondence of the hodograph and a domain in the physical plane, and by using an inverse expansion for θ in powers of ψ in the triply covered region...
>
> Cherry considerably improved and simplified the use of the hodograph method in the channel design problem. Firstly, he ingeniously sidesteps the difficulties in the triply covered region by changing the independent

variables from q, θ to q, ϕ, where

$$\theta = \phi - 2\alpha \arctan \frac{q \sin \phi}{1 - q \cos \phi}, \qquad 2\alpha = \left(\frac{\gamma + 1}{\gamma - 1}\right)^{1/2} - 1, \qquad q = \lambda a_*.$$

The Legendre potential Ω (and also the stream function) is a single valued function of q, ϕ so that a one-to-one correspondence exists between the physical plane and the transformed q, ϕ-plane. Cherry is able to construct special solutions of $\Omega, \ldots,$ in terms of hypergeometric functions and superpose these to produce a variety of desirable channel contours. Basically, he takes a linear combination of the fundamental potentials Ω_T, Ω_R, Ω_U corresponding to three types of flow: transonic, radial and uniform, respectively. Each potential is dominant in the region associated with it so that the complete flow field converts a subsonic flow which is approximately radial into a supersonic, asymptotically uniform flow by acceleration through a transonic convergent–divergent nozzle. The formulation is very general and gives the nozzle designer freedom to provide for a wide variety of requirements.

In spite of its high state of development, the hodograph method is quite difficult to apply to axially symmetric flow since the presence of the radial term in the continuity equation destroys the linear character of the equations in the hodograph plane (Cole 1983; Cook *et al.* 1999).

Norton (1972), Pirumov and Roslyakov (1990) developed a numerical method for solving the Cauchy problem with initial data on the centreline (in case of axisymmetric flow) or on the zero streamline (in case of plane flow). If the data are analytic, the Cauchy–Kowalewsky Theorem establishes the existence of the solution in the neighbourhood of the zero streamline or centreline. However, for elliptic equations and arbitrarily specified initial data, the solution of the Cauchy problem may not depend continuously on the data, or numerical instabilities may arise. In order to avoid the instabilities, Norton used the change of variables $(r, x) \rightarrow (\psi(r, x), \xi(x))$, where $\xi = 1 + \tanh(bx)$ is a stretching function, and transformed the governing equations into a form which puts any geometry into a rectangular shape and which spaces the mesh points more finely in regions of the greatest gradients of the flow field. Numerical integration was initiated at the centreline and was continued radially. On the centreline, i.e. x-axis, Norton employed the velocity distribution as follows:

$$\lambda(x, 0) = 1 + \tfrac{1}{2}A[\tanh(Bx) + \tanh(Cx)].$$

Variation of the parameters A, B and C makes it possible to control the key features of the nozzle such as the contraction ratio, the radius of curvature at the throat, and the angle of divergence of the supersonic part at the outlet. However, the contours generated are analytic. Hence, they cannot in general match precisely an *a priori* given contour. Figure 2.8 displays three solutions of the inverse problem with Cauchy data on the centreline at fixed A and different values of the parameters B and C.

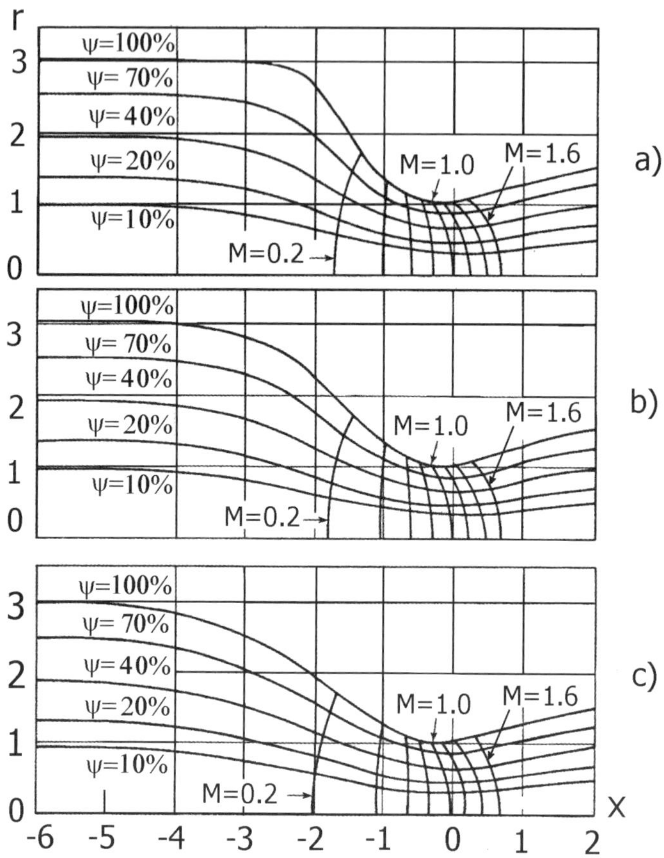

Figure 2.8 Solutions of the inverse problem with initial data on the axis of symmetry. $A = 0.931$, $B + C = 1.4$. (a) $B/C = 1.0$, (b) $B/C = 1.5$, (c) $B/C = 2.0$. Computations of Norton (1972). Copyright © 1972 AIAA. Reproduced with permission.

Pirumov and Roslyakov (1990) represented governing equations in Mises' variables (ξ, ψ) and employed a special finite-difference method for integration of the equations from the axis in the radial direction. The method was based on a specific approximation for the derivatives involved into the equations and special choice of mesh cell sizes. A two-parametric distribution of the flow velocity coefficient was considered along the x-axis

$$\lambda(x, 0) = 1 + \frac{(1 - \lambda_\infty)(\bar{\lambda}_\infty - 1)(e^{bx} - 1)}{(1 - \lambda_\infty)e^{bx} + (\bar{\lambda}_\infty - 1)},$$

where λ_∞, $\bar{\lambda}_\infty$ are the asymptotic values of the velocity λ at infinity upstream and downstream of the origin, respectively, $b > 0$ is a parameter. Analysis of the computed flow fields at various sets of the parameters enabled Pirumov and Roslyakov to make

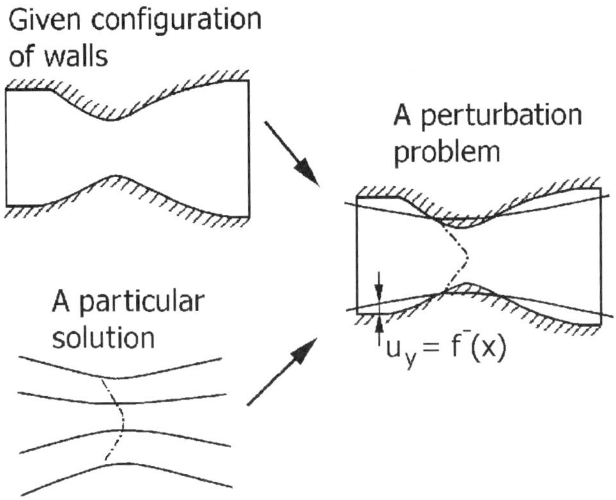

Figure 2.9 Perturbation approach for solving the direct problem.

a few qualitative conclusions and to observe local regions of flow deceleration that appeared under certain choices of λ_∞ and $\bar{\lambda}_\infty$.

2.2.2 A perturbation problem

The available particular solutions to the von Kármán and Chaplygin equations make it possible to employ the perturbation approach for solving the direct problem as follows. For a given configuration of the walls of a channel/nozzle, one can choose an appropriate particular solution $\varphi(x, y)$ with two streamlines located close enough to the given walls (Figure 2.9). Then one can calculate the deviation of the streamlines from the walls and find the boundary condition with respect to the perturbation $u(x, y)$ of $\varphi(x, y)$. This enables one to fix the types of the equation, which are determined by $\varphi(x, y)$ if $u(x, y)$ is small, at the ends of the channel. The perturbation approach was used for a transonic flow first by Taylor (1930a), who considered a potential vortex with concentric circumferential streamlines. Taylor superimposed a harmonic perturbation on the vortex and obtained an ordinary differential equation for the change of the amplitude in radial direction.

As pointed out in the previous section, the direct problem for a channel with prescribed configuration of walls can be reformulated in the nearsonic approximation as the problem of finding a solution $\varphi(x, y)$ to the von Kármán Equation (2.10) in the rectangle $G = \{(x, y) \in \mathbb{R}^2;\ 0 < x < l,\ -1 < y < 1\}$ endowed with the boundary conditions (2.13).

Let $\varphi(x, y)$ be a given particular solution to Equation (2.10). We denote by $\hat{\varphi}(x, y)$ the sought-after solution of Problem (2.10), (2.13). Both $\varphi(x, y)$ and $\hat{\varphi}(x, y)$ satisfy

Equation (2.10), whence the difference

$$u(x, y) = \varphi(x, y) - \hat{\varphi}(x, y)$$

is governed by the equation

$$(1 - \varphi_x + u_x)u_{xx} - \varphi_{xx}u_x + \frac{1}{\gamma + 1}u_{yy} = 0, \tag{2.20}$$

the same as (2.15). The derivative $\hat{\varphi}_y$ on the upper and lower sides $y = \pm 1, 0 < x < l$, of the rectangle is specified by the boundary conditions (2.13) written with respect to $\hat{\varphi}$, while the derivative φ_y of the particular solution $\varphi(x, y)$ is known. Therefore $u_y(x, \pm 1) = \varphi_y(x, \pm 1) - \hat{\varphi}_y(x, \pm 1)$ is known, and the conditions with respect to the perturbation $u(x, y)$ read

$$\left.\begin{aligned} u &= 0 && \text{at the inlet } x = 0, \\ u_y &= f^{\pm}(x) && \text{at } y = \pm 1. \end{aligned}\right\} \tag{2.21}$$

The functions $f^{\pm}$ are perturbations of the angles $\theta^{\pm}(x)$ of the velocity vector inclination to the x-axis on the upper and lower walls of the channel.

Along with (2.20), let us consider the linear equation

$$L_g u := (1 - \varphi_x + g)u_{xx} - \varphi_{xx}u_x + \frac{1}{\gamma + 1}u_{yy} = 0, \tag{2.22}$$

where $g(x, y)$ is a given function. At $g \equiv 0$, Equation (2.22) reduces to

$$L_0 u := [(1 - \varphi_x)u_x]_x + \frac{1}{\gamma + 1}u_{yy} = 0. \tag{2.23}$$

Theorem 2.3. *Let $\varphi \in W^{2,2}(G), \varphi_x(0, y) < 1, \varphi_x(l, y) > 1$, in addition, $\varphi_{xx}(x, y) > 0$ in the domain G, that is the acceleration of the given flow be positive in the nozzle. Then the linearized Problem (2.21), (2.23) can have at most one solution $u \in W^{2,2}(G)$.*

The inequality $\varphi_{xx}(x, y) > 0$ is implied to be true in the sense of 'almost everywhere' in the domain G, as the derivative φ_{xx} is an element of $L^2(G)$. The validity of Theorem 2.3 immediately follows from Theorem 1.7.

Theorem 2.4. *Assume that $\varphi(x, y) \in W^{6,2}(G)$ is a given function such that $\varphi_x(0, y) < 1, \varphi_x(l, y) > 1$, in addition, $\varphi_{xx} \geqslant \delta > 0$ in G. Then for any perturbation $f^{\pm}(x)$ vanishing in the small vicinity of the point $x = 0$ and having sufficiently small norm $\|f^{\pm}\|_{W^{3,2}(0,l)}$, there exists a solution $u \in W^{4,2}(G)$ of the nonlinear perturbation Problem (2.20), (2.21). That solution is unique in the class of functions $\|u\|_{W^{3,2}(G)} < r$, where r depends on $\varphi(x, y)$ and G.*

Proof. The solution will be constructed through the following iteration scheme

$$\left.\begin{aligned} (1 - \varphi_x + g_i)u_{xx}^{(i+1)} - \varphi_{xx}u_x^{(i+1)} + \frac{1}{\gamma + 1}u_{yy}^{(i+1)} = 0, \\ g_i = u_x^{(i)}, \quad i = 0, 1, \ldots, \qquad g_0 \equiv 0, \end{aligned}\right\} \tag{2.24}$$

under the boundary conditions (2.21). The proof is split into four steps.

Step 1. The linear Problem (2.21), (2.22) easily reduces to the problem for the non-homogeneous equation

$$L_g u := (1 - \varphi_x + g)u_{xx} - \varphi_{xx}u_x + \frac{1}{\gamma + 1}u_{yy} = f \in W^{3,2}(G), \tag{2.25}$$

endowed with the homogeneous boundary conditions

$$\left.\begin{aligned} u &= 0 \qquad \text{at the inlet } x = 0, \\ u_y &= 0 \qquad \text{at } y = \pm 1. \end{aligned}\right\} \tag{2.26}$$

Indeed, we can obtain this by substituting $u + \tilde{u}$ for u in (2.21), (2.22), where

$$\tilde{u} = \tfrac{1}{4}[f^+(x)(1 + y)^2 - \tfrac{1}{4}f^-(x)(1 - y)^2] \tag{2.27}$$

is a function satisfying the boundary conditions (2.21).

Equation (2.25) coincides with (1.1) in the special case $k = 1 - \varphi_x + g$, $\alpha = \varphi_{xx}$, $a = 1/(\gamma + 1)$, $c \equiv 0$. The expressions

$$\left.\begin{aligned} 2\alpha + k_x &= \varphi_{xx} + g_x, & 2\alpha - k_x &= 3\varphi_{xx} - g_x, \\ 2\alpha - 3k_x &= 5\varphi_{xx} - 3g_x, & 2\alpha - 5k_x &= 7\varphi_{xx} - 5g_x, \end{aligned}\right\} \tag{2.28}$$

are positive at sufficiently small $|g|$ because $\varphi_{xx} \geqslant \delta > 0$ in G. The condition $\varphi \in W^{6,2}(G)$ yields $\varphi_x \in C^3(\overline{G})$ due to the embedding theorems for the two-dimensional domain (Adams 1975).

Hence, at least for $g \equiv 0$ the coefficients of (2.25) are smooth enough, and all the inequalities required in Theorem 1.7 are satisfied. Consequently, there exists a unique solution $u \in W^{2,2}(G) \cap W^{4,2}(G_{\text{in}})$ to Problem (2.25), (2.26), where $G_{\text{in}} = \{(x, y) \in \mathbb{R}^2;\ \varepsilon < x < l,\ -1 < y < 1\}$, $\varepsilon > 0$ is small enough.

Step 2. The solution $u \in W^{2,2}(G) \cap W^{4,2}(G_{\text{in}})$ obtained at $g \equiv 0$ may in general have singularities at the corner points $x = 0$, $y = \pm 1$ of the domain G. However, the condition of vanishing $f^{\pm}(x)$ in the vicinity of $x = 0$ yields $f(x, y) \equiv 0$ at sufficiently small x and provides $u \in W^{4,2}(G)$. In order to prove the latter, we use the even continuation of $u \in W^{2,2}(G)$ and smooth continuation of the coefficients of (2.25) across the segments $y = \pm 1$, $0 \leqslant x \leqslant \sigma$, where σ is small enough, into domains $\{(x, y) \in \mathbb{R}^2;\ 1 < |y| < 1 + \sigma_1,\ 0 < x < \sigma\}$; see Figure 2.10. By formal differentiation of (2.25) with respect to y, we find that $u_y \in W^{1,2}$ is a weak solution of the Dirichlet problem for the equation of elliptic type $L_0 u_y = \bar{f}$ in the strip $0 < x < \sigma$, $|y| < 1 + \sigma_1$, where $\bar{f}$ is square-integrable. A weak solution of such a problem is necessarily regular (Ladyzhenskaya 1985). Hence, $u_y \in W^{2,2}(G_\sigma)$, $G_\sigma = \{(x, y) \in \mathbb{R}^2;\ 0 < x < \sigma,\ |y| < 1\}$.

Similarly, the second differentiation of (2.25) shows that $u_{yy} \in W^{2,2}(G_\sigma)$. Then Equation (2.25) yields $u_{xx} \in W^{2,2}(G_\sigma)$, consequently, $u \in W^{4,2}(G_\sigma)$. Taking into consideration that $u \in W^{4,2}(G_{\text{in}})$, where G_{in} and G_σ can overlap, we obtain $u \in W^{4,2}(G)$.

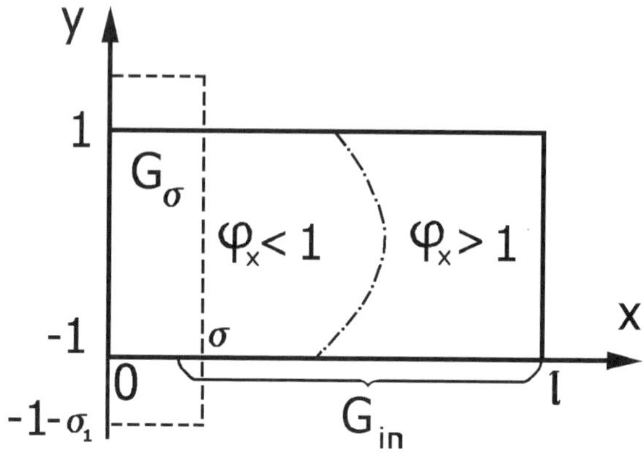

Figure 2.10 An extension of the given domain G used in the proof of the smoothness of $u(x, y)$ in the vicinities of the points $x = 0$, $y = \pm 1$.

Step 3. In the case $g \not\equiv 0$, $g \in W^{3,2}(G)$, which is crucial to the validity of scheme (2.24), the coefficient in front of u_{xx} in (2.25) does not belong to $C^3(\overline{G})$. However, $g \in C^1(\overline{G})$ due to embedding theorems (Adams 1975). The assertion of Theorem 1.7 in the part of the existence of the solution $u \in W^{2,2}(G)$ remains valid at $k \in C^1(\overline{G})$, since estimate (1.23) remains true. The latter is seen from inequality (1.40) which does not involve higher-order derivatives of k. Hence, there exists the solution $u \in W^{2,2}(G)$ of Problem (2.25), (2.26) owing to the arguments of Step 1.

In order to prove that the solution belongs to $W^{4,2}(G)$ in the case $g \in W^{3,2}(G)$, we first validate *a priori* estimates

$$\|u\|_{W^{p,2}(G)} \leqslant m\|L_g u\|_{W^{p-1,2}(G)}, \qquad p = 1, 2, 3, 4, \quad m > 0, \tag{2.29}$$

for sufficiently small $\|g\|_{W^{3,2}(G)}$ and any function $u \in W^{p,2}(G)$ satisfying boundary conditions (2.26) and such that $L_g u \in W^{p-1,2}(G)$, $L_g u \equiv 0$ in the vicinity of $x = 0$. The validity of (2.29) at $p = 1$ follows from the analysis of the integral $\int_G L_g u \cdot u_x e^{\mu x}\, dG$ similar to (1.8). Let us prove that (2.29) is true at $p = 4$ if it holds at $p = 1, 2, 3$. The condition $\varphi_x < 1$ at $x = 0$ provides $\varphi_x < 1$ in the strip $0 \leqslant x \leqslant 4\sigma$, $|y| \leqslant 1$ at sufficiently small $\sigma > 0$. Consider a nonnegative cut-off function $\eta(x) \in C^\infty[0, l]$ vanishing at $x \leqslant \sigma$ and equal to unity at $x \geqslant 2\sigma$. An inspection shows that the function $\tilde{u} = \eta u$ satisfies the equation

$$L_g\tilde{u} = \tilde{f} := \eta L_g u + (1 - \varphi_x + g)(\eta_{xx}u + 2\eta_x u_x) - \varphi_{xx}\eta_x u \in W^{3,2}(G).$$

By differentiating $L_g\tilde{u} = \tilde{f}$ twice with respect to x, we obtain

$$\begin{aligned}(1 - \varphi_x + g)(\tilde{u}_{xx})_{xx} - 3\varphi_{xx}(\tilde{u}_{xx})_x + (\tilde{u}_{xx})_{yy}/(\gamma + 1)\\ = \tilde{f}_{xx} - g_{xx}\tilde{u}_{xx} - 2g_x\tilde{u}_{xxx} + 3\varphi_{xxx}\tilde{u}_{xx} + \varphi_{xxxx}\tilde{u}_x.\end{aligned} \tag{2.30}$$

Hence, $\tilde{u}_{xx}$ can be considered as a solution of the second-order equation that only differs from (2.25) by the multiplier 3 in front of φ_{xx}, which does not influence the validity of (2.29). In addition, $\tilde{u}_{xx}$ satisfies the boundary conditions (2.26), because $\tilde{u}_{xx} = 0$ at $x \leqslant \sigma$. That is why one can use (2.29) with $p = 2$ in order to estimate $\|\tilde{u}_{xx}\|_{W^{2,2}(G)}$ by the right-hand side of (2.30),

$$\|\tilde{u}_{xx}\|_2 \leqslant m\|\tilde{f}_{xx} - g_{xx}\tilde{u}_{xx} - 2g_x\tilde{u}_{xxx} + 3\varphi_{xxx}\tilde{u}_{xx} + \varphi_{xxxx}\tilde{u}_x\|_1, \tag{2.31}$$

where $\|\cdot\|_p$ denotes $\|\cdot\|_{W^{p,2}(G)}$ for brevity. Due to Remark 2.5 below, the second and third terms on the right-hand side of (2.31) can be estimated as follows,

$$\|g_{xx}\tilde{u}_{xx}\|_1 \leqslant m\|g_{xx}\|_1\|\tilde{u}_{xx}\|_2 \leqslant m\|g\|_3\|\tilde{u}_{xx}\|_2$$

and

$$\|g_x\tilde{u}_{xxx}\|_1 \leqslant m\|g_x\|_2\|\tilde{u}_{xxx}\|_1 \leqslant m\|g\|_3\|\tilde{u}_{xx}\|_2,$$

where $m > 0$ may as usual be different in different formulae, meanwhile, it does not depend on $u(x, y)$. Therefore, under sufficiently small $\|g\|_3$, inequality (2.31) reduces to

$$\|\tilde{u}_{xx}\|_2 \leqslant m\|\tilde{f}_{xx} + 3\varphi_{xxx}\tilde{u}_{xx} + \varphi_{xxxx}\tilde{u}_x\|_1. \tag{2.32}$$

Since $\varphi_{xxx} \in C^1(\overline{G})$, the second and third terms on the right-hand side of (2.32) can be estimated by (2.29) with $p = 3$ and $p = 2$. Then the equation $L_g\tilde{u} = \tilde{f}$ makes it possible to estimate $\|\tilde{u}_{yy}\|_2$ by $\|\tilde{u}_{xx}\|_2$, and we arrive at $\|\tilde{u}\|_4 \leqslant m\|\tilde{f}\|_3$. Recalling that $\eta = 1$ and $\tilde{u} = u$ at $x \geqslant 2\sigma$, we obtain

$$\|u\|_{W^{4,2}(G\cap(x>2\sigma))} \leqslant m\|\tilde{f}\|_3.$$

A similar inequality

$$\|u\|_{W^{4,2}(G\cap(x<3\sigma)} \leqslant m\|L_g u\|_3$$

holds due to the elliptic type of the operator $L_g u$ at $0 \leqslant x \leqslant 4\sigma$ and the considerations of Step 2. By combining the two inequalities, we get (2.29) with $p = 4$.

Remark 2.5. The estimate $\|uv\|_1 \leqslant m\|u\|_2\|v\|_1$ is true if $u \in W^{2,2}(G)$, $v \in W^{1,2}(G)$.

Indeed,

$$\begin{aligned}
\|uv\|_1^2 &= \int_G [(u_x v + uv_x)^2 + (u_y v + uv_y)^2 + u^2v^2]\,\mathrm{d}G \\
&\leqslant 2\int_G (u_x^2 v^2 + u^2 v_x^2 + u_y^2 v^2 + u^2 v_y^2)\,\mathrm{d}G + \int_G u^2v^2\,\mathrm{d}G \\
&\leqslant 2\int_G (u_x^2 + u_y^2)v^2\,\mathrm{d}G + 2(\max_G u)^2\|v\|_1^2 \\
&\leqslant 2\|u_x^2 + u_y^2\|_0\|v^2\|_0 + m\|u\|_2^2\|v\|_1^2 \leqslant m\|u\|_2^2\|v\|_1^2,
\end{aligned}$$

owing to multiplicative inequalities (Adams 1975; Ladyzhenskaya 1985) which yield for a rectangle G:

$$\|v^2\|_0 = \|v\|^2_{L^4(G)} \leqslant m\|v\|_1\|v\|_0 \leqslant m\|v\|^2_1.$$

Consequently,

$$\|u_x^2 + u_y^2\|_0 \leqslant m\|u\|_2^2,$$

where the constants $m > 0$ are independent of u and v.

Now, we use (2.29) in order to prove that the solution $u \in W^{2,2}(G)$ of Problem (2.25), (2.26) belongs to $W^{4,2}(G)$. Let us approximate g in the norm $\|\cdot\|_{W^{3,2}(G)}$ by a sequence $g_j \in C^3(\overline{G})$, $j = 1, 2, \ldots$. The equations

$$(1 - \varphi_x + g_j)u_{xx} - \varphi_{xx}u_x + \frac{1}{\gamma+1}u_{yy} = f, \tag{2.33}$$

endowed with the boundary conditions (2.26) have solutions $u_j \in W^{4,2}(G)$ due to the result of Step 2. These solutions are bounded in the norm $\|\cdot\|_4$ owing to (2.29). Hence, there exists a subsequence with a weak limit $\hat{u} \in W^{4,2}(G)$. On the other hand, the sequence u_j converges to the solution $u \in W^{2,2}(G)$, because by subtracting (2.33) from (2.25), we obtain $L_g(u - u_j) = (g_j - g)u_{jxx}$. Therefore,

$$\|u - u_j\|_2 \leqslant m\|L_g(u - u_j)\|_1 = m\|(g_j - g)u_{jxx}\|_1 \to 0$$

as $j \to \infty$. That is why $u = \hat{u} \in W^{4,2}(G)$.

Step 4. The scheme (2.24) implies solving a sequence of Problems (2.24), (2.21) for

$$g_i \in B_r = \{g \in W^{3,2}(G);\ \|g\|_3 < r\}.$$

On account of Step 3, the solutions $u^{(i+1)}$ exist if the radius r of the ball B_r is small enough. Then the scheme (2.24) can be represented in the form $g_{i+1} = Tg_i$, in which the operator Tg is defined in the following way: $\forall g \in B_r$, $Tg = u_x$, where u is the solution of equation $L_g u = 0$ under the boundary conditions (2.21). Consequently, the nonlinear Problem (2.21), (2.22) can be rewritten as $g = Tg$.

Under sufficiently small $\|f^{\pm}\|$, the operator Tg maps the ball B_r into B_r, because

$$\begin{aligned}\|Tg\|_3 &= \|u_x\|_3 \leqslant \|u_x - \tilde{u}_x\|_3 + \|\tilde{u}_x\|_3\\ &\leqslant m\|f\|_3 + \|\tilde{u}_x\|_3 \leqslant m\||f^+| + |f^-|\|_{W^{3,2}(0,l)} < r.\end{aligned} \tag{2.34}$$

where $\tilde{u}$ is defined by (2.27). Let us prove that the operator Tg is a contracting one in the norm $\|\cdot\|_1$ if the radius r is small enough. Due to (2.29) with $p = 2$, we obtain for two elements g_1 and g_2:

$$\|Tg_1 - Tg_2\|_1 = \|u_{1x} - u_{2x}\|_1 \leqslant \|u_1 - u_2\|_2 \leqslant m\|L_{g_1}(u_1 - u_2)\|_1. \tag{2.35}$$

From Equation (2.22) we find $L_{g_1}u_1 - L_{g_2}u_2 = L_{g_1}(u_1 - u_2) + (g_1 - g_2)u_{2xx} = 0$. Therefore, the right-hand side of (2.35) can be estimated by

$$\begin{aligned}m\|(g_1 - g_2)u_{2xx}\|_1 &\leqslant m\|g_1 - g_2\|_1 \cdot \|u_{2xx}\|_2\\ &\leqslant m\|g_1 - g_2\|_1 \cdot \|u_{2x}\|_3 \leqslant mr\|g_1 - g_2\|_1\end{aligned}$$

owing to (2.34). Hence, the operator Tg is a contracting one at $r < 1/m$. Then the principle of contracting mappings shows that the successive approximations $g_{i+1} = Tg_i$, $i = 0, 1, \ldots$, converge to the unique solution $g \in W^{1,2}(G)$ of the equation $g = Tg$. On the other hand, owing to the weak compactness of B_r, there exists a subsequence which weakly converges to $\hat{g} \in W^{3,2}(G)$. Due to the uniqueness, one gets $g = \hat{g} \in W^{3,2}(G)$. Since $g = u_x$, the solution of Problem (2.21), (2.22) reads $u = \int_0^x g\,\mathrm{d}x$. □

2.3 The Full-Potential and the Chaplygin Equations

2.3.1 Formulation of a problem

System (2.1) governing steady, two-dimensional, isentropic, potential flow of inviscid gas reduces to the classical full-potential equation as follows. The derivatives ρ_x and ρ_y of the flow density can be expressed in terms of the velocity components by differentiation of the Bernoulli Equation (2.2):

$$\rho_x = -\rho^{2-\gamma}(uu_x + vv_x), \qquad \rho_y = -\rho^{2-\gamma}(uu_y + vv_y). \tag{2.36}$$

Substitution of (2.36) into the continuity equation $\rho(u_x + v_y) + \rho_x u + \rho_y v = 0$ yields

$$\begin{aligned} \rho^{\gamma-1}(u_x + v_y) - u(uu_x + vv_x) - v(uu_y + vv_y) = 0, \\ (\rho^{\gamma-1} - u^2)u_x - uv(u_y + v_x) + (\rho^{\gamma-1} - v^2) = 0, \end{aligned} \tag{2.37}$$

where $\rho^{\gamma-1} = a^2$ is the squared ratio of the speed of sound to the critical speed a_*. Since $u = \varphi_x$ and $v = \varphi_y$, from (2.37) we obtain the full-potential equation

$$(a^2 - \varphi_x^2)\varphi_{xx} - 2\varphi_x\varphi_y\varphi_{xy} + (a^2 - \varphi_y^2)\varphi_{yy} = 0. \tag{2.38}$$

The presence of the mixed derivative φ_{xy} would not complicate the mathematical study significantly due to arguments of Section 1.5. However, Equation (2.38) is difficult for the analysis of boundary-value problems because of complications caused by the curved boundaries of domain of interest in the physical plane (x, y).

In order to avoid treatment of the curved boundaries, one can map the given domain onto a rectangle by introducing orthogonal coordinates (ξ, η) in such a way that lines $\eta = \eta_1$ and $\eta = \eta_2$ correspond to the lower and upper walls of a channel, respectively. Then the continuity equation becomes

$$\left(\frac{\rho E_1}{J}\varphi_\xi\right)_\xi + \left(\frac{\rho E_3}{J}\varphi_\eta\right)_\eta = 0, \tag{2.39}$$

where

$$E_1 = \xi_x^2 + \xi_y^2, \qquad E_3 = \eta_x^2 + \eta_y^2, \qquad J = \xi_x\eta_y - \xi_y\eta_x,$$
$$\rho = [1 + \tfrac{1}{2}(\gamma - 1)(1 - E_1\varphi_\xi^2 - E_3\varphi_\eta^2)]^{1/(\gamma-1)}.$$

We refer the reader to Fletcher (1991, Section 14.3) or Holst (1999) for details. However, the drawback of Equation (2.39) is that second-order derivatives of φ are hidden in ρ_ξ and ρ_ξ in addition to those represented explicitly. An attempt to collect all the terms with the second-order derivatives, which determine the type of the equation, and to transform (2.39) to a form of the full-potential Equation (2.38) results in a complex equation which appears quite difficult to study mathematically.

The above-mentioned difficulties do not arise if one employs the streamline coordinates (φ, ψ). By eliminating the angle $\theta = \arctan(\varphi_y/\varphi_x)$ from the system of Chaplygin Equations (2.7), we arrive at the second-order equation

$$\left[\frac{(\lambda\rho)_\lambda}{\lambda\rho^2}\lambda_\varphi\right]_\varphi + \left[\frac{\rho}{\lambda}\lambda_\psi\right]_\psi = 0, \tag{2.40}$$

where the dependence of ρ on λ is given by the Bernoulli law (2.2):

$$\rho(\lambda) = [1 + \tfrac{1}{2}(\gamma - 1)(1 - \lambda^2)]^{1/(\gamma-1)}.$$

We notice that the coordinates (φ, ψ) or (x, ψ) were used in many numerical studies of transonic flow. Employment of these coordinates makes it possible to avoid the grid generation step in computations (An and Barron 1995; Latypov 1993; Verhoff 1999).

Equation (2.40) can be treated in a very simple domain, a rectangle in the plane of the streamline coordinates, if one studies, for example, a flow in a channel. However, Equation (2.40) with respect to $\lambda(\varphi, \psi)$ is also inconvenient for the mathematical analysis because of the boundary condition $\lambda_\varphi = 0$ prescribed at the inlet $\varphi = \varphi_1$, which follows from $\theta(\varphi_1, \psi) = 0$ due to the continuity equation in the form represented by the first equation of the system (2.7): $\theta_\psi + \lambda_\varphi(\lambda\rho)_\lambda/(\lambda\rho^2) = 0$. The Neumann condition at the inlet does not permit one to use results of Chapter 1 on the *unique* solvability of boundary-value problems, as they were obtained for the case when the Dirichlet condition was imposed on the left-hand side of the rectangle.

In order to gain the Dirichlet condition at $\varphi = \varphi_1$, we pass to a function $\Phi(\varphi, \psi)$, a generalized velocity potential in the plane (φ, ψ), defined with the relations

$$\Phi_\varphi = 1 + \int_1^\lambda \frac{\rho(\lambda)}{\lambda}\,\mathrm{d}\lambda, \qquad \Phi_\psi = \theta. \tag{2.41}$$

This definition is evidently correct as $\Phi_{\varphi\psi} = \Phi_{\psi\varphi}$ due to the second equation of system (2.7). There is a one-to-one correspondence between Φ_φ and λ, as sketched in Figure 2.11a. Owing to (2.41), the first equation of system (2.7) yields the equation

$$k(\Phi_\varphi)\Phi_{\varphi\varphi} + \Phi_{\psi\psi} = 0, \tag{2.42}$$

where

$$k(\Phi_\varphi) = \tfrac{1}{2}(\gamma + 1)(1 - \lambda^2)\rho^{-(\gamma+1)}. \tag{2.43}$$

Clearly, $k(1) = 0$, and

$$\frac{\mathrm{d}k}{\mathrm{d}\Phi_\varphi} = \frac{\mathrm{d}k}{\mathrm{d}\lambda}\frac{\mathrm{d}\lambda}{\mathrm{d}\Phi_\varphi} = -(\gamma + 1)\lambda^4\rho^{-2\gamma-1} < 0; \tag{2.44}$$

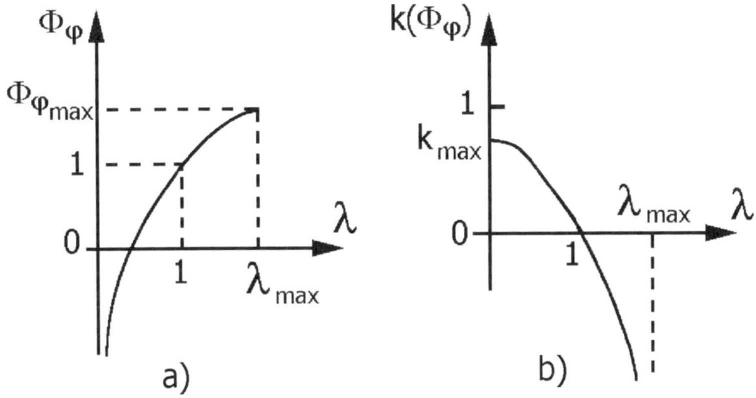

Figure 2.11 Plots of functions defined by (2.41) and (2.43): $\lambda_{\max} = ((\gamma + 1)/(\gamma - 1))^{1/2}$, $k_{\max} = (2/(\gamma + 1))^{2/(\gamma - 1)}$.

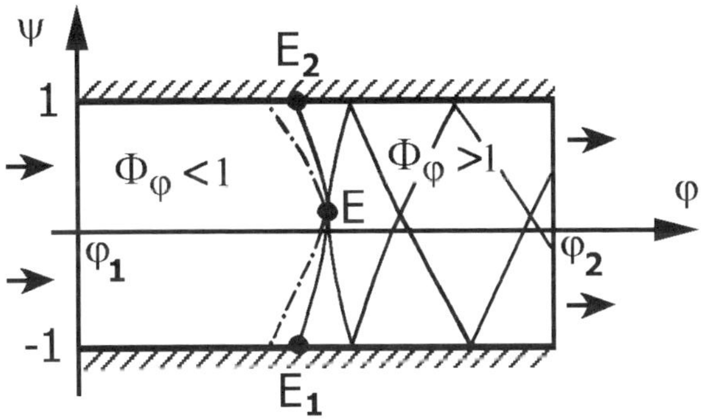

Figure 2.12 A domain in which the boundary-value problem (2.42), (2.46) is formulated.

the monotonous dependence of k on λ is illustrated in Figure 2.11b. Equation (2.42) is of hyperbolic type in the subdomain of supersonic flow, where $\lambda > 1$, $\Phi_\varphi > 1$, $k(\Phi_\varphi) < 0$, and of elliptic type in the subdomains of subsonic flow, where $\lambda < 1$, $\Phi_\varphi < 1$, $k(\Phi_\varphi) > 0$. In the nearsonic approximation $\lambda - 1 = o(1)$, we obtain $\Phi_\varphi \simeq \lambda$, $k \simeq (\gamma + 1)(1 - \Phi_\varphi)$. Consequently, (2.42) reduces to the von Kármán Equation (2.10) with respect to $\Phi(\varphi, \psi)$.

Another form of Equation (2.42) is

$$[K(\Phi_\varphi)]_\varphi + \Phi_{\psi\psi} = 0, \quad \text{where } K = \int_1^{\Phi_\varphi} k(\Phi_\varphi)\, \mathrm{d}\Phi_\varphi. \tag{2.45}$$

In the nearsonic approximation, we obtain $K(\Phi_\varphi) \simeq -(\gamma + 1)(1 - \Phi_\varphi)^2/2$.

Formulation of the problem for transonic flow through a nozzle in the plane (φ, ψ) is as follows. Find a solution $\Phi(\varphi, \psi)$ of Equation (2.42) or (2.45) in the rectangle

$G = \{(\varphi, \psi) \in \mathbb{R}^2;\ \varphi_1 < \varphi < \varphi_2,\ -1 < \psi < 1\}$ (Figure 2.12) under the boundary conditions:

$$\left.\begin{aligned} \Phi = h(\psi),\ \Phi_\varphi < 1 \qquad & \text{at the inlet } \varphi = \varphi_1, \\ \Phi_\psi = \theta^\pm(\varphi) \qquad & \text{on the walls } \psi = \pm 1, \\ \Phi_\varphi > 1 \qquad & \text{at the outlet } \varphi = \varphi_2. \end{aligned}\right\} \tag{2.46}$$

Since we consider shock-free flow, no Rankine–Hugoniot conditions are to be fulfilled in the flow field.

The Neumann condition prescribed on the walls defines *a posteriori* the shape of the walls in the plane (x, y). On the upper wall $\psi = 1$, for example, the dependence of x on φ is determined by the relation:

$$\frac{\mathrm{d}\varphi}{\mathrm{d}x} = \varphi_x + \varphi_y \frac{\mathrm{d}y}{\mathrm{d}x} = \lambda\cos\theta + \lambda\sin\theta\tan\theta$$
$$= \frac{\lambda}{\cos\theta} \quad \Rightarrow \quad x(\varphi) = \int_{\varphi_1}^{\varphi} \frac{\cos\theta^+(\varphi)}{\lambda(\varphi, 1)}\,\mathrm{d}\varphi, \qquad \theta^+(\varphi) = \theta(\varphi, 1).$$

Similarly, the dependence of y on x is given by

$$y(\varphi) = 1 + \int_{\varphi_1}^{\varphi} \frac{\sin\theta^+(\varphi)}{\lambda(\varphi, 1)}\,\mathrm{d}\varphi.$$

In order to study Problem (2.42), (2.46) for flow in a nozzle, we employ the perturbation approach as in Section 2.2. We assume that $\Phi(\varphi, \psi)$ is a known solution of Problem (2.42), (2.46) and shall prove the existence of a solution $\hat{\Phi}(\varphi, \psi)$ under a given small perturbation $f^\pm(\varphi)$ of the angle $\theta^\pm(\varphi) = \Phi_\psi(\varphi, \pm 1)$, which determines the shape of walls in the physical plane.

The perturbation

$$u = \Phi - \hat{\Phi}$$

of the flow field is governed by the equation obtained by subtracting Equation (2.42) written in terms of $\hat{\Phi}$ from that written for Φ:

$$k(\Phi_\varphi)\Phi_{\varphi\varphi} - k(\Phi_\varphi - u_\varphi)\cdot(\Phi_{\varphi\varphi} - u_{\varphi\varphi}) + u_{\psi\psi} = 0. \tag{2.47}$$

In view of (2.46), the boundary conditions with respect to the perturbation $u(\varphi, \psi)$ are

$$\left.\begin{aligned} u = 0 \qquad & \text{at} \quad \varphi = \varphi_1, \quad -1 \leqslant \psi \leqslant 1, \\ u_\psi = f^\pm(\varphi) \quad & \text{at} \quad \psi = \pm 1, \quad \varphi_1 \leqslant \varphi \leqslant \varphi_2. \end{aligned}\right\} \tag{2.48}$$

2.3.2 Solvability of the nonlinear perturbation problem

Theorem 2.6. *If there exists a solution $\Phi(\varphi, \psi) \in W^{6,2}(G)$ of Problem (2.42), (2.46) such that $\Phi_{\varphi\varphi} \geqslant \delta > 0$, i.e. the flow acceleration is positive in the nozzle, and*

$$\max_G\{\lambda(\varphi, \psi)\} \leqslant \lambda_{\max} - \varepsilon, \quad \textit{where } \lambda_{\max} = \sqrt{\frac{\gamma+1}{\gamma-1}}, \qquad \varepsilon > 0,$$

then for any perturbation $f^{\pm}(\varphi)$ of $\theta^{\pm}(\varphi)$ vanishing in the small vicinity of $\varphi = \varphi_1$ and having sufficiently small norm $\|f^{\pm}\|_{W^{3,2}(\varphi_1,\varphi_2)}$ there exists a unique solution $u(\varphi,\psi) \in W^{4,2}(G)$ of Problem (2.47), (2.48).

Proof. Let Φ be the solution pointed out in Theorem 2.6.

Step 1. The expression

$$\begin{aligned} k'(\Phi_\varphi, u_\varphi) &= [k(\Phi_\varphi - u_\varphi) - k(\Phi_\varphi)]/u_\varphi, \quad && \text{for } u_\varphi \neq 0, \\ &= -\frac{\mathrm{d}k}{\mathrm{d}\Phi_\varphi}, && \text{for } u_\varphi = 0, \end{aligned}$$

is a continuous function of u_φ since $\mathrm{d}k/\mathrm{d}\Phi_\varphi$ is finite at $0 \leqslant \lambda < \lambda_{\max} - \varepsilon$ (Figure 2.11b). Using $k'(\Phi_\varphi, u_\varphi)$, one can represent Equation (2.47) in the form

$$k(\Phi_\varphi)\Phi_{\varphi\varphi} - [k(\Phi_\varphi) + k'(\Phi_\varphi, u_\varphi)\cdot u_\varphi](\Phi_{\varphi\varphi} - u_{\varphi\varphi}) + u_{\psi\psi} = 0,$$

i.e.

$$[k(\Phi_\varphi) + k'(\Phi_\varphi, u_\varphi)\cdot u_\varphi]u_{\varphi\varphi} - k'(\Phi_\varphi, u_\varphi)\Phi_{\varphi\varphi}u_\varphi + u_{\psi\psi} = 0. \tag{2.49}$$

Along with this equation, let us consider the linear equation

$$L_g u := [k(\Phi_\varphi) + k'(\Phi_\varphi, g)\cdot g]u_{\varphi\varphi} - k'(\Phi_\varphi, g)\Phi_{\varphi\varphi}u_\varphi + u_{\psi\psi} = 0, \tag{2.50}$$

where $g(\varphi,\psi)$ is a given function. We note that $k' \simeq (\gamma + 1)$ in the nearsonic approximation, cf. (2.22).

The linear Problem (2.48), (2.50) can be easily reduced to the problem

$$L_g u = f, \tag{2.51}$$

$$\left.\begin{aligned} u &= 0 \quad \text{at} \quad \varphi = \varphi_1, \quad && -1 \leqslant \psi \leqslant 1, \\ u_\psi &= 0 \quad \text{at} \quad \psi = \pm 1, && \varphi_1 \leqslant \varphi \leqslant \varphi_2, \end{aligned}\right\} \tag{2.52}$$

with the homogeneous boundary conditions and the right-hand side $f \in W^{3,2}(G)$. Since $\Phi \in W^{6,2}(G)$, we obtain $\Phi_\varphi, k(\Phi_\varphi) \in W^{5,2}(G)$. Therefore, $\Phi_\varphi, k(\Phi_\varphi) \in C^3(\overline{G})$ due to the embedding theorems for the two-dimensional domain.

Equation (2.51) coincides with (1.1) if one employs appropriate notation for the coefficients of the equation. Then condition $2\alpha \pm k_x \geqslant \delta > 0$ (see (1.51)) reads

$$2k'(\Phi_\varphi, g)\Phi_{\varphi\varphi} \pm [k(\Phi_\varphi) - k'(\Phi_\varphi, g)g]_\varphi \geqslant \delta > 0.$$

The left-hand side of this inequality is arbitrarily close to $2k'(\Phi_\varphi, 0) \pm k_\varphi(\Phi_\varphi)$ if $|g|$ and $|g_\varphi|$ are small enough. The latter expression is positive, since

$$k'(\Phi_\varphi, 0) = -k_\varphi(\Phi_\varphi) = -\frac{\mathrm{d}k}{\mathrm{d}\Phi_\varphi}\Phi_{\varphi\varphi} > 0$$

due to (2.44) and the condition of positive acceleration $\Phi_{\varphi\varphi} > 0$. Therefore, if $g \in C^3(\overline{G})$ is sufficiently small, then Theorem 1.7 establishes the existence of the unique solution $u \in W^{2,2}(G) \cap W^{4,2}(G_{\text{in}})$ to the linear Problem (2.51), (2.52).

Step 2. The condition of vanishing $f^{\pm}(\varphi)$ in the vicinity of $\varphi = \varphi_1$ ensures that the solution $u(\varphi, \psi)$ belongs to $W^{4,2}$ over vicinities of the corner points $\varphi = \varphi_1$, $\psi = \pm 1$ of the domain G. This can be proved by continuation of the equation into an extended domain across the segments $\varphi_1 \leqslant \varphi \leqslant \varphi_1 + \sigma$, $\psi = \pm 1$ and by treating a boundary-value problem for an elliptic equation in the same way as in Section 2.2.

Step 3. If $g \in W^{3,2}(G)$ then there exists a solution $u(\varphi, \psi) \in W^{2,2}(G)$ owing to the arguments of Step 1. In order to prove that $u(\varphi, \psi)$ belongs to $W^{4,2}(G)$ we have to validate estimates (2.29). As in Section 2.2, this can be done using the relation

$$\begin{aligned}(k + k'g) \cdot (u_{\varphi\varphi})_{\varphi\varphi} &+ [2(k + k'g)_\varphi - k'\Phi_{\varphi\varphi}] \cdot (u_{\varphi\varphi})_\varphi + (u_{\varphi\varphi})_{\psi\psi} \\ &= -k_{\varphi\varphi} u_{\varphi\varphi} + (k'g)_{\varphi\varphi} u_{\varphi\varphi} + \cdots + f_{\varphi\varphi}, \end{aligned} \qquad (2.53)$$

obtained by differentiating (2.51) twice with respect to φ. At sufficiently small $|g|$ and $|g_\varphi|$, the coefficient $-2(k + k'g)_\varphi + k'\Phi_{\varphi\varphi}$ is arbitrarily close to

$$-2k_\varphi + k'\Phi_{\varphi\varphi} = \left[-2\frac{\mathrm{d}k}{\mathrm{d}\Phi_\varphi} + k'\right]\Phi_{\varphi\varphi} > 0.$$

If $k'g \in W^{3,2}(G)$, then the $W^{1,2}$-norm of the term $(k'g)_{\varphi\varphi}u_{\varphi\varphi}$ on the right-hand side of (2.53) can be estimated by $\|k'g\|_3 \cdot \|u_{\varphi\varphi}\|_2$, where $\|u\|_n$ denotes $\|u\|_{W^{n,2}(G)}$ for brevity. This enables one to estimate the norm $\|u_{\varphi\varphi}\|_2$ of the solution of (2.53) by the $W^{1,2}$-norm of the right-hand side due to (2.29) at $p = 2$.

Step 4. We construct the solution of the nonlinear Problem (2.42), (2.46) through the iteration scheme

$$\left.\begin{aligned}[k(\Phi_\varphi) + k'(\Phi_\varphi, g_i) \cdot g_i]u_{\varphi\varphi}^{(i+1)} - k'(\Phi_\varphi, g_i)\Phi_{\varphi\varphi}u_\varphi^{(i+1)} + u_{\psi\psi}^{(i+1)} = 0,\\ g_i = u_x^{(i)}, \quad i = 0, 1, \ldots, \qquad g_0 \equiv 0.\end{aligned}\right\} \qquad (2.54)$$

On account of Step 3, the solutions $u^{(i+1)}$ exist if $\|g_i\|_{W^{3,2}(G)}$ is small enough. The scheme (2.54) can be represented in the form $g_{i+1} = Tg_i$, in which the operator Tg from $W^{3,2}(G)$ into $W^{3,2}(G)$ is defined in the following way: $Tg = u_\varphi$, where u is the solution of equation $L_g u = 0$ under conditions (2.48). Therefore, the nonlinear Problem (2.48), (2.49) can be rewritten in the form $Tg = g$.

The operator Tg is a contracting one in the norm of $W^{1,2}(G)$ under sufficiently small $\|g\|_3 \leqslant r$, as Equation (2.50) yields

$$\begin{aligned}&L_{g_1}u_1 - L_{g_2}u_2 \\ &\quad = L_{g_1}(u_1 - u_2) + [k'(g_1)g_1 - k'(g_2)g_2]u_{2\varphi\varphi} + [k'(g_2) - k'(g_1)]\Phi_{\varphi\varphi}u_{2\varphi} \\ &\quad = 0,\end{aligned}$$

hence,

$$\begin{aligned}\|Tg_1 - Tg_2\|_1 &= \|u_{1\varphi} - u_{2\varphi}\|_1 \leqslant \|u_1 - u_2\|_2 \leqslant c\|L_{g_1}(u_1 - u_2)\|_1 \\ &= c\|[k'(g_1)g_1 - k'(g_2)g_2]u_{2\varphi\varphi} + [k'(g_2) - k'(g_1)]\Phi_{\varphi\varphi}u_{2\varphi}\|_1 \\ &\leqslant c\|u_{2\varphi\varphi}\|_2 \cdot \|g_1 - g_2\|_1 \leqslant cr\|g_1 - g_2\|_1,\end{aligned}$$

where the positive constants c are independent of u. In addition, if $f^{\pm}$ is sufficiently small, then the operator T maps the ball $B_r \subset W^{3,2}(G)$ of the radius r into B_r, since from (2.29) at $p = 4$ one finds

$$\|u_\varphi\|_3 \leqslant cr\| |f^+| + |f^-| \|_{W^{3,2}(\varphi_1,\varphi_2)} < r.$$

Therefore, the principle of contracting mappings establishes that the successive approximations $g_{i+1} = Tg_i, i = 0, 1, \ldots,$ converge to the unique solution $g \in W^{1,2}(G)$ of the equation $g = Tg$. On the other hand, due to the weak compactness of the ball B_r, there exists a subsequence which weakly converges to $\hat{g} \in W^{3,2}(G)$. Owing to the uniqueness, one gets $g = \hat{g} \in W^{3,2}(G)$. Then $u(\varphi, \psi) = \int_{\varphi_1}^{\varphi} g \, \mathrm{d}\varphi$ is the solution of Problem (2.47), (2.48). □

We note that at $g = 0$ the linear Equation (2.50) reduces to

$$k(\Phi_\varphi)u_{\varphi\varphi} + k'(\Phi_\varphi, 0)\Phi_{\varphi\varphi}u_\varphi + u_{\psi\psi} = 0,$$

that is,

$$[k(\Phi_\varphi)u_\varphi]_\varphi + u_{\psi\psi} = 0, \tag{2.55}$$

where $k(\Phi_\varphi)$ is defined by (2.43). Equation (2.55) with respect to the perturbation $u(\varphi, \psi)$ of the potential $\Phi(\varphi, \psi)$ was actually obtained by formal linearization of (2.42). This is not quite correct from the physical point of view because the quantities $\varphi(x, y)$ and $\psi(x, y)$, which play the role of independent variables in Equation (2.42), are subjected to small perturbations in the plane (x, y) as well as $\Phi(x, y)$.

A physically consistent linear equation governing perturbation $u(\varphi, \psi)$ of the velocity potential in the streamline coordinates determined by a given particular solution of system (2.5) reads

$$\left[\frac{(\rho\lambda)_\lambda}{\rho}u_\varphi\right]_\varphi + (\rho^2 u_\psi)_\psi = 0. \tag{2.56}$$

This equation was obtained by Nikolsky (1948) as follows. From the first and the second equations of system (2.5), replacing φ by $\varphi + \varphi'$, λ by $\lambda + \lambda'$, θ by $\theta + \theta'$, where the prime denotes small perturbations, we obtain

$$\begin{aligned}
\varphi_x + \varphi'_x &= (\lambda + \lambda')\cos(\theta + \theta') \\
&= \lambda\cos\theta + \lambda'\cos\theta - \theta'\lambda\sin\theta + o(|\lambda'| + |\theta'|), \\
\varphi_y + \varphi'_y &= (\lambda + \lambda')\sin(\theta + \theta') \\
&= \lambda\sin\theta + \lambda'\sin\theta + \theta'\lambda\cos\theta + o(|\lambda'| + |\theta'|).
\end{aligned}$$

Omitting the small terms of higher order, we arrive at the linearized relations:

$$\left.\begin{aligned}
\varphi'_x &= \lambda'\cos\theta - \theta'\lambda\sin\theta, \\
\varphi'_y &= \lambda'\sin\theta + \theta'\lambda\cos\theta.
\end{aligned}\right\} \tag{2.57}$$

Elimination of θ' yields

$$\lambda' = \varphi'_x\cos\theta + \varphi'_y\sin\theta. \tag{2.58}$$

Now consider the perturbations as functions of the unperturbed potential φ and stream function ψ and use the relations

$$\frac{\partial}{\partial x} = \varphi_x \frac{\partial}{\partial \varphi} + \psi_x \frac{\partial}{\partial \psi} = \lambda \cos\theta \frac{\partial}{\partial \varphi} - \rho\lambda \sin\theta \frac{\partial}{\partial \psi}, \tag{2.59}$$

$$\frac{\partial}{\partial y} = \varphi_y \frac{\partial}{\partial \varphi} + \psi_y \frac{\partial}{\partial \psi} = \lambda \sin\theta \frac{\partial}{\partial \varphi} + \rho\lambda \cos\theta \frac{\partial}{\partial \psi}. \tag{2.60}$$

Then (2.58) becomes

$$\lambda' = \lambda \varphi'_{\varphi}. \tag{2.61}$$

On the other hand, eliminating λ' from (2.57) we obtain

$$\theta'\lambda = -\varphi'_x \sin\theta + \varphi'_y \cos\theta,$$

i.e.

$$\theta' = \rho \varphi'_{\psi}. \tag{2.62}$$

Similarly, replacing ψ by $\psi + \psi'$, λ by $\lambda + \lambda'$, θ by $\theta + \theta'$, from the third and the fourth equations of system (2.39) we arrive at the linearized relations

$$\left.\begin{aligned} \psi'_x &= -\lambda'(\rho\lambda)_\lambda \sin\theta - \theta'\rho\lambda\cos\theta, \\ \psi'_y &= \lambda'(\rho\lambda)_\lambda \cos\theta - \theta'\rho\lambda\sin\theta, \end{aligned}\right\} \tag{2.63}$$

where $(\rho\lambda)_\lambda = \mathrm{d}(\rho\lambda)/\mathrm{d}\lambda$. Elimination of θ' yields

$$\lambda'(\rho\lambda)_\lambda = -\psi'_x \sin\theta + \psi'_y \cos\theta,$$

$$\lambda'(\rho\lambda)_\lambda = \rho\lambda\psi'_{\psi}. \tag{2.64}$$

On the other hand, eliminating λ' from (2.63) we obtain

$$\theta'\rho\lambda = -\psi'_x \cos\theta - \psi'_y \sin\theta,$$

$$\theta'\rho = -\psi'_{\varphi}. \tag{2.65}$$

A comparison of (2.61) and (2.64) shows that

$$\varphi'_{\varphi} = \frac{\rho}{(\rho\lambda)_\lambda} \psi'_{\psi}. \tag{2.66}$$

Similarly, elimination of θ' from (2.62) and (2.65) yields

$$\rho^2 \varphi'_{\psi} = -\psi'_{\varphi}. \tag{2.67}$$

The system (2.66), (2.67) of first-order equations easily reduces to the second-order Equation (2.56).

The obtained linear model (2.48), (2.56) is of the same order of approximation as that in classical linearized problems for purely subsonic or supersonic flow (Oswatitsch 1976). Meanwhile, the linearization is made with respect to a given two-dimensional solution $\lambda(\varphi, \psi)$, not a uniform flow.

The notation

$$\tilde{k}(\varphi, \psi) = \frac{(\rho\lambda)_\lambda}{\rho}, \qquad \alpha(\varphi, \psi) = -\left[\frac{(\rho\lambda)_\lambda}{\rho}\right]_\varphi,$$

transforms Equation (2.56) into the form

$$\tilde{k}(\varphi, \psi)u_{\varphi\varphi} + [\rho^2(\varphi, \psi)u_\psi]_\psi - \alpha(\varphi, \psi)u_\varphi = 0. \qquad (2.68)$$

Solvability of the linear Problem (2.48), (2.68) depends on the validity of the conditions

$$\alpha \geqslant \delta > 0, \qquad (\rho^2)_\varphi \leqslant 0, \qquad (2.69)$$

as seen from Theorem 1.7. Using the relation $\rho_\lambda = -\lambda\rho^{2-\gamma}$, we obtain

$$\alpha = -\left(\frac{\rho_\lambda\lambda + \rho}{\rho}\right)_\varphi = -\left(\frac{\rho_\lambda\lambda}{\rho}\right)_\lambda \lambda_\varphi = \left(\frac{\lambda^2}{\rho^{\gamma-1}}\right)_\lambda \lambda_\varphi = (\gamma+1)\lambda\rho^{2-2\gamma}\lambda_\varphi,$$

$$(\rho^2)_\varphi = 2\rho\rho_\lambda\lambda_\varphi = -2\lambda\rho^{3-\gamma}\lambda_\varphi.$$

Hence, the conditions (2.69) are true if the streamwise acceleration $\lambda_\varphi(\varphi, \psi)$ is positive in the nozzle.

The condition $\lambda_\varphi(\varphi, \psi) > 0$ may be required only in the region bounded by the walls, the inlet section $\varphi = \varphi_1$, and the characteristic line E_1EE_2 which contacts the sonic line at a point E (see Figure 2.7). Analysis of the solvability of the boundary-value problem in this case is based on considerations similar to those given in the end of Section 2.1. If the solution is obtained in the above domain, one can construct it downstream of the characteristic E_1EE_2 by solving a sequence of Goursat's problems.

We emphasize that it was the angle θ of the velocity vector inclination to the x-axis, i.e. the streamline deflection angle, not the velocity magnitude λ, which was prescribed at the inlet in all the boundary conditions (2.12), (2.13), (2.46) formulated above. Theorems 2.1 and 2.6 rigorously establish correctness of the direct problem for transonic flow under the imposed angle θ at the inlet and on the walls of the nozzle/channel. Having solved the problem, one gets the velocity field throughout the nozzle including the inlet section. Thus, the velocity field depends on the shape of walls and cannot be prescribed at the inlet and outlet *a priori*. This is in agreement with the elementary one-dimensional approach to the problem of flow in the Laval nozzle (Oswatitsch 1976). The latter approach assumes λ to depend only on the x-coordinate. As a consequence, the continuity equation yields the relation which expresses $\lambda(x)$ in terms of the cross-sectional area $S(x)$ of the nozzle,

$$\frac{S_*}{S(x)} = \lambda(x)[1 + \tfrac{1}{2}(\gamma-1)(1-\lambda^2(x))]^{1/(\gamma-1)},$$

where S_* is the area of the critical section. Therefore, $\lambda(x_1)$ at the inlet is determined by the ratio $S_*/S(x_1)$.

In two-dimensional flow, the angle θ prescribed at the inlet in the formulations (2.12), (2.13) is typically small and does not influence significantly the velocity field.

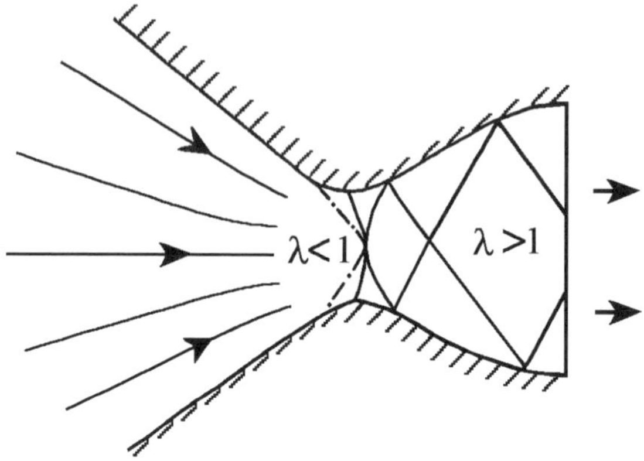

Figure 2.13 Flow in a Laval nozzle of the infinite length.

Therefore, the imposing of the angle is justified from both mathematical and physical points of view. At the same time, if the configuration of walls is fixed while the angle θ at the inlet slightly varies, then the flow field in general changes. Consequently, the location of the sonic line and the gas flux through the nozzle will change under the variation of θ at the inlet. This is accounted for by the formulation of the problem for a nozzle of the finite length which assumes a part of the boundary of the domain (the inlet section $\varphi = \varphi_1$) to pass across subsonic flow. In fact, data at the subsonic inlet section just simulate flow conditions upstream of the inlet. An alternative formulation of the Laval nozzle problem would admit either an infinite domain (Figure 2.13) or a finite domain bounded by a rigid wall at $\varphi = \varphi_1$ instead of the inlet section. In the latter case, one should examine either an unsteady flow or a steady flow through the nozzle under a given mass source inside of it (actually, in a combustion chamber to which the nozzle is attached).

2.3.3 Transonic flow through an orifice

Plane transonic flow originating from an orifice was examined in classical works of Frankl (1945) and Guderley (1947), who constructed particular solutions and studied their behaviour in case of an infinite reservoir that was bounded by two straight walls symmetric with respect to the x-axis. Flow patterns were shown to depend on the pressure in the outer media as follows. If the difference between the stagnation pressure in the reservoir p_0 and the static pressure in the media p_{out} is small, then a subsonic jet emerges from the orifice, so that the pressure on the jet boundary equals p_{out}. The flow velocity on the free boundary is related to p_{out} via the isentropic relation

$$\frac{p}{p_0} = \left(1 - \frac{\gamma - 1}{\gamma + 1}\lambda^2\right)^{\gamma/(\gamma-1)}.$$

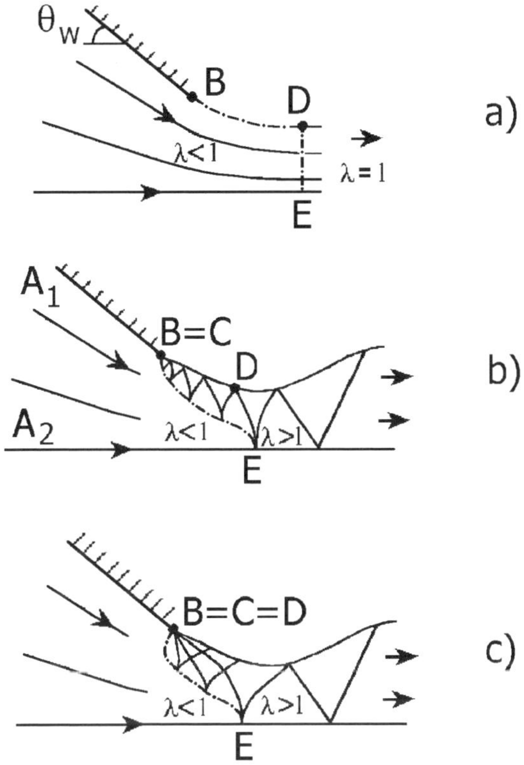

Figure 2.14 Qualitative patterns of flow emerging from an orifice.

The static pressure p is variable at sections x = const. of the jet, and it tends to p_{out} as $x \to \infty$, while the tangents to the streamlines become parallel to the x-axis.

If the outer pressure p_{out} reduces to

$$p_* = p_0 \cdot \left(\frac{2}{\gamma + 1}\right)^{\gamma/(\gamma-1)},$$

then the velocity on the free boundary becomes sonic, and asymptotic expansions of the Chaplygin functions show that the cross-section with the sonic velocity occurs at a finite distance from the orifice (Figure 2.14a).

If the pressure p_{out} drops slightly below p_*, then a sonic line begins at the edge B of the orifice (Figure 2.14b). The flow expands at the edge, so that the pressure jumps from p_* to the value p_{out}. The edge is a corner point of the wall streamline, and the jump of the tangents to the streamline $\psi = \psi_{\text{B}}$ is determined by the ratio p_*/p_{out}.

Further decrease of the outer pressure causes an upstream shift of the characteristic line ED. The point D moves to the edge B and reaches it at a sufficiently small p_{out}. In all cases, if the flow field is found in the mixed subsonic/supersonic region upstream

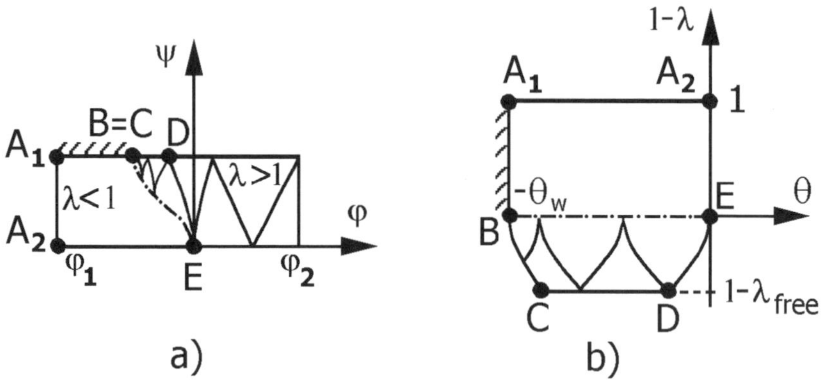

Figure 2.15 Domains in the plane of streamline coordinates and in the hodograph plane in which boundary-value problems are formulated.

of the characteristic line ED, then in the region downstream of that line the flow can be obtained by solving a sequence of Goursat problems with data on two characteristic lines or a characteristic line and the jet boundary.

Instead of studying the free boundary problem in the physical plane, one can use the streamline coordinates (φ, ψ) and treat the problem in the domain determined by the inequalities $\{(\varphi, \psi) \in \mathbb{R}^2;\ \varphi_1 < \varphi < \varphi_2,\ 0 < \psi < 1\}$, where the stream function is normalized in such a way that $\psi = 1$ on the wall $A_1 B$ (Figure 2.15a). Formulation of boundary conditions differs from (2.48) by the Dirichlet condition, not Neumann's one, given on the segment $\psi = 1$, $\varphi_B < \varphi < \varphi_2$ that corresponds to the free boundary of the jet. Inviscid flow asymptotics at large distances from a mass sink is given by classical expressions

$$\varphi \simeq \frac{Q}{\pi}\log(x^2 + y^2), \qquad \lambda \simeq \frac{|Q|}{2\pi}(x^2 + y^2)^{-1/2}, \qquad \psi \simeq \frac{Q}{2\pi}\theta, \qquad Q < 0,$$

which yield the linear dependence of θ on ψ at $\varphi = \varphi_1$. Therefore, the full set of boundary conditions for flow originating from the orifice can be written in the form

$$\left.\begin{array}{llll} \Phi_\psi = -\theta_{\rm w}\psi,\ \Phi_\varphi < 1 & \text{at} & \varphi = \varphi_1, & 0 < \psi < 1, \\ \Phi_\psi = -\theta_{\rm w} & \text{on the rigid wall} & \psi = 1, & \varphi_1 < \varphi < \varphi_{\rm B}, \\ \Phi = h(\varphi) & \text{on the jet boundary} & \psi = 1, & \varphi_{\rm B} < \varphi < \varphi_2, \\ \Phi_\varphi > 1 & \text{at a downstream section} & \varphi = \varphi_2, & \\ \Phi_\psi = 0 & \text{on the axis of symmetry} & \psi = 0, & \end{array}\right\} \tag{2.70}$$

where $h(\varphi)$ is determined by the pressure in the outer media via the relation (2.41), and $\theta_{\rm w}$ is the angle made by the reservoir wall with the inverse direction of the x-axis. The replacement of the Neumann condition by the Dirichlet one on the part of the boundary does not influence the validity of the existence and uniqueness theorems proved in Chapter 1. However, difficulties can arise in the analysis of Problem (2.42),

(2.70) because of the discontinuity of the velocity field at the edge B of the orifice. The discontinuity violates the condition of smooth coefficients of the equation governing transonic flow and requires development of new approaches for the proof of the solvability.

That is why the problem for flow emerging from a reservoir constituted by two straight walls appears worthwhile to study in the hodograph plane (λ, θ) rather than in the streamline coordinates. The flow regime depicted in Figure 2.14b corresponds to the domain of the hodograph plane shown in Figure 2.15b where λ_{free} is determined by the pressure in the outer media. The above considerations yield the boundary conditions

$$\left.\begin{array}{ll} \psi = -\theta/\theta_{\text{w}} & \text{at } \lambda = 0,\ -\theta_{\text{w}} < \theta < 0, \\ \psi = 0 & \text{on the segment } \mathrm{A_2E} \text{ of the axis } \theta = 0, \\ \psi = 1 & \text{on the wall } \mathrm{A_1B} \text{ and free boundary BCD.} \end{array}\right\} \tag{2.71}$$

The Chaplygin equations governing the stream function in the hodograph plane can be obtained from (2.7) by assigning λ and θ for the independent variables:

$$\varphi_\lambda + \psi_\theta \frac{(\lambda\rho)_\lambda}{\lambda\rho^2} = 0, \qquad \frac{\varphi_\theta}{\lambda} - \frac{\psi_\lambda}{\rho} = 0.$$

Elimination of the potential φ yields

$$(1 - M^2)\psi_{\theta\theta} + \lambda^2\psi_{\lambda\lambda} + \lambda(1 + M^2)\psi_\lambda = 0. \tag{2.72}$$

Examples of numerical simulation of Problem (2.71), (2.72) will be discussed in Section 4.5.

2.4 Finite-Element Solutions

In order to analyse finite-element solutions, one needs a weak formulation of the problem for transonic flow in a nozzle. Let V be the subspace of functions from $W^{1,2}(G)$ which vanish at the inlet, while $\hat{V}$ be the subspace of functions vanishing at both the inlet and the outlet:

$$V = \{\Phi \in W^{1,2}(G);\ \Phi(\varphi_1, \psi) = 0\},$$
$$\hat{V} = \{\chi \in W^{1,2}(G);\ \chi(\varphi_1, \psi) = \chi(\varphi_2, \psi) = 0\}.$$

We denote by V_{ad} the class of admissible functions,

$$\begin{aligned} V_{\text{ad}} = \{\Phi \in V;\ &\Phi_\varphi \geqslant 1 + \varepsilon \text{ a.e. in a vicinity of } \varphi = \varphi_2; \\ &\Phi_\varphi \leqslant 1 - \varepsilon \text{ a.e. in a vicinity of } \varphi = \varphi_1;\ \varepsilon > 0\}, \end{aligned}$$

which requires the velocity to be supersonic at the outlet and subsonic at the inlet; the notation 'a.e.' as usual stands for 'almost everywhere'. Obviously, V_{ad} is a closed and convex subset of V.

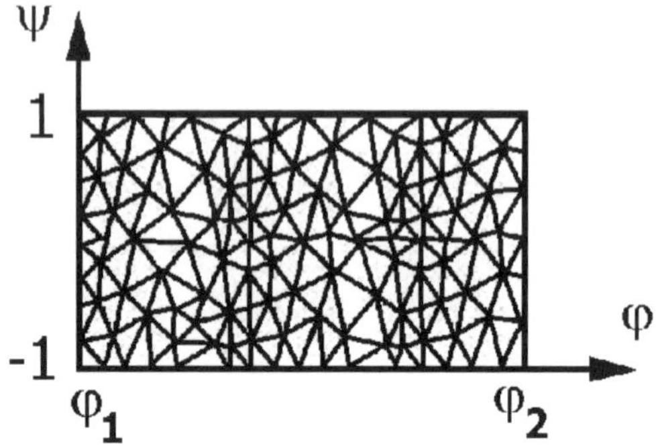

Figure 2.16 A scheme of the triangulation of the domain under consideration.

A weak formulation of Problem (2.45), (2.46) is as follows. Find a function $\Phi \in V_{\text{ad}}$ such that

$$\int_G [K(\Phi_\varphi)\chi_\varphi + \Phi_\psi \chi_\psi]\,\mathrm{d}G = \int_{\varphi_1}^{\varphi_2} [-\theta^- \chi|_{\psi=-1} + \theta^+ \chi|_{\psi=1}]\,\mathrm{d}\varphi \quad \text{for all } \chi \in \hat{V}. \tag{2.73}$$

If $\Phi(\varphi, \psi)$ is a given solution of Problem (2.73), then Theorem 2.6 yields existence of the solution $\hat{\Phi}(\varphi, \psi) = \Phi(\varphi, \psi) - u(\varphi, \psi)$ to the perturbation problem in the form (2.73) with $\theta^\pm(\varphi) - f^\pm(\varphi)$ instead of $\theta^\pm(\varphi)$ in the right-hand side.

Denote by $\mathcal{T}_h$ a decomposition of $\overline{G}$ into closed triangular subsets (Figure 2.16), where h is the discretization parameter, and assume that the triangulations $\mathcal{T}_h$ are *regular* and *quasi-uniform*. Let $V_h \subset V$, $\hat{V}_h \subset \hat{V}$ be the finite-dimensional spaces of continuous functions which are linear on every triangle $\tau_i \in \mathcal{T}_h$. Formulation of the discretized problem corresponding to (2.73) is as follows: find $\Phi_h(\varphi, \psi) \in V_h \cap V_{\text{ad}}$ such that

$$a(\Phi_h, \chi_h) = l(\chi_h), \quad \text{for all } \chi_h \in \hat{V}_h, \tag{2.74}$$

where

$$a(\Phi_h, \chi_h) = \int_G [K(\Phi_{h\varphi})\chi_{h\varphi} + \Phi_{h\psi}\chi_{h\psi}]\,\mathrm{d}G,$$

$$l(\chi_h) = \int_{\varphi_1}^{\varphi_2} [-\theta^- \chi_h|_{\psi=-1} + \theta^+ \chi_h|_{\psi=1}]\,\mathrm{d}\varphi.$$

We follow the idea suggested by Glowinski and Pironneau (1978) and Glowinski (1984) and reformulate Problem (2.74) as a discrete minimization problem for the functional

$$\mathcal{J}_h(\phi_h) = \tfrac{1}{2}\|\nabla \xi_h(\phi_h)\|^2_{L^2(G)}, \tag{2.75}$$

where $\xi_h(\phi_h) \in \hat{V}_h$ is the solution to the discrete weak Poisson problem

$$\int_G \nabla \xi_h \nabla \chi_h \,\mathrm{d}G = a(\phi_h, \chi_h) - l(\chi_h), \qquad \forall \chi_h \in \hat{V}_h. \tag{2.76}$$

Let Φ_h be a solution of the minimization problem

$$\mathcal{J}_h(\Phi_h) = \min_{\phi_h \in V_h \cap V_{\mathrm{ad}}} \mathcal{J}_h(\phi_h) = \min_{\phi_h \in V_h \cap V_{\mathrm{ad}}} \tfrac{1}{2}\|\nabla \xi_h(\phi_h)\|^2_{L^2(G)}. \tag{2.77}$$

If the minimum vanishes, i.e. $\mathcal{J}_h(\Phi_h) = 0$, then $\nabla \xi_h(\Phi_h) = 0$. Therefore, the right-hand side of (2.76) vanishes, and Φ_h is a solution of (2.74).

The functional $\mathcal{J}_h$ is continuous on $\mathbb{R}^n$, and its minimum over $V_h \cap V_{\mathrm{ad}}$ is attained in at least one element because we minimize the continuous functional over a bounded, convex and nonempty subset of a finite-dimensional space.

Theorem 2.7. *Under the assumptions of Theorem 2.6 a sequence $\{\Phi_h\}$ of solutions of the discrete minimization Problem (2.77) converges to a solution of the continuous problem in the space V as $h \to 0$.*

Proof.

Step 1. By the finite-element interpolation theory (Ciarlet 1978, p. 70), for every $\Phi \in W^{2,2}(G)$ there exists an interpolant $\Phi_h^I \in V_h$ of Φ satisfying the inequality $\|\Phi - \Phi_h^I\|_{W^{1,2}(G)} \leqslant ch\|\Phi\|_{W^{2,2}(G)}$, where c is a constant independent of Φ and h. In particular, such an interpolant $\hat{\Phi}_h \in V_h \cap V_{\mathrm{ad}}$ exists for the solution $\hat{\Phi} \in W^{6,2}(G) \cap V_{\mathrm{ad}}$ of the continuous Problem (2.73).

If Φ_h is a solution of the minimization Problem (2.77), then it satisfies the inequality

$$0 \leqslant \mathcal{J}_h(\Phi_h) = \min_{\phi_h \in V_h \cap V_{\mathrm{ad}}} \mathcal{J}_h(\phi_h) \leqslant \mathcal{J}_h(\hat{\Phi}_h).$$

Due to definition (2.75), (2.76) we get

$$\begin{aligned}
2\mathcal{J}_h(\hat{\Phi}_h) &= \|\nabla \xi_h(\hat{\Phi}_h)\|^2_{L^2} = a(\hat{\Phi}_h, \xi_h) - l(\xi_h) \\
&= a(\hat{\Phi}_h, \xi_h) - a(\hat{\Phi}, \xi_h) \\
&= \int_G [(\tilde{K}(\hat{\Phi}_{h\varphi}) - \tilde{K}(\hat{\Phi}_\varphi))\xi_{h\varphi} + (\hat{\Phi}_{h\psi} - \hat{\Phi}_\psi)\xi_{h\psi}]\,\mathrm{d}G \\
&\leqslant c\|\hat{\Phi} - \hat{\Phi}_h\|_{W^{1,2}}\|\xi_h\|_{W^{1,2}} \to 0 \quad \text{as } h \to 0,
\end{aligned}$$

therefore,

$$\mathcal{J}_h(\Phi_h) \to 0 \quad \text{as } h \to 0. \tag{2.78}$$

Step 2. Since the sequence $\{\Phi_h\}_{h \in (0,h_0)}$ is bounded in $V \cap V_{\mathrm{ad}}$, the weak compactness yields the existence of a subsequence $\{\Phi_h\}$ and a function $\Phi^* \in V \cap V_{\mathrm{ad}}$ such that $\Phi_h \rightharpoonup \Phi^*$ weakly in $W^{1,p}(G)$ for every $p \in (1, \infty)$ (Berger 1989). Then the strong convergence

$$\|\Phi_h - \Phi^*\|_{W^{1,2}(G)} \to 0 \tag{2.79}$$

can be established in the way suggested by Berger (1989), using (2.78) and the auxiliary functionals

$$\mathcal{I}_h(\eta) := -\int_G \nabla\Phi_h \cdot \nabla \Pi_h \eta \,\mathrm{d}G - M\int_G \Pi_h \eta \,\mathrm{d}G,$$

where Π_h is the interpolation operator.

Step 3. Let us prove that the function Φ^* is a solution of the continuous Problem (2.73). For an arbitrary but fixed $\chi \in \hat{V}$, due to (2.78) we find

$$\begin{aligned}
&|a(\Phi^*, \chi) - l(\chi)| \\
&\qquad = \lim_{h\to 0} |a(\Phi_h, \chi) - l(\chi)| \\
&\qquad = \lim_{h\to 0} |a(\Phi_h, \chi - R_h\chi) - l(\chi - R_h\chi) + a(\Phi_h, R_h\chi) - l(R_h\chi)|, \quad (2.80)
\end{aligned}$$

where $R_h\chi \in \hat{V}_h$ is the Ritz projection of χ defined by

$$\int_G \nabla R_h\chi \cdot \nabla \chi_h \,\mathrm{d}G = \int_G \nabla\chi \cdot \nabla\chi_h \,\mathrm{d}G \quad \text{for all } \chi_h \in \hat{V}_h.$$

Now, we replace the last two terms in (2.80) by the integral of $\nabla\xi_h \cdot \nabla R_h\chi$ according to (2.76):

$$\begin{aligned}
&= \lim_{h\to 0}\left| a(\Phi_h, \chi - R_h\chi) - l(\chi - R_h\chi) + \int_G \nabla\xi_h(\Phi_h)\cdot\nabla R_h\chi \,\mathrm{d}G \right| \\
&\leqslant \lim_{h\to 0}\Bigg(c_1\|\Phi_h\|_V \cdot \|\chi - R_h\chi\|_V + c_2\|f^\pm + \theta^\pm\|_{L^\infty(G)} \cdot \|\chi - R_h\chi\|_V \\
&\qquad\qquad + \left|\int_G \nabla\xi_h(\Phi_h)\cdot\nabla R_h\chi \,\mathrm{d}G\right| \Bigg) \\
&\leqslant \lim_{h\to 0} c\left(\|\chi - R_h\chi\|_V + \left|\int_G \nabla\xi_h(\Phi_h)\cdot\nabla R_h\chi \,\mathrm{d}G\right| \right) \\
&\leqslant \lim_{h\to 0} c(\|\chi - R_h\chi\|_V + \|\nabla\xi_h(\Phi_h)\|_{L^2(G)} \cdot \|\nabla R_h\chi\|_{L_2(G)}) = 0,
\end{aligned}$$

$\forall\chi \in \hat{V}$, since the first term tends to zero due to properties of the Ritz projection, and the second term does due to (2.75), (2.78). Hence, Φ^* is a solution of (2.73). □

2.5 Unsteady Inviscid Flow

Unsteady, isentropic, two-dimensional, irrotational flow of compressible gas is governed by the time-dependent full-potential equation (Cole and Cook 1986),

$$(a^2 - \varphi_x^2)\varphi_{xx} - 2\varphi_x\varphi_y\varphi_{xy} + (a^2 - \varphi_y^2)\varphi_{yy} - 2\varphi_x\varphi_{xt} - 2\varphi_y\varphi_{yt} - \varphi_{tt} = 0, \quad (2.81)$$

where $\varphi(x, y, t)$ is the nondimensional velocity potential, and $a(x, y, t)$ is the nondimensional speed of sound given by the Cauchy–Lagrange equation

$$a^2 = \tfrac{1}{2}(\gamma + 1) - \tfrac{1}{2}(\gamma - 1)(\varphi_x^2 + \varphi_y^2 + 2\varphi_t),$$

which is a generalization of Bernoulli's law. For simplicity, in this section we consider the nearsonic approximation and assume that flow variation in time is slow. Then Equation (2.81) reduces to

$$(\gamma + 1)(1 - \varphi_x)\varphi_{xx} + \varphi_{yy} - 2\varphi_{xt} = 0 \quad (2.82)$$

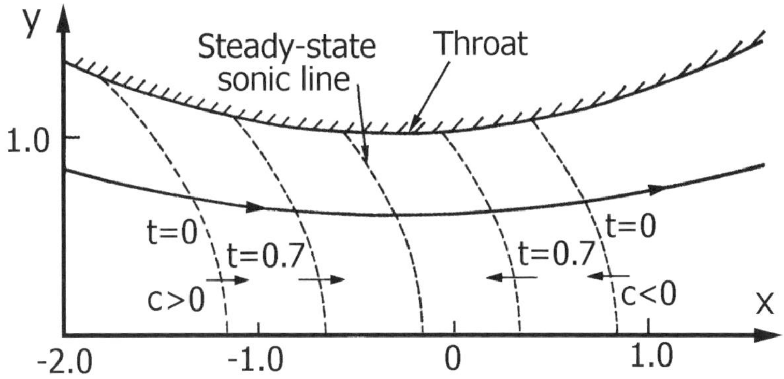

Figure 2.17 Locations of the sonic line in unsteady flow with stationary walls of the channel at different instants as described by the solution (2.83). Arrows indicate motion of the sonic line as time increases. Reproduced from Adamson (1972) by permission of Cambridge University Press.

(see Cole and Cook 1986; Lin *et al.* 1948). Differentiating Equation (2.82) with respect to x, we obtain the equation governing the velocity component $\bar{u} = \varphi_x$:

$$(\gamma + 1)(1 - \bar{u})\bar{u}_{xx} + \bar{u}_{yy} - (\gamma + 1)\bar{u}_x^2 - 2\bar{u}_{xt} = 0.$$

A particular solution to this equation is (Adamson 1972)

$$\bar{u} = 1 + \frac{4b}{\gamma + 1}[x - x_0 + 2by^2 + 2ce^{-2bt}], \tag{2.83}$$

the corresponding velocity component $\bar{v} = \varphi_y$ obtained from the condition of irrotationality is

$$\bar{v} = \frac{16b^2 y}{\gamma + 1}[x - x_0 + 2by^2/3].$$

The sign of b determines whether the flow is accelerating ($b > 0$) or decelerating ($b < 0$). For $b > 0$, the x-coordinate of the sonic point on the axis tends to x_0 as $t \to \infty$. In this limit (2.83) reduces to the proper steady-state solution. Meanwhile, for $b < 0$ the unsteady term increases without limit as t increases; this may be an indication that a shock wave must form in the decelerating flow.

The solution (2.83) for $b = 1/2, \gamma = 1.4$, and $c = \pm 0.8068$ is shown in Figure 2.17. The flow is subsonic upstream and supersonic downstream of the sonic line in each case. It can be seen that for $c > 0$ the solution represents a flow returning to a steady-state condition as the upstream pressure recovers from a decrease in its steady-state value. For $c < 0$ the solution represents a flow returning to the steady state as the upstream pressure recovers from an increase over its steady-state value. The remarkable feature is that the sonic line is initially located away from the throat. Hence, for $c > 0$ and $t = 0$, for example, the flow is supersonic in an area smaller than that corresponding to the position of the sonic line.

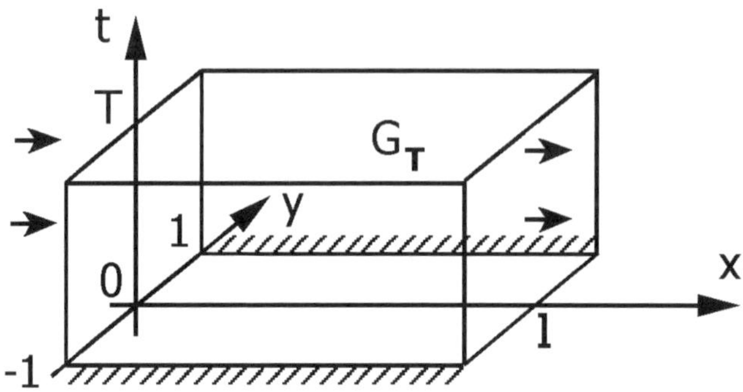

Figure 2.18 A three-dimensional domain in which problem (2.82), (2.84) is formulated.

A direct problem for unsteady transonic flow in a channel of slightly variable cross-section can be reduced to a problem for Equation (2.82) in a box $G_T = \{(x, y, t) \in \mathbb{R}^3;\ 0 < x < l,\ -1 < y < 1, 0 < t < T\}$ with the same technique as used in Section 2.1. The initial–boundary conditions are (Figure 2.18)

$$\left.\begin{array}{llll} \varphi = 0,\ \varphi_x < 1 & \text{at the inlet} & x = 0, & -1 < y < 1, \\ \varphi_x > 1 & \text{at the outlet} & x = l, & -1 < y < 1, \\ \varphi_y = \theta^{\pm}(x, t) & \text{at} & y = \pm 1, & 0 < x < l, \\ \varphi(x, y, 0) = \varphi_0(x, y) & \text{at} & t = 0, & -1 < y < 1,\ 0 < x < l, \end{array}\right\} \tag{2.84}$$

where $\theta^{\pm}$, φ_0 are given functions.

Below we treat the linear stability of the solution of Problem (2.82), (2.84) with respect to unsteady perturbation of the initial–boundary conditions. Let $\varphi(x, y, t)$ be a given solution to Problem (2.82), (2.84). We represent the sought-after velocity potential in the form $\varphi(x, y, t) + u(x, y, t)$, where u is the small perturbation. By linearizing Equation (2.82), we obtain

$$(\gamma + 1)(1 - \varphi_x)u_{xx} + u_{yy} - (\gamma + 1)\varphi_{xx}u_x - 2u_{tx} = 0. \tag{2.85}$$

The initial–boundary conditions with respect to $u(x, y, t)$ read

$$\left.\begin{array}{llll} u = 0, & \text{at} & x = 0, & -1 < y < 1, \\ u_y = h^{\pm}(x, t) & \text{at} & y = \pm 1, & 0 < x < l, \\ u(x, y, 0) = u_0(x, y) & \text{at} & t = 0, & -1 < y < 1,\ 0 < x < l, \end{array}\right\} \tag{2.86}$$

where $h^{\pm}$, u_0 are given perturbations of $\theta^{\pm}$ and φ_0, respectively.

As in Section 2.1, the first and the second conditions in (2.86) mean that the angle of the velocity vector inclination to the x-axis is prescribed at the inlet and on the walls of the channel.

Problem (2.85), (2.86) can be reduced in a standard way to the one for the nonhomogeneous equation endowed with the homogeneous boundary conditions. For the latter problem, Larkin (1991) established the unique solvability in the functional class $u \in L^\infty(0, T; W^{2,2}(G))$, $u_{xt} \in L^\infty(0, T; L^2(G))$, $0 \leqslant t \leqslant T$, under the condition that the given solution $\varphi(x, y, t)$ is smooth enough. Here $L^\infty(0, T; X)$ is the Banach space of measurable functions on the segment $0 \leqslant t \leqslant T$ with values in the space X and the norm $\|u\|_\infty = \text{vrai sup}_{0\leqslant t\leqslant T} \|u(t)\|_X$.

One does not need the condition of positive acceleration of the given flow in order to prove the solvability of the problem with respect to unsteady perturbation at a finite interval $0 \leqslant t \leqslant T$. For any sign of the acceleration, one can gain sufficiently large values of the coefficient in front of u_x in (2.85) by the substitution $u = e^{\mu t} v$, where $\mu = \text{const.} > 0$. However, for analysis of the behaviour of the perturbation at infinitely increasing time $t \to \infty$, the condition of positive acceleration is of principal importance. If it holds, then the solution $u(x, y, t)$ tends to the solution $u^s(x, y)$ of the related problem for steady perturbation in such a way that

$$\int_G |u_x(x, y, t) - u_x^s(x, y)|^2 \, dx \, dy \to 0 \quad \text{as } t \to \infty$$

(Larkin 1991). This means that unsteady nearsonic flow in a channel is linearly stable with respect to small perturbation of the shape of walls if the acceleration of the given flow is positive.

We note that in a number of papers the stability of *one-dimensional* steady transonic flow of inviscid gas was studied by examining the unsteady nonlinear propagation of sound impulses of a finite extent (Emmons 1958; Prasad 1973; Karabutov and Sapozhnikov 1986). It was shown that such impulses are decaying in the vicinity of the point where the velocity is sonic and the flow acceleration is positive. At the same time the shape of the impulses steepens in the vicinity of the point where the velocity is sonic and the flow is decelerating. This indicates that a flow which decelerates smoothly from supersonic to subsonic velocities must give rise to a shock wave. Thus, the condition of the positive acceleration at the sonic point is of principal importance for the stability of the unsteady one-dimensional flow, as well as of the two-dimensional flow model given by Problem (2.85), (2.86).

Theorem 2.8. *Let a classical solution $\varphi(x, y, t) \in C^1(\overline{G}_T) \cap C^2(\overline{G}_T)$ or a solution from the Sobolev space $W^{2,2}(G_T)$ of the nonlinear Problem (2.82), (2.84) exist, and the flow acceleration be positive over the nozzle: $\varphi_{xx}(x, y, t) > 0$ at $0 \leqslant t \leqslant T$. Then the solution φ is unique in the above-mentioned classes of functions.*

Proof. Suppose there are two solutions $\varphi(x, y)$ and $\hat{\varphi}(x, y)$ of Problem (2.82), (2.84). By subtracting Equation (2.82) written with respect to $\hat{\varphi}$ from that written with respect to φ, we obtain

$$(\gamma + 1)[(1 - \varphi_x)\varphi_{xx} - (1 - \hat{\varphi}_x)\hat{\varphi}_{xx}] + \varphi_{yy} - \hat{\varphi}_{yy} - 2(\varphi_{xt} - \hat{\varphi}_{xt}) = 0. \quad (2.87)$$

As in Section 2.1, we use simple algebra in order to rearrange the expression in the square brackets as follows,

$$(1 - \varphi_x)\varphi_{xx} - (1 - \hat{\varphi}_x)\hat{\varphi}_{xx} = -u_x\varphi_{xx} + (1 - \hat{\varphi}_x)u_{xx},$$

where $u = \varphi - \hat{\varphi}$. Then Equation (2.87) becomes

$$(1 - \hat{\varphi}_x)u_{xx} - \varphi_{xx}u_x + \frac{1}{\gamma + 1}(u_{yy} - 2u_{xt}) = 0. \tag{2.88}$$

By adding and subtracting the term $pu_x u_{xx}$ with a parameter p, $0 \leqslant p \leqslant 1$, we obtain the equation

$$[1 - p\varphi_x - (1 - p)\hat{\varphi}_x]u_{xx} - [(1 - p)\varphi_{xx} + p\hat{\varphi}_{xx}]u_x + \frac{1}{\gamma + 1}(u_{yy} - 2u_{xt}) = 0. \tag{2.89}$$

Both φ and $\hat{\varphi}$ satisfy boundary conditions (2.84). Therefore, the difference u satisfies the homogeneous conditions:

$$\left.\begin{aligned} u &= 0 \quad \text{at} \quad x = 0, \quad -1 < y < 1, \\ u_y &= 0 \quad \text{at} \quad y = \pm 1, \quad 0 < x < l. \end{aligned}\right\} \tag{2.90}$$

Using the notation

$$\begin{aligned} k(x, y, t) &= 1 - p\varphi_x - (1 - p)\hat{\varphi}_x, \\ \alpha(x, y, t) &= (1 - p)\varphi_{xx} + p\hat{\varphi}_{xx}, \end{aligned}$$

we represent Equation (2.89) in the form

$$k(x, y, t)u_{xx} - \alpha(x, y, t)u_x + \frac{1}{\gamma + 1}(u_{yy} - 2u_{xt}) = 0. \tag{2.91}$$

Now we multiply Equation (2.91) by u_x and integrate over the three-dimensional domain G_{T}:

$$\int_{G_{\mathrm{T}}} \left[ku_{xx}u_x - \alpha u_x^2 + \frac{1}{\gamma + 1}(u_{yy}u_x - 2u_{xt}u_x)\right] \mathrm{d}G_{\mathrm{T}} = 0.$$

Simple transformations of the integrand yield

$$\int_{G_{\mathrm{T}}} \left[-(\alpha + \tfrac{1}{2}k_x)u_x^2 + \tfrac{1}{2}(ku_x^2)_x + \frac{1}{\gamma + 1}[(u_y u_x)_y - \tfrac{1}{2}(u_y^2)_x - (u_x^2)_t]\right] \mathrm{d}G_{\mathrm{T}} = 0.$$

By applying the divergence theorem to the left-hand side of the equation, we get

$$\begin{aligned} &-\int_{G_{\mathrm{T}}} (\alpha + \tfrac{1}{2}k_x)u_x^2 \,\mathrm{d}G_{\mathrm{T}} \\ &\quad + \int_{S_{\mathrm{T}}} \left[\tfrac{1}{2}ku_x^2 \cos(\boldsymbol{n}, x) + \frac{1}{\gamma + 1}(u_x u_y \cos(\boldsymbol{n}, y) \right. \\ &\qquad \left. - \tfrac{1}{2}u_y^2 \cos(\boldsymbol{n}, x) - u_x^2 \cos(\boldsymbol{n}, t))\right] \mathrm{d}S_{\mathrm{T}} = 0, \end{aligned} \tag{2.92}$$

where S_T is the boundary of G_T and $\boldsymbol{n}$ is the outward normal vector to S_T. One can easily verify that the surface integral is nonpositive due to the boundary conditions (2.90) and the inequalities $k > 0$ at $x = 0$, $k < 0$ at $x = l$. The nonpositiveness becomes clear if we represent k in the form $k = p(1 - \varphi_x) + (1 - p)(1 - \hat{\varphi}_x)$ and recall conditions (2.84) with respect to φ_x and $\hat{\varphi}_x$ at the inlet and outlet. Since $2\alpha + k_x \equiv (2 - 3p)\varphi_{xx} + (3p - 1)\hat{\varphi}_{xx}$, setting $p = 1/3$ we obtain the expression $2\alpha + k_x \equiv \varphi_{xx}$, which is positive by the assumptions of Theorem 2.8. Therefore, all the terms on the left-hand side of (2.92) are nonpositive. Hence, $u \equiv 0$ and $\varphi(x, y, t) \equiv \hat{\varphi}(x, y, t)$ in the domain G_T. □

2.6 Finite-Difference Schemes

An approximate solution to the perturbation Problem (2.20), (2.21) for steady, plane, isentropic flow of compressible inviscid gas can be found, in principle, by the iteration scheme (2.24) where the solution $u^{(i+1)}$ is a limit of a subsequence of the Galerkin sums

$$u^{N,\varepsilon}(x, y) = \sum_{l=1}^{N} X_l^{N,\varepsilon}(x) Y_l(y)$$

as $\varepsilon \to 0$, $N \to \infty$. The functions $X_l^{N,\varepsilon}(x)$ are to be found by integrating the system of third-order ordinary differential equations (1.16), while the functions $Y_l(y)$, $l = 1, 2, \ldots$, constitute a fundamental system in $W^{2,2}(-1, 1)$ (see (1.15)). However, this way of calculating the solution to Problem (2.20), (2.21) is quite complicated and rather inconvenient for practical use.

In a pioneering paper, Murman and Cole (1971) gave rise to a rapid development of numerical algorithms for simulation of transonic flow problems. Their original idea was that a finite-difference scheme must depend on the type of the governing equation at the mesh node under consideration. In case of the von Kármán equation

$$(\gamma + 1)(1 - \varphi_x)\varphi_{xx} + \varphi_{yy} = 0, \tag{2.93}$$

if the node lies in the subsonic region, then the type is elliptic, and the second-order derivatives are approximated by the standard central differences,

$$(\gamma + 1)(1 - \varphi_x)\frac{\varphi_{i+1,j} - 2\varphi_{i,j} + \varphi_{i-1,j}}{h_1^2} + \frac{\varphi_{i,j+1} - 2\varphi_{i,j} + \varphi_{i,j-1}}{h_2^2} = 0, \tag{2.94}$$

where h_1 and h_2 are the mesh cell sizes in the directions of the x- and y-axes. In the supersonic region, the finite-difference approximation for φ_{xx} must be upwinded:

$$(\gamma + 1)(1 - \varphi_x)\frac{\varphi_{i,j} - 2\varphi_{i-1,j} + \varphi_{i-2,j}}{h_1^2} + \frac{\varphi_{i,j+1} - 2\varphi_{i,j} + \varphi_{i,j-1}}{h_2^2} = 0. \tag{2.95}$$

In the proximity of the sonic line, through which the flow accelerates, the scheme reduces to

$$\frac{\varphi_{i,j+1} - 2\varphi_{i,j} + \varphi_{i,j-1}}{h_2^2} = 0. \tag{2.96}$$

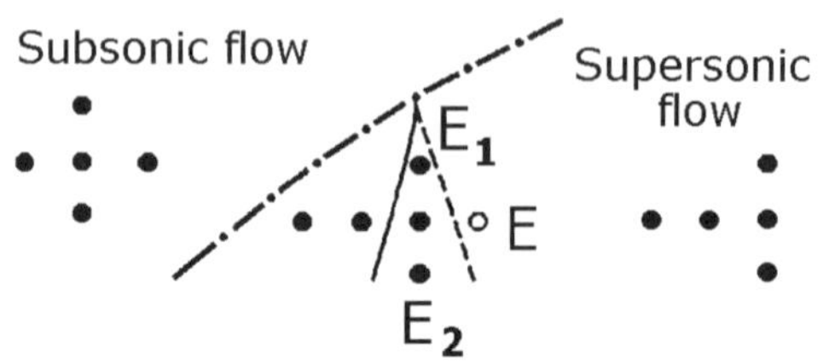

Figure 2.19 Different stencils at subsonic and supersonic points.

The necessity of upwinding in the supersonic region is accounted for by the fact that the central differences, which work in explicit schemes for purely supersonic flow, would violate the Courant–Friedrichs–Lewy (CFL) stability condition near the sonic line in the case of transonic flow. Indeed, the CFL condition is known to require a sufficiently small mesh size in the marching direction, so that the downstream node E of the mesh would belong to the domain of dependence of the vertical segment E_1E_2 (see Figure 2.19). The domain of dependence is bounded by two characteristic lines

$$\frac{dx}{dy} = \pm\sqrt{(\gamma+1)(\varphi_x - 1)}$$

emanating from points E_1 and E_2. In the vicinity of the sonic line, the difference $\varphi_x - 1$ turns out to be very small. Therefore, the characteristic lines are located very close to the segment E_1E_2. Hence, it is difficult in practice to choose a sufficiently small step size in the x-direction and to ensure that the node E belongs to the domain of dependence. Meanwhile, if the node E drops outside the domain of dependence, then the scheme (2.94) becomes unstable. As to the implicit scheme (2.95), it is known to be stable at any ratio of mesh sizes in the x- and y-directions, even though the error in the approximation for φ_{xx} is of the order $O(h_1)$, as compared with the second-order approximation by the central differences (see, for example, Fletcher 1991).

By inspection, one can verify the identity

$$\frac{\varphi_{i,j} - 2\varphi_{i-1,j} + \varphi_{i-2,j}}{h_1^2} = \frac{\varphi_{i+1,j} - 2\varphi_{i,j} + \varphi_{i-1,j}}{h_1^2} - h_1\bar{\varphi}_{xxx},$$

where

$$\bar{\varphi}_{xxx} = (\varphi_{i+1,j} - 3\varphi_{i,j} + 3\varphi_{i-1,j} - \varphi_{i-2,j})/h_1^3$$

approximates the third-order derivative φ_{xxx}. Hence, the upwind differencing suggested by Murman and Cole can be interpreted as the addition of the artificial viscosity term $-h_1\bar{\varphi}_{xxx}$ to the central differenced second-order derivative. One of two techniques can therefore be used in computations: either the streamwise derivatives are upwind differenced in the supersonic regions or all derivatives are centrally differenced everywhere but an artificial viscosity term is added to the equations in the supersonic region. In the latter approach, if one employs the nonconservative form

(2.94)–(2.96), the scheme reads

$$(\gamma+1)(1-\varphi_x)\frac{\varphi_{i+1,j}-2\varphi_{i,j}+\varphi_{i-1,j}}{h_1^2}+\frac{\varphi_{i,j+1}-2\varphi_{i,j}+\varphi_{i,j-1}}{h_2^2}$$
$$=-h_1(\gamma+1)(\varphi_x-1)\varphi_{xxx} \quad \text{for } M>1,$$
$$=0 \quad \text{for } M<1.$$

As mentioned in Section 1.5, Romanova (1995) used a finite-difference technique to study a boundary-value problem for the third-order equation

$$k(x,y)u_{xx}+u_{yy}-\alpha(x,y)u_x+c(x,y)u=-\varepsilon u_{xxx}+f(x,y), \tag{2.97}$$

and proved convergence of the solution $u^\varepsilon(x,y)$ to the solution of the second-order equation as $\varepsilon\to 0$. The distinction of (2.97) from the previous scheme is that Romanova employed the singular perturbation term $-\varepsilon u_{xxx}$ in both hyperbolic and elliptic regions. Also, Equation (2.97) in contrast to (2.93) contains the term $c(x,y)u$ with negative c which must have sufficiently large absolute value in order to provide the validity of the existence theorem. Finally, Equation (2.97) is linear. Therefore, it could only be employed to establish the existence of the solution to the linearized system of finite-difference equations at each iteration step (cf. (2.101) below).

The first-order derivative φ_x in (2.94), (2.95) is supposed to be approximated by the central difference $(\varphi_{i+1,j}-\varphi_{i-1,j})/(2h_1)$. To provide an adequate treatment of shock waves, it is worthwhile to put the von Kármán Equation (2.93) into the conservation form

$$[K(\varphi_x)]_x+\varphi_{yy}=0, \tag{2.98}$$

where

$$K(\varphi_x)=-\tfrac{1}{2}(\gamma+1)(\varphi_x-1)^2.$$

Then the upwinding in the supersonic region yields the expression

$$(\gamma+1)\left(1-\frac{\varphi_{i,j}-\varphi_{i-2,j}}{2h_1}\right)\frac{\varphi_{i,j}-2\varphi_{i-1,j}+\varphi_{i-2,j}}{h_1^2}$$
$$+\frac{\varphi_{i,j+1}-2\varphi_{i,j}+\varphi_{i,j-1}}{h_2^2}=0. \tag{2.99}$$

The scheme (2.94), (2.96), (2.99) can be represented over the flow field in a compact form as follows:

$$-D_+K(u_{i,j})+\sigma_{i,j}D_+K(u_{i,j})-\sigma_{i-1,j}D_+K(u_{i-1,j})$$
$$+\frac{\varphi_{i,j+1}-2\varphi_{i,j}+\varphi_{i,j-1}}{h_2^2}=0, \tag{2.100}$$

where

$$u_{i,j}=(\varphi_{i,j}-\varphi_{i-1,j})/h_1, \qquad u_{i+1/2,j}=(u_{i,j}+u_{i+1,j})/2$$

are approximations to the velocity component in the direction of the x-axis,

$$\sigma_{i,j} = 1 \quad \text{if } u_{i+1/2,j} \geqslant 1, \qquad \sigma_{i,j} = 0 \quad \text{if } u_{i+1/2,j} < 1.$$

Here and in the sequel we use the notation

$$\begin{aligned} D_+K(u_{i,j}) &= [K(u_{i+1,j}) - K(u_{i,j})]/h_1, \\ D_-K(u_{i,j}) &= [K(u_{i,j}) - K(u_{i-1,j})]/h_1. \end{aligned}$$

In the supersonic region, both $\sigma_{i,j}$ and $\sigma_{i-1,j}$ are equal to unity. Therefore, the first and second terms in (2.100) are cancelled, and (2.100) yields (2.99). In the subsonic region, both $\sigma_{i,j}$ and $\sigma_{i-1,j}$ vanish, therefore, (2.100) reduces to (2.94) with the central difference for the first-order derivative. At the shock points, expression (2.100) yields a difference equation which approximates the derivative $[K(\varphi_x)]_x$ by the sum of the central and upwind difference operators (Murman 1974).

In order to solve the system of nonlinear algebraic Equations (2.100), Murman and Cole used an iterative approach in which the coefficient approximating φ_x is taken from the previous iteration. For instance, the iterative scheme for Equation (2.99) reads

$$(\gamma+1)\left(1 - \frac{\varphi_{i,j}^{(n-1)} - \varphi_{i-2,j}^{(n-1)}}{2h_1}\right)\frac{\varphi_{i,j}^{(n)} - 2\varphi_{i-1,j}^{(n)} + \varphi_{i-2,j}^{(n)}}{h_1^2} + \frac{\varphi_{i,j+1}^{(n)} - 2\varphi_{i,j}^{(n)} + \varphi_{i,j-1}^{(n)}}{h_2^2} = 0. \tag{2.101}$$

The resulting large system of linear equations was solved by a successive line over-relaxation algorithm. Each vertical line was relaxed successively, proceeding in the positive x-direction. At each stage of the iteration the local velocity was tested to select the elliptic, hyperbolic, sonic, or shock point approximation.

Though the scheme (2.100) is linearly stable, Ballhaus and Goorjian (1977), Krupp and Cole (1976) reported nonlinear instabilities which occurred in computations. Furthermore, Jameson (1976) pointed out that the scheme admits entropy violating shocks as solutions.

In order to eliminate entropy-violating shocks, Engquist and Osher (1980) modified the Murman–Cole scheme near the sonic lines and shock waves by introducing the functions K_1, K_2:

$$\begin{aligned} K_1(u) &= K(u) \quad \text{if } u > 1, \qquad & K_1(u) &= 0 \quad \text{if } u \leqslant 1, \\ K_2(u) &= K(u) \quad \text{if } u < 1, \qquad & K_2(u) &= 0 \quad \text{if } u \geqslant 1, \end{aligned}$$

where $u = \varphi_x$ is the velocity component in the x-direction, so that $u = 1$ on the sonic line. The finite-difference approximations for $[K(\varphi_x)]_x$ in the vicinity of the sonic lines and shock waves are listed below. The Engquist–Osher scheme is marked by the subscript 'E–O', and the Murman–Cole scheme by 'M–C'. For brevity, we drop the second subscript in the notation $u_{i,j}$.

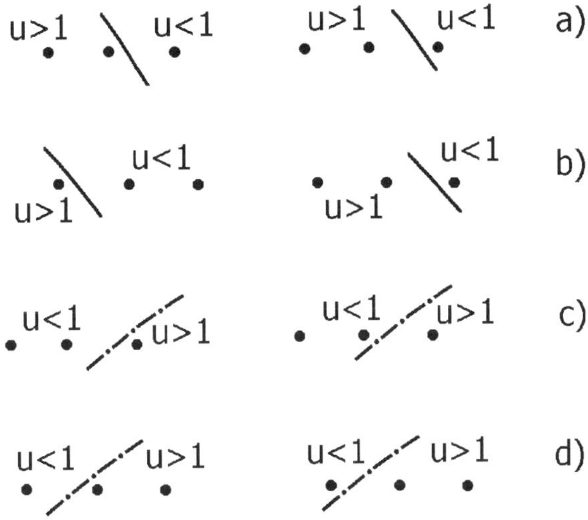

Figure 2.20 Different locations of the mesh with respect to the sonic line/shock wave.

(a) $u_{i-1}, u_i > 1 > u_{i+1}$ (shock or sonic line through which the flow decelerates, see Figure 2.20a):

$$[K(\varphi_x)]_x|_{\text{E–O}} \approx D_- K(u_i) + D_+ K_2(u_i),$$
$$[K(\varphi_x)]_x|_{\text{M–C}} \approx 2D_0 K(u_i) \quad \text{if } u_{i+1/2} < 1,$$
$$\approx D_- K(u_i) \quad \text{if } u_{i+1/2} \geqslant 1,$$

where $D_0 K(u_{i,j}) = [K(u_{i+1,j}) - K(u_{i-1,j})]/(2h_1)$;

(b) $u_{i-1} > 1 > u_i, u_{i+1}$ (shock or sonic line through which the flow decelerates, see Figure 2.20b):

$$[K(\varphi_x)]_x|_{\text{E–O}} \approx D_- K_1(u_i) + D_+ K(u_i),$$
$$[K(\varphi_x)]_x|_{\text{M–C}} \approx D_+ K(u_i) \quad \text{if } u_{i-1/2} < 1,$$
$$\approx 2D_0 K(u_i) \quad \text{if } u_{i-1/2} \geqslant 1;$$

(c) $u_{i-1}, u_i < 1 < u_{i+1}$ (sonic line through which the flow accelerates, see Figure 2.20c):

$$[K(\varphi_x)]_x|_{\text{E–O}} \approx D_+ K_2(u_i),$$
$$[K(\varphi_x)]_x|_{\text{M–C}} \approx D_+ K(u_i) \quad \text{if } u_{i+1/2} < 1,$$
$$\approx 0 \quad \text{if } u_{i+1/2} \geqslant 1;$$

(d) $u_{i-1} < 1 < u_i$, u_{i+1} (sonic line through which the flow accelerates, see Figure 2.20d):

$$
\begin{aligned}
[K(\varphi_x)]_x|_{\text{E–O}} &\approx D_- K_1(u_i),\\
[K(\varphi_x)]_x|_{\text{M–C}} &\approx 0 \quad \text{if } u_{i+1/2} \leqslant 1,\\
&\approx D_- K(u_i) \quad \text{if } u_{i-1/2} > 1.
\end{aligned}
$$

In particular, the E–O representation in case (c) reads

$$D_+ K_2(u_i) = [K_2(u_{i+1}) - K_2(u_i)]/h_1 = [0 - K(u_i)]/h_1,$$

while in case (d) it is

$$D_- K_1(u_i) = [K_1(u_i) - K_1(u_{i-1})]/h_1 = [K(u_i) - 0]/h_1.$$

The Murman–Cole scheme, e.g. in case (d), $u_{i-1/2} > 1$ reads

$$
\begin{aligned}
D_- K(u_i) &= [K(u_i) - K(u_{i-1})]/h_1\\
&= -\frac{\gamma+1}{2h_1}[(u_i - 1)^2 - (u_{i-1} - 1)^2]\\
&= -\frac{\gamma+1}{2h_1}(u_i - u_{i-1})(u_i + u_{i-1} - 2)\\
&= -(\gamma+1)\left(\frac{\varphi_i - \varphi_{i-2}}{2h_1} - 1\right)\frac{\varphi_i - 2\varphi_{i-1} + \varphi_{i-2}}{h_1^2},
\end{aligned}
$$

which coincides with the first term on the left-hand side of Equation (2.99).

The E–O scheme is first order accurate in the supersonic region, and second-order accurate in the subsonic region. The theorem presented below establishes the validity of an entropy inequality in terms of the transonic small disturbance equation

$$[K_{\text{sim}}\phi_x - \tfrac{1}{2}(\gamma+1)\phi_x^2]_x + \phi_{yy} = 0, \qquad (2.102)$$

where $K_{\text{sim}} = (1 - M_\infty^2)/\delta^{2/3}$ is the transonic similarity parameter, δ is the thickness of an airfoil or the maximum deviation of the nozzle walls from a straight segment (Cole and Cook 1986). The potential $\phi(x, y)$ in (2.102) is normalized by the freestream velocity instead of the critical one as in the von Kármán Equation (2.98). Equation (2.102) reduces to (2.98) by the substitution

$$\phi = \varphi + \left(\frac{K_{\text{sim}}}{\gamma+1} - 1\right)x.$$

Theorem 2.9 (Engquist and Osher 1980). *Suppose that $\varphi_{i,j}$, $D_-^x\varphi_{i,j}$ and $D_-^y\varphi_{i,j} = (\varphi_{i,j} - \varphi_{i,j-1})/h_2$ converge boundedly a.e. to φ, φ_x, φ_y, respectively, as $h_1 \to 0$, $h_2 \to 0$. Then φ is a weak solution of Equation (2.93) which satisfies the entropy inequality*

$$\iint [\eta_x F(\phi_x, \phi_y) - \eta_y \phi_x \phi_y \,\mathrm{d}x\,\mathrm{d}y] \geqslant 0, \qquad \phi = \varphi + \left(\frac{K_{\text{sim}}}{\gamma+1} - 1\right)x, \quad (2.103)$$

for any $\eta \in C^1(G)$ *with* $\eta \geqslant 0$ *vanishing on the boundary of the domain* G, *where*

$$F(\phi_x, \phi_y) = \tfrac{1}{2}\phi_y^2 - \tfrac{1}{2}K_{\text{sim}}\phi_x^2 + (\gamma + 1)\tfrac{1}{3}\phi_x^3.$$

The entropy inequality in the weak formulation (2.103) comes from the inequality

$$(\tfrac{1}{2}\phi_y^2 - \tfrac{1}{2}K_{\text{sim}}\phi_x^2 + (\gamma + 1)\tfrac{1}{3}\phi_x^3)_x - (\phi_x\phi_y)_y \leqslant 0,$$

which is equivalent in the piecewise-smooth flow to the inequality $\phi_{xx} < 0$ eliminating the expansion shocks (Mock 1980).

Alternatively to the E–O scheme, the switch function $\sigma_{i,j}$ can be modified in the Murman–Cole scheme (2.100) so that limit solutions of the difference equation must satisfy the entropy inequality. Engquist and Osher suggested the modification

$$\sigma_{i,j}\left(\frac{1}{2}\left[\frac{\partial K(u_{i,j})}{\partial u} + \frac{\partial K(u_{i+1,j})}{\partial u}\right]\right)$$
$$= \frac{2}{3}\frac{\partial K(u_{i+1,j})}{\partial u} + \frac{1}{3}\frac{\partial K(u_{i,j})}{\partial u} + a\left(\left|\frac{\partial K(u_{i,j})}{\partial u}\right| + \left|\frac{\partial K(u_{i+1,j})}{\partial u}\right|\right),$$

for some $a > 0$ when $u_{i,j} < 1 < u_{i+1,j}$ or $u_{i,j} > 1 > u_{i+1,j}$. Since

$$\sigma_{i,j}\Delta_+K(u_{i,j}) = (\Delta_+u_{i,j})\sigma_{i,j}\left(\frac{1}{2}\left[\frac{\partial K(u_{i,j})}{\partial u} + \frac{\partial K(u_{i+1,j})}{\partial u}\right]\right),$$

where $\Delta_+K(u_{i,j}) = K(u_{i+1,j}) - K(u_{i,j})$, the resulting scheme is well defined and first order accurate, although the coefficients are not continuous.

The *time-dependent* transonic Equation (2.82) in the conservation form reads

$$[K(\varphi_x)]_x + \varphi_{yy} - 2\varphi_{tx} = 0, \qquad K(\varphi_x) = -\tfrac{1}{2}(\gamma + 1)(\varphi_x - 1)^2. \qquad (2.104)$$

For this equation, one can remove the factor of 2 by normalizing the t-variable. Then a half step $\tau/2$ with respect to time can be performed via the difference scheme

$$u_{i,j}^{(n+1/2)} = u_{i,j}^{(n)} - \frac{\tau}{h_1}\Delta_+^x K(\min(u_{i,j}^{(n)}, 1)) - \frac{\tau}{h_1}\Delta_-^x K(\max(u_{i,j}^{(n)}, 1)), \qquad (2.105)$$

under appropriate boundary conditions. Next, the potential can be recovered via the relation $D_-^x\varphi_{i,j}^{(n+1/2)} = u_{i,j}^{(n+1/2)}$. Then one performs a second half step via the linear Crank–Nicolson type differencing:

$$u_{i,j}^{(n+1)} = u_{i,j}^{(n+1/2)} - \frac{\tau}{2h_2}\Delta_+^y(v_{i,j}^{(n+1)} + v_{i,j}^{(n+1/2)}). \qquad (2.106)$$

The above scheme with the Courant–Friedrichs–Lewy condition

$$\max_{u_{i,j}^{(n)}} \frac{\tau}{h_1}\left|\frac{\partial K(u_{i,j}^{(n)})}{\partial u}\right| = \varepsilon_0 < \tfrac{1}{3},$$

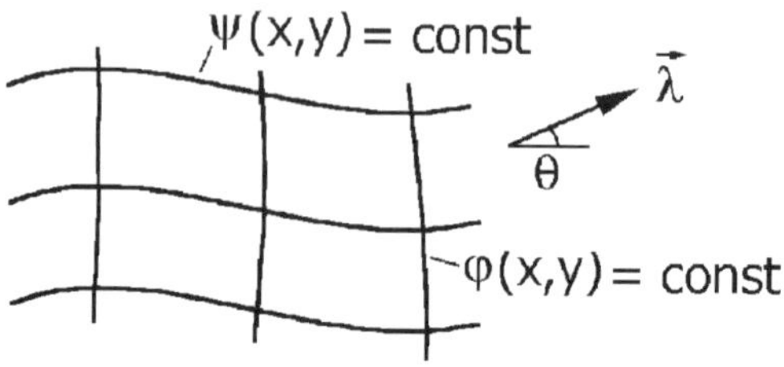

Figure 2.21 The streamline coordinates (φ, ψ).

is stable with respect to the norm,

$$\|u\|_0^2 = \sum_{i=0}^{N}\sum_{j=0}^{M} u_{i,j}^2 h_1 h_2.$$

Expressions (2.105) and (2.106) are first order accurate as operators on $\varphi_{i,j}^{(n)}$. If there is convergence to a time-independent solution, then (2.105) and (2.106) are second-order accurate in subsonic regions and first-order accurate in supersonic regions. Engquist and Osher analysed this scheme in the case when the velocity is supersonic at the inlet and subsonic at the outlet, e.g. for decelerating flow in a channel. A study of the solvability of a boundary-value problem for the decelerating flow will be presented in Section 3.5. We shall discuss further developments of the E–O scheme in Section 3.6.

2.7 Acoustical Oscillations in Transonic Flow

2.7.1 Equations governing acoustical oscillations in moving media

The potential of acoustical waves in steady mean flow must satisfy an equation obtained by the linearization of the time-dependent full-potential Equation (2.81). The resulting linear equation, however, appears to be very complex and inconvenient for mathematical study. Alternatively, equations that describe propagation of acoustical waves can be obtained by substituting the asymptotic expansions for the flow variables into the continuity and momentum equations, subtracting the corresponding steady equations, and neglecting terms of higher than first order in the amplitude of oscillations. This procedure yields a system of linearized equations, which reduces to a second-order equation in the case when the flow is isentropic and irrotational (Verdon 1989),

$$\frac{\mathrm{D}}{\mathrm{D}t}\left(a^{-2}\frac{\mathrm{D}\varphi}{\mathrm{D}t}\right) - \rho^{-1}\nabla\cdot(\rho\nabla\varphi) = 0, \tag{2.107}$$

where $a(x, y, z)$, $\rho(x, y, z)$, $\boldsymbol{v}(x, y, z)$ are the speed of sound, the density, and the velocity field of the mean flow, respectively,

$$\frac{\mathrm{D}}{\mathrm{D}t} \equiv \frac{\partial}{\partial t} + \boldsymbol{v} \cdot \nabla = 0$$

is a convective derivative operator. The linear Equation (2.107), however, is inconvenient for mathematical study because of the curved boundaries of the domain under consideration in the Cartesian coordinates. That is why below we derive a convected wave equation for two-dimensional acoustical oscillations in the streamline coordinates (Figure 2.21), which will permit us to get a few conclusions about eigenfrequencies of acoustical oscillations in transonic flow.

As in Section 2.5, we neglect the entropy and vorticity perturbations which only cause an effect of the second-order smallness (Stolyarov 1983). Then unsteady two-dimensional flow is governed by the time-dependent continuity equation

$$\tilde{\rho}_t + (\tilde{\rho}\tilde{\lambda}\cos\tilde{\theta})_x + (\tilde{\rho}\tilde{\lambda}\sin\tilde{\theta})_y = 0, \tag{2.108}$$

and the condition of irrotationality

$$(\tilde{\lambda}\cos\tilde{\theta})_y - (\lambda\sin\tilde{\theta})_x = 0, \tag{2.109}$$

which makes it possible to define the velocity potential $\tilde{\varphi}(x, y, t)$ by the relations $\tilde{\varphi}_x = \tilde{\lambda}\cos\tilde{\theta}$, $\tilde{\varphi}_y = \tilde{\lambda}\sin\tilde{\theta}$. The density is now expressed in terms of the potential by the Cauchy–Lagrange equation:

$$\tilde{\rho}(x, y, t) = [1 + \tfrac{1}{2}(\gamma - 1)(1 - \tilde{\lambda}^2 - 2\tilde{\varphi}_t)]^{1/(\gamma-1)}. \tag{2.110}$$

As usual, $\tilde{\lambda}(x, y, t)$ denotes the ratio of the flow velocity to the critical speed a_* of the given steady flow, $\tilde{\rho}(x, y, t)$ is the ratio of the flow density to the critical density ρ_*.

We replace $\tilde{\rho}$, $\tilde{\lambda}$, θ, $\tilde{\varphi}$ in (2.108)–(2.110) by

$$\begin{aligned} &\rho(x, y) + \rho'(x, y, t), \qquad && \lambda(x, y) + \lambda'(x, y, t), \\ &\theta(x, y) + \theta'(x, y, t), && \varphi(x, y) + \varphi'(x, y, t), \end{aligned}$$

correspondingly, where the first terms refer to the mean steady flow while the second ones marked by a prime denote small unsteady perturbations. By linearizing the continuity Equation (2.108), we get

$$\rho_t' + [-\rho\lambda\theta'\sin\theta + (\rho\lambda)'\cos\theta]_x + [\rho\lambda\theta'\cos\theta + (\rho\lambda)'\sin\theta]_y = 0. \tag{2.111}$$

From (2.110) we obtain

$$(\rho\lambda)' = (\rho\lambda)_\lambda\lambda' - \lambda\rho^{2-\gamma}\varphi_t'. \tag{2.112}$$

Substitution of the latter into (2.111) yields

$$\begin{aligned} \rho_t' + [-\rho\lambda\theta'\sin\theta + [(\rho\lambda)_\lambda\lambda' - \lambda\rho^{2-\gamma}\varphi_t']\cos\theta]_x& \\ + [\rho\lambda\theta'\cos\theta + [(\rho\lambda)_\lambda\lambda' - \lambda\rho^{2-\gamma}\varphi_t']\sin\theta]_y = 0,& \end{aligned} \tag{2.113}$$

which can be represented in the form

$$\Sigma_1 + \Sigma_2 = 0,$$

where

$$\Sigma_1 = [-\rho\lambda\theta' \sin\theta + (\rho\lambda)_\lambda \lambda' \cos\theta]_x + [\rho\lambda\theta' \cos\theta + (\rho\lambda)_\lambda \lambda' \sin\theta]_y, \quad (2.114)$$

$$\Sigma_2 = \rho_t' - [\lambda\rho^{2-\gamma}\varphi_t' \cos\theta]_x - [\lambda\rho^{2-\gamma}\varphi_t' \sin\theta]_y. \quad (2.115)$$

It should be recalled that $\rho(x, y)$, $\lambda(x, y)$, $\theta(x, y)$ occurring in these expressions correspond to steady flow, so that ρ is expressed in terms of λ by the Bernoulli law (2.2).

Let us transform Σ_1 into the form of the left-hand side of Equation (2.56) governing steady perturbation. By linearizing the condition of irrotationality (2.109), we gain the same relations as (2.61) and (2.62),

$$\lambda' = \lambda(x, y)u_\varphi, \qquad \theta' = \rho(x, y)u_\psi, \quad (2.116)$$

where the notation $u = \varphi'$ for the perturbation of the velocity potential is employed. Substitution of (2.116) into (2.114) yields

$$\begin{aligned}\Sigma_1 &= [-\rho^2\lambda \sin\theta\, u_\psi + (\rho\lambda)_\lambda \lambda u_\varphi \cos\theta]_x + [\rho^2\lambda\cos\theta\, u_\psi + (\rho\lambda)_\lambda \lambda u_\varphi \sin\theta]_y \\ &= [-\rho^2\lambda\sin\theta\, u_\psi]_x + [\rho^2\lambda\cos\theta u_\psi]_y \\ &\quad + [(\rho\lambda)_\lambda \lambda u_\varphi \cos\theta]_x + [(\rho\lambda)_\lambda \lambda u_\varphi \sin\theta]_y. \end{aligned} \quad (2.117)$$

The sum of the first and second terms in the right-hand side can be represented as follows,

$$\begin{aligned}[-\rho^2\lambda\sin\theta u_\psi]_x + [\rho^2\lambda\cos\theta u_\psi]_y &= -[\rho^2 u_\psi]_x \lambda\sin\theta - \rho^2 u_\psi[\lambda\sin\theta]_x, \\ +[\rho^2 u_\psi]_y \lambda\cos\theta + \rho^2 u_\psi[\lambda\cos\theta\,]_y &= -[\rho^2 u_\psi]_x\lambda\sin\theta + [\rho^2 u_\psi]_y \lambda\cos\theta,\end{aligned}$$

due to the condition (2.109). The third and fourth terms in (2.117) can be rearranged using the continuity equation $(\rho\lambda\cos\theta)_x + (\rho\lambda\sin\theta)_y = 0$, since ρ, λ, θ correspond to the steady flow:

$$\begin{aligned}&[(\rho\lambda)_\lambda \lambda u_\varphi \cos\theta]_x + [(\rho\lambda)_\lambda \lambda u_\varphi \sin\theta]_y \\ &\qquad = \left[\frac{(\rho\lambda)_\lambda}{\rho}\rho\lambda u_\varphi\cos\theta\right]_x + \left[\frac{(\rho\lambda)_\lambda}{\rho}\rho\lambda u_\varphi\sin\theta\right]_y \\ &\qquad = \left[\frac{(\rho\lambda)_\lambda}{\rho}u_\varphi\right]_x \rho\lambda\cos\theta + \left[\frac{(\rho\lambda)_\lambda}{\rho}u_\varphi\right]_y \rho\lambda\sin\theta.\end{aligned}$$

Therefore, (2.117) can be rewritten in the form

$$\begin{aligned}\Sigma_1 = &-\lambda\sin\theta\,[\rho^2 u_\psi]_x + \lambda\cos\theta[\rho^2 u_\psi]_y \\ &+ \rho\lambda\cos\theta\left[\frac{(\rho\lambda)_\lambda}{\rho}u_\varphi\right]_x + \rho\lambda\sin\theta\left[\frac{(\rho\lambda)_\lambda}{\rho}u_\varphi\right]_y.\end{aligned}$$

The derivatives $\partial/\partial x$, $\partial/\partial x$ can be represented in terms of $\partial/\partial\varphi$ and $\partial/\partial\psi$ by means of (2.59), (2.60):

$$\Sigma_1 = \rho\lambda^2(\rho^2 u_\psi)_\psi + \rho\lambda^2\left[\frac{(\rho\lambda)_\lambda}{\rho}u_\varphi\right]_\varphi. \tag{2.118}$$

Now, we turn to the sum Σ_2 given by (2.115). Using the continuity equation, we represent it as follows:

$$\begin{aligned}\Sigma_2 &= \rho'_t - \rho\lambda\cos\theta[\rho^{1-\gamma}u_t]_x - \rho\lambda\sin\theta[\rho^{1-\gamma}u_t]_y \\ &= \rho'_t - \rho\lambda^2[\rho^{1-\gamma}u_t]_\varphi = \rho'_t - \rho\lambda^2[\rho^{1-\gamma}u_{t\varphi} + (\gamma-1)\lambda\rho^{2-2\gamma}\lambda_\varphi u_t].\end{aligned} \tag{2.119}$$

Linearization of (2.110) yields

$$\rho' = -\rho^{2-\gamma}(\lambda\lambda' + \varphi'_t) = -\rho^{2-\gamma}(\lambda^2 u_\varphi + u_t). \tag{2.120}$$

By substituting this into (2.119), we obtain

$$\begin{aligned}\Sigma_2 &= -\rho^{2-\gamma}(\lambda^2 u_\varphi + u_t)_t - \rho\lambda^2[\rho^{1-\gamma}u_{t\varphi} + (\gamma-1)\lambda\rho^{2-2\gamma}\lambda_\varphi u_t] \\ &= -\rho\lambda^2[\rho^{1-\gamma}\lambda^{-2}u_{tt} + 2\rho^{1-\gamma}u_{t\varphi} + (\gamma-1)\lambda\rho^{2-2\gamma}\lambda_\varphi u_t].\end{aligned} \tag{2.121}$$

Finally, by summing up (2.118) and (2.121), we turn (2.113) into the *convected wave equation*

$$\left[\frac{(\rho\lambda)_\lambda}{\rho}u_\varphi\right]_\varphi + (\rho^2 u_\psi)_\psi - \rho^{1-\gamma}[\lambda^{-2}u_{tt} + 2u_{t\varphi} + (\gamma-1)\lambda\rho^{1-\gamma}\lambda_\varphi u_t] = 0, \tag{2.122}$$

where $\lambda(\varphi, \psi)$ is the velocity of the mean flow in the streamline coordinates, and

$$\rho^{1-\gamma} = \frac{2}{\gamma + 1 - (\gamma-1)\lambda^2}.$$

A similar equation governing acoustical perturbations was derived by Lebedinsky *et al.* (1982). However, they used the assumption that λ depends only on the x-coordinate.

In the nearsonic approximation, Equation (2.122) becomes

$$(\gamma+1)(1-\lambda)u_{\varphi\varphi} + u_{\psi\psi} - (\gamma+1)\lambda_\varphi u_\varphi - 2u_{t\varphi} - u_{tt} - (\gamma-1)\lambda_\varphi u_t = 0. \tag{2.123}$$

The presence of the terms $2u_{t\varphi}$ and $(\gamma+1)\lambda u_{\varphi\varphi}$ is virtually associated with the convection of acoustical perturbations at the flow velocity $\lambda(\varphi, \psi)$, which is close to the speed of sound. The terms with λ_φ show influence of the nonuniformity of the flow which causes diffraction and attenuation of sound waves.

If the flow velocity is small, $0 < \lambda(\varphi, \psi) \ll 1$, then multiplying Equation (2.122) by λ^2 and omitting the terms of the order $O(\lambda)$, one gets

$$\lambda^2(u_{\varphi\varphi} + \rho_0^2 u_{\psi\psi}) - \frac{2}{\gamma+1}u_{tt} = 0,$$

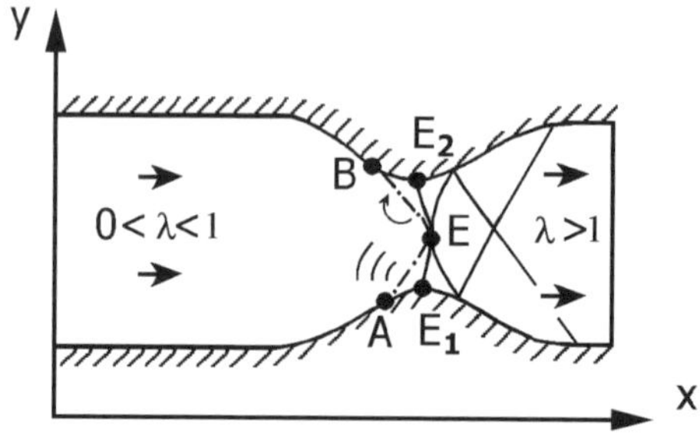

Figure 2.22 A domain of the plane (x, y) in which acoustical oscillations are considered.

where $\rho_0 = (\frac{1}{2}(\gamma+1))^{1/(\gamma-1)}$ is the stagnation density. Since

$$\lambda\frac{\partial}{\partial\varphi} = \cos\theta\frac{\partial}{\partial x} + \sin\theta\frac{\partial}{\partial y}, \qquad \lambda\rho_0\frac{\partial}{\partial\psi} = -\sin\theta\frac{\partial}{\partial x} + \cos\theta\frac{\partial}{\partial y},$$

the equation can be represented as follows,

$$\lambda\left(\cos\theta\frac{\partial u_\varphi}{\partial x} + \sin\theta\frac{\partial u_\varphi}{\partial y}\right) + \lambda\rho_0\left(\cos\theta\frac{\partial u_\psi}{\partial y} - \sin\theta\frac{\partial u_\psi}{\partial x}\right) - \frac{2}{\gamma+1}u_{tt} = 0,$$

$$\frac{\partial\lambda\cos\theta u_\varphi}{\partial x} + \frac{\partial\lambda\sin\theta u_\varphi}{\partial y} + \frac{\partial\lambda\rho_0\cos\theta u_\psi}{\partial y} - \frac{\partial\lambda\rho_0\sin\theta u_\psi}{\partial x} - \frac{2}{\gamma+1}u_{tt} = 0,$$

$$\frac{\partial}{\partial x}(\lambda\cos\theta u_\varphi - \lambda\rho_0\sin\theta u_\psi) + \frac{\partial}{\partial y}(\lambda\sin\theta u_\varphi + \lambda\rho_0\cos\theta u_\psi) - \frac{2}{\gamma+1}u_{tt} = 0,$$

$$u_{xx} + u_{yy} - \frac{2}{\gamma+1}u_{tt} = 0,$$

where $\frac{1}{2}(\gamma+1)$ is the nondimensional speed of sound at the stagnation state.

2.7.2 Fredholm solvability of a boundary-value problem

In the sequel, we consider harmonic oscillations, i.e. solutions of the form $u(\varphi,\psi,t) = \mathrm{Re}[u_\mathrm{a}(\varphi,\psi)\mathrm{e}^{\mathrm{i}\omega t}]$. Substituting this into Equation (2.122) and omitting the subscript 'a' in the notation of amplitude $u_\mathrm{a}(\varphi,\psi)$, we obtain

$$\left[\frac{(\rho\lambda)_\lambda}{\rho}u_\varphi\right]_\varphi + (\rho^2 u_\psi)_\psi - \rho^{1-\gamma}[2\mathrm{i}\omega u_\varphi - \omega^2\lambda^{-2}u + \mathrm{i}\omega(\gamma-1)\lambda\rho^{1-\gamma}\lambda_\varphi u] = 0, \tag{2.124}$$

where $u(\varphi,\psi)$ is the complex amplitude and ω is the complex frequency of oscillations. Let the rectangle $G = \{(\varphi,\psi)\in\mathbb{R}^2;\ \varphi_1<\varphi<\varphi_2,\ \psi_1<\psi<\psi_2\}$ in the streamline coordinates correspond to a nozzle in the plane (x, y). We prescribe the boundary conditions with respect to the amplitude of harmonic oscillations as follows:

$$\left.\begin{array}{lll} u_\psi = 0 & \text{at} & \psi = \psi_1 \text{ and } \psi = \psi_2, \\ u = g(\psi) & \text{at} & \varphi = \varphi_1. \end{array}\right\} \tag{2.125}$$

The Neumann condition on the sides $\psi = \psi_1$ and $\psi = \psi_2$ is a classical condition on acoustically rigid walls. The nonhomogeneous Dirichlet condition at $\varphi = \varphi_1$ is associated with forced oscillations under consideration. It means prescription of the acoustical energy flux at the inlet, as seen from the formula (2.133) discussed at the end of this section. The 'nozzle' can refer either to a Laval nozzle itself or to the nozzle along with the attached reservoir, see Figure 2.22. Though the velocity of the mean flow in the reservoir is small, it cannot be neglected since the nonzero velocity provides the correctness of using the mapping $(\varphi, \psi) \leftrightarrow (x, y)$.

Theorem 2.10. *Suppose that the flow velocity is subsonic at the nozzle inlet, $\lambda(\varphi_1, \psi) < 1$, and supersonic at the outlet, $\lambda(\varphi_1, \psi) > 1$. Let the flow acceleration be positive on the sonic line, $\lambda_\varphi > 0$ at $\lambda = 1$, and the complex frequency of the forced acoustical oscillations satisfy the inequality $4\,\mathrm{Im}\,\omega < (\gamma + 1)\lambda_\varphi|_{\lambda=1}$. Then Problem (2.124), (2.125) is Fredholm solvable in the class of functions $C^1(\overline{G}) \cap C^2(G)$, i.e. the following Fredholm alternative is true:*

(i) *either there exists a unique solution of Problem (2.124), (2.125) for every function $g(\psi)$ which is smooth enough and satisfies consistency conditions at the corner points $\psi = \psi_1, \varphi = \varphi_1$ and $\psi = \psi_2, \varphi = \varphi_1$ of the domain G, or*

(ii) *the function $g(\psi)$ is to satisfy l integral conditions, where $l \geqslant 1$, in order to provide the solvability of this problem; in the latter case the homogeneous problem (i.e. with $g(\psi) \equiv 0$) has l linearly independent solutions.*

Theorem 2.10 shows that if there exist free harmonic oscillations in transonic flow through the nozzle, which decay slowly, $0 \leqslant 4\,\mathrm{Im}\,\omega < (\gamma + 1)\lambda_\varphi|_{\lambda=1}$, or increase, $\mathrm{Im}\,\omega < 0$, then Problem (2.124), (2.125) with $g(\psi) \equiv 0$ can have only a finite number of linearly independent solutions representing different acoustical fields at this frequency ω.

A proof of Theorem 2.10 is based on consideration of Equation (1.1) with respect to the complex function $u = u_1 + \mathrm{i}u_2$,

$$Lu \equiv k(x, y)u_{xx} + [a(x, y)u_y]_y - \alpha(x, y)u_x + c(x, y)u = f(x, y),$$

where k, a are real-valued functions, while $\alpha = \alpha_1 + \mathrm{i}\alpha_2$, $c = c_1 + \mathrm{i}c_2$, $f = f_1 + \mathrm{i}f_2$ are complex-valued ones. An analysis of the integral

$$\int_G \mathrm{Re}[Lu \cdot (u_1 - \mathrm{i}u_2)_x]\,\mathrm{d}G, \tag{2.126}$$

makes it possible to obtain necessary *a priori* estimates and to prove the Fredholm solvability of Problem (2.124), (2.125), as described in Kuz'min (1992a).

One cannot infer from Theorem 2.10 the discreteness of the frequency spectrum for free acoustical oscillations in transonic flow in the nozzle even at

$$4\,\mathrm{Im}\,\omega < (\gamma + 1)\lambda_{\varphi}|_{\lambda=1}.$$

Indeed, Equation (2.124) can obviously be represented in the form

$$L_{\omega}u + \omega^2 u = 0,$$

where $L_{\omega}u$ is a compact operator. If the eigenvalues μ_j of this operator do not coincide with $-\omega^2$, then Problem (2.124), (2.125) is uniquely solvable, and we arrive at the case (i) pointed out in Theorem 2.10. On the contrary, if an eigenvalue μ_j coincides with $-\omega^2$, then the case (ii) occurs. The values of μ_j, in general, vary if ω varies. Therefore, in the second case one cannot rule out the situation in which the equality $\mu_j = -\omega^2$ holds for values of ω which constitute a curve or a domain in the complex frequency plane. Such frequencies ω would represent a continuous spectrum of Problem (2.124), (2.125).

Note that a problem of free acoustical oscillations in the supersonic boundary layer has both discrete and continuous spectra (Gaponov 1977). Theorem 2.10 provides no estimates for the spectrum of ω. Probably, free acoustical oscillations in transonic flow that accelerates from subsonic to supersonic velocities are always decaying (this is rigorously proved below for the one-dimensional flow) as opposed to free oscillations in subsonic flow which are neutrally stable. Then the inequality $\mathrm{Im}\,\omega > 0$ must hold for the spectral values of ω. Hence, frequencies ω with $\mathrm{Im}\,\omega = 0$ appear not to be spectral. This means that Problem (2.124), (2.125) with $g \equiv 0$ and real-valued ω has no solutions except for the identical zero. Since the condition of Fredholm solvability mentioned in Theorem 2.10 holds for real ω, the uniqueness of the solution to Problem (2.124), (2.125) on forced vibrations is followed by its existence.

The possibility of free oscillations with $\mathrm{Im}\,\omega > 0$ can be physically accounted for by the existence of receding waves which propagate in the upstream direction through the nozzle and interact with the waves moving downstream. The reflection in the upstream direction occurs on the walls in the convergent part of the nozzle, and also inside the flow due to the nonuniformity caused by the nonzero velocity gradient. In the one-dimensional approach, the critical section of the nozzle is acoustically equivalent to a solid boundary partially absorbing sound waves (Crocco and Chen 1956).

In two-dimensional flow, a pattern of the sound propagation is as follows. In the domains G' and G'' bounded by the walls, the sonic line AEB, and the characteristic line E_1EE_2, acoustical perturbation, being convected by the supersonic flow, partially penetrates into the subsonic region through the sonic line, as sketched in Figure 2.22. In the subsonic region, it propagates in all directions including the upstream one. A portion of the perturbation penetrates back into G', G'' and is convected by the supersonic flow again. Meanwhile, the portion of the perturbation which crosses the characteristic line E_1EE_2 cannot propagate upstream in the supersonic flow and does not influence the phenomena in the domains G', G'' and in the subsonic region.

It should be noted that an analysis of acoustical perturbations in the nozzle on the basis of linearized equations may be incorrect if one studies propagation of sound impulses of high amplitudes (Culick 1994; Ferrand *et al.* 1998). However, if the intensity of acoustical oscillations is not too large, and the flow acceleration is positive in the supersonic region up to the sonic line, then nonlinear phenomena play a minor role as estimates show. Consequently, in order to find the acoustical field, one can use the linear equations. Furthermore, the knowledge of the spectrum of Problem (2.124), (2.125) is of interest for studies of the linear stability of transonic flow. A nearsonic approximation to Equation (2.124) reads

$$(\gamma+1)(1-\lambda)u_{\varphi\varphi}+u_{\psi\psi}-[(\gamma+1)\lambda_\varphi+2\mathrm{i}\omega]u_\varphi-[(\gamma-1)\lambda_\varphi\mathrm{i}\omega-\omega^2]u=0. \tag{2.127}$$

It turns out that estimates of spectral values of ω on the basis of the integral (2.126) are inefficient. They only enable one to prove the absence of the spectrum below the parabola $\omega_2 = c_1 + c_2\omega_1^2$ in the plane of complex frequency $\omega = \omega_1 + \mathrm{i}\omega_2$, where $c_1 < 0$ and $c_2 < 0$ are defined by the given velocity of the mean flow. The inefficiency of the estimates is due to the absence of the imaginary part of $\alpha(x, y)$ in (2.126). Meanwhile it is the term $2\mathrm{i}\,\mathrm{Re}\,\omega$ that plays a substantial role in Equation (2.127).

2.7.3 A one-dimensional model for acoustical oscillations

Before 1990s, mathematical studies of acoustical oscillations in a Laval nozzle were based on the one-dimensional approximation, in which parameters of the mean flow and the acoustical field depend only on the longitudinal coordinate x (Natanzon 1986). Such an approximation is physically justified for nozzles with smooth shape of walls and slightly varying cross-sectional area. If one applies Equation (2.124) to this case, it reduces to the ordinary differential equation

$$\left[\frac{(\rho\lambda)_\lambda}{\rho}u_\varphi\right]_\varphi-\rho^{1-\gamma}[2\mathrm{i}\omega u_\varphi-\omega^2\lambda^{-2}u+\mathrm{i}\omega(\gamma-1)\lambda\rho^{1-\gamma}\lambda_\varphi u]=0. \tag{2.128}$$

Let us prove that, if the boundary condition $u(\varphi_1) = 0$ holds at the nozzle inlet, then exponentially increasing free oscillations in the transonic flow are impossible. Assume that $0 < \lambda < 1$ at $\varphi_1 \leqslant \varphi < 0$, $1 < \lambda$ at $0 < \varphi \leqslant \varphi_2$, $\lambda = 1$ at $\varphi = 0$. In addition, let $u(\varphi) \in C^1[\varphi_1, \varphi_2] \cap C^2(\varphi_1, \varphi_2)$. We multiply Equation (2.128) by $\bar{\omega}\bar{u} = (\omega_1 - \mathrm{i}\omega_2)(u_1 - \mathrm{i}u_2)$, and integrate the result from φ_1 to 0:

$$\int_{\varphi_1}^{0}\left\{\bar{\omega}\left[\frac{(\rho\lambda)_\lambda}{\rho}u_\varphi\right]_\varphi\bar{u}\right.$$
$$\left.-\rho^{1-\gamma}[2\mathrm{i}|\omega|^2u_\varphi\bar{u}-\omega^2\bar{\omega}\lambda^{-2}|u|^2+\mathrm{i}|\omega|^2(\gamma-1)\rho^{1-\gamma}\lambda\lambda_\varphi|u|^2]\right\}\mathrm{d}\varphi=0.$$

Then we integrate by parts the first term, use the condition $u(\varphi_1) = 0$, and pick out the imaginary part of the equation:

$$\int_{\varphi_1}^{0} \left\{ \omega_2 \frac{(\rho\lambda)_\lambda}{\rho} |u_\varphi|^2 - \rho^{1-\gamma}[2|\omega|^2(|u|^2)_\varphi - \omega_2|\omega|^2\lambda^{-2}|u|^2 + |\omega|^2(\gamma-1)\rho^{1-\gamma}\lambda\lambda_\varphi|u|^2] \right\} \mathrm{d}\varphi = 0.$$

Finally, integrating by parts the term that contains $(|u|^2)_\varphi$, we obtain

$$-|\omega|^2|u(0)|^2 + \omega_2 \int_{\varphi_1}^{0} \left[\frac{(\rho\lambda)_\lambda}{\rho} |u_\varphi|^2 + \rho^{1-\gamma}|\omega|^2\lambda^{-2}|u|^2 \right] \mathrm{d}\varphi = 0. \qquad (2.129)$$

If $\omega_2 < 0$, then both terms in the left-hand side are nonpositive, hence, $u(\varphi) \equiv 0$ at $\varphi_1 \leqslant \varphi \leqslant 0$. This enables one to consider the interval $0 < \varphi \leqslant \varphi_2$, in which the flow is supersonic, and to treat the solution as follows. Multiplying Equation (2.128) by $\bar{u}_\varphi$ and integrating the result from 0 to $\varepsilon > 0$, we find that $u(\varphi) \equiv 0$ at $0 \leqslant \varphi \leqslant \varepsilon$ if ε is sufficiently small and $\lambda_\varphi(0) > 0$. Then solving the differential equation under the vanishing initial data at $\varphi = \varepsilon$, we obtain $u(\varphi) \equiv 0$ at $\varepsilon \leqslant \varphi \leqslant \varphi_2$ because the flow is purely supersonic here and the differential equation has no singular points. Thus we see that complex values of the frequency ω with negative imaginary parts are not spectral. The real-valued ω are not spectral as well. Indeed, in the class of functions $C^1[\varphi_1, \varphi_2]$ a unique solution of the ordinary differential Equation (2.128) can be written as $u(\varphi) = F(\varphi) \cdot \text{const.}$, where $F(\varphi)$ is the hypergeometric function, $F(0) \neq 0$. On the other hand, setting $\omega_2 = \operatorname{Im}\omega = 0$ in (2.129), we conclude that $u(0) = 0$. Hence, const. $= 0$ and $u(\varphi) \equiv 0$.

Therefore, spectral values of ω must satisfy the inequality $\operatorname{Im}\omega > 0$. It means that free acoustical oscillations in Laval nozzles are decaying. The condition of positive acceleration was required at the critical section $\varphi = 0$ actually in order to prove that $u \equiv 0$ in the half-vicinity $0 \leqslant \varphi \leqslant \varepsilon$ of the singular point $\varphi = 0$ of the equation.

For $\operatorname{Im}\omega > \frac{1}{2}(\gamma + 1)\lambda_\varphi(0)$, any value of ω is a spectral one as follows from an analysis of the boundary-value problem for the ordinary differential Equation (2.128) endowed with an extra condition at the outlet $\varphi = \varphi_2$. Therefore, the discrete spectrum of free oscillations in the Laval nozzle can only lie in the strip $0 < \operatorname{Im}\omega < \frac{1}{2}(\gamma+1)\lambda_\varphi(0)$ (see Figure 2.23). Meanwhile, there are no spectral values in this strip at large values of $|\operatorname{Re}\omega|$; this follows from a high-frequency asymptotics for the general solution of Equation (2.128) (King and Karamcheti 1973). More accurate spectral estimates in the one-dimensional approach can probably be obtained using available results for nonself-adjoint ordinary differential equations (Pivovarchik 1987).

Landahl (1964) studied one-dimensional propagation of harmonic waves in the nearsonic flow on the basis of an equation similar to (2.123). He concluded that, if the source is placed in the subsonic region, then the receding wave is very much different in the two cases:

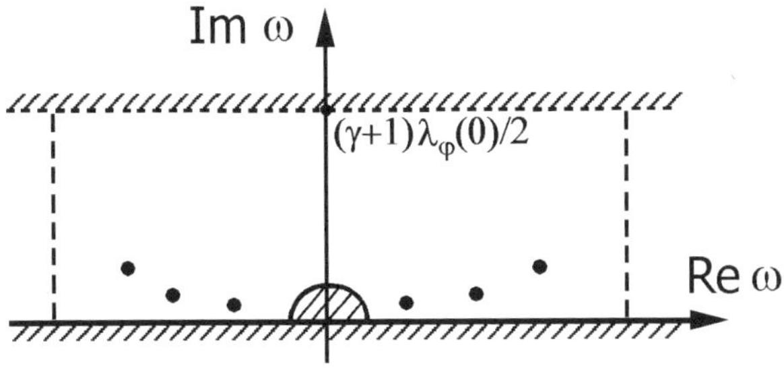

Figure 2.23 A domain of the complex plane ω in which the spectrum of free acoustical oscillations is located. Reproduced from Kuz'min (1992a) by permission of Birkhäuser.

In the accelerated flow its amplitude decreases and wavelength increases away from the source. The amplitude decreases faster away from the source than for the case of a constant Mach number. In the decelerated flow, on the other hand, the amplitude goes to infinity and wave length to zero as the sonic line is approached. This behaviour is to be expected for physical reasons since the energy transported by the receding wave gets trapped at the sonic line. This would indicate that a flow that decelerates smoothly from supersonic to subsonic flow is unstable and must eventually develop a compression shock. However, care must be exercised in applying the present result to general two- and three-dimensional flows since curvature of the flow and the sonic line or surface may allow the disturbance energy to escape sideways.

2.7.4 Two-dimensional subsonic flow in a channel

Assume the velocity $\lambda(\varphi, \psi)$ involved in the coefficients of Equation (2.124) governing the amplitude of oscillations to be subsonic $0 < \lambda(\varphi, \psi) < 1$ over the domain $\overline{G} = \{(\varphi, \psi) \in \mathbb{R}^2;\ \varphi_1 \leqslant \varphi \leqslant \varphi_2,\ \psi_1 \leqslant \psi \leqslant \psi_2\}$. Consider the homogeneous boundary conditions

$$\left.\begin{aligned} u_\psi &= 0 \quad \text{at} \quad \psi = \psi_1 \text{ and } \psi = \psi_2, \\ u &= 0 \quad \text{at} \quad \varphi = \varphi_1 \text{ and } \varphi = \varphi_2. \end{aligned}\right\} \tag{2.130}$$

Below we prove that spectral values of ω for which there are nonzero solutions of Problem (2.124), (2.130) can only be real-valued ones. Let us multiply Equation (2.124) by $\bar{\omega}\bar{u}$ and integrate the result over the domain G. After that we integrate by parts the first term as in the previous subsection, then use the condition of vanishing u_ψ at $\psi = \psi_1$, $\psi = \psi_2$, and pick out the imaginary part of the equation; in this way we obtain instead of (2.129)

$$\omega_2 \int_G [(\rho\lambda)_\lambda \rho^{-1}|u_\varphi|^2 + \rho^2|u_\psi|^2 + \rho^{1-\gamma}|\omega|^2\lambda^{-2}|u|^2]\,\mathrm{d}G$$
$$- \int_{\psi_1}^{\psi_2} [\rho^{1-\gamma}|\omega|^2|u|^2 + (\rho\lambda)_\lambda\rho^{-1}\,\mathrm{Im}[\omega u(u_1 - \mathrm{i}u_2)_\varphi]]\Big|_{\varphi=\varphi_1}^{\varphi=\varphi_2} \mathrm{d}\psi = 0. \quad (2.131)$$

The second integral vanishes due to the boundary conditions (2.130). Hence, the integral over G vanishes if $\omega_2 = \mathrm{Im}\,\omega \neq 0$. The expression $(\rho\lambda)_\lambda$ is positive due to $0 < \lambda < 1$, hence, the integrand is nonnegative. Therefore, $u \equiv 0$ if $\omega_2 \neq 0$, i.e. the real-valued frequencies are not spectral. This means that free acoustical oscillations in subsonic potential flow are neutrally stable.

We note that Problem (2.124), (2.130) for subsonic flow is self-adjoint at a fixed real-valued ω. In order to prove this assertion, we denote by $M_\omega u$ the differential operator on the left-hand side of (2.124)

$$M_\omega u = [(\rho\lambda)_\lambda \rho^{-1} u_\varphi]_\varphi + (\rho^2 u_\psi)_\psi + 2\varrho u_\varphi + \varrho_\varphi u + \mathrm{i}\omega\varrho\lambda^{-2} u,$$

where the notation $\varrho = -\mathrm{i}\omega\rho^{1-\gamma}$ is introduced for brevity. For a complex-valued function $\bar{v} = v_1 - \mathrm{i}v_2$, using Green's formula, we obtain

$$\int_G M_\omega u \bar{v}\,\mathrm{d}G$$
$$= \int_G [(\rho\lambda)_\lambda \rho^{-1}\bar{v}_\varphi]_\varphi u + (\rho^2 \bar{v}_\psi)_\psi u - 2\varrho\bar{v}_\varphi u - \varrho_\varphi \bar{v} u + \mathrm{i}\omega\varrho\lambda^{-2}\bar{u}]\,\mathrm{d}G$$
$$= \int_G \overline{M_\omega v \bar{u}}\,\mathrm{d}G,$$

since $\bar{\varrho} = -\varrho$ for the real-valued ω. The obtained relation means that the problem is self-adjoint. The operator $M_\omega u$ is a compact one. Therefore the equation $M_\omega u - \mu u = 0$ under boundary conditions (2.130) has at most a countable set of real eigenvalues $\mu_j(\omega)$ for any real ω. If any of μ_j vanishes, then the chosen ω is a spectral frequency for Problem (2.124), (2.130).

The amplitude $p_\mathrm{a}(\varphi, \psi)$ of harmonic oscillations of the pressure normalized by the critical pressure p_* is expressed in terms of the amplitude $u(\varphi, \psi)$ of oscillations of the velocity potential through the formula

$$p_\mathrm{a} = -\gamma\rho(\lambda^2 u_\varphi + \mathrm{i}\omega u), \quad (2.132)$$

which follows from (2.120) if we replace ρ' by $\rho_\mathrm{a}\mathrm{e}^{\mathrm{i}\omega t}$ and u by $u\mathrm{e}^{\mathrm{i}\omega t}$, and after that use the well-known relations $p = \rho^\gamma$, $p_\mathrm{a}/p = \gamma\rho_\mathrm{a}/\rho$.

If $u = 0$ at the inlet or outlet of the channel, then (2.132) yields $p_\mathrm{a} = -\gamma\rho\lambda^2 u_\varphi = -\gamma\rho\lambda\lambda_\mathrm{a}$. Substitution of this relation into the expression for the density of acoustical energy flux written in the nondimensional form (Morfey 1971),

$$N = \frac{\lambda}{\gamma^2\rho a^2} p_\mathrm{a}^2 + \left(1 + \frac{\lambda^2}{a^2}\right)\frac{p_\mathrm{a}\lambda_\mathrm{a}}{\gamma} + \rho\lambda\lambda_\mathrm{a}^2, \quad (2.133)$$

where a is the ratio of the sound speed to a_*, shows that $N = 0$. Hence, the condition $u = 0$ at the inlet or outlet section of the channel implies that the acoustical energy flux through that section vanishes. The above established neutral stability of acoustical oscillations in subsonic potential flow coincides with the conclusion inferred from the equation of acoustical energy balance by Morfey (1971). It is remarkable that the result on the neutral stability differs from the known Rayleigh theorem for incompressible fluid, though it has been proved in a similar way. The Rayleigh theorem establishes a necessary condition for instability of parallel (vortex) flow of inviscid incompressible fluid; that condition is the existence of an inflexion point of the velocity profile at which the curvature changes the sign (Swinney and Gollub 1981).

In the classical acoustics of immovable media, the usual condition at the open ends of a channel is the vanishing amplitude p_a of acoustical oscillations. For $\lambda = 0$ it is equivalent to the condition of vanishing u as seen from (2.132). Meanwhile, in the moving media, the condition $p_a = 0$ yields the relation $\lambda^2 u_\varphi + i\omega u = 0$. In this case, by analysing (2.131), one can prove the two assertions as follows. If $u = 0$ at the inlet, $p_a = 0$ at the outlet, and ω is an eigenfrequency of free oscillations, then $\omega_2 > 0$, i.e. the oscillations are decaying. If $p_a = 0$ at the inlet and $u = 0$ at the outlet, then $\omega_2 < 0$, i.e. the oscillations are increasing. The latter result can be accounted for by the presence of a fictitious source of acoustical energy if $p_a = 0$ is imposed at the inlet (this follows from an analysis of acoustical oscillations in the uniform flow through a channel of constant cross-sectional area (Raushenbakh 1961)). At the same time, the prescription of $p_a = 0$ at the outlet admits losses of acoustical energy due to the effect of its convection into the outer media.

Both the condition $p_a = 0$ and the condition $u = 0$ at open ends of a channel with moving media are approximate because they do not take into account two-dimensional phenomena in the vicinity of the edges of the channel. However, the use of the streamline coordinates (φ, ψ) enables one to reduce the inaccuracy of these boundary conditions. Indeed, in order to take into account the two-dimensional phenomena at the edges, one can extend the domain under consideration by enclosing a near field of the outer media, as pointed out in Figure 2.24. Since the extended domain maps again onto a rectangle of the plane (φ, ψ), we can assume from the very beginning of our analysis that the rectangle G corresponds to the extended domain. As the distance from the outer portions of the boundary to the ends of the channel increases, their influence on the acoustical field is reduced. In other words, it becomes weaker as the length of the rectangle G increases.

From the physical point of view, it would be more accurate to pose a condition of radiation on the outer portions of the boundary $\varphi = \varphi_1$, $\varphi = \varphi_2$, as compared with $p_a = 0$ or $u = 0$. Let us consider the influence of the condition of radiation in the case of axisymmetric acoustical oscillations in axisymmetric subsonic flow. We choose the streamline $\psi = 0$ to coincide with the x-axis of flow symmetry and denote by

$$G_\varepsilon = \{\varphi, \psi) : \varphi_1 < \varphi_2, \ \varepsilon < \psi < \psi_2\}$$

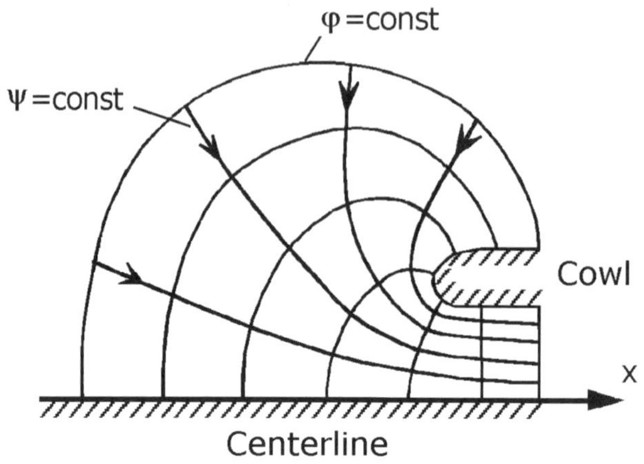

Figure 2.24 An extended domain for treatment of the sound radiation into the outer media.

the rectangle separated from the axis of symmetry by a strip of the width ε. The condition of radiation can be written in the form (Osipov and Shirkovsky 1988):

$$\lambda u_\varphi + \left(\mp \mathrm{i}\omega\sqrt{\frac{2}{\gamma+1}} + \frac{1}{\sqrt{x^2+y^2}}\right)u = 0$$
$$\text{on the portions } \varphi = \varphi_1 \text{ and } \varphi = \varphi_2,\ 0 < \psi < \psi_2, \text{ of the boundary,} \quad (2.134)$$

where $x = x(\varphi, \psi)$ is the longitudinal coordinate, $y = y(\varphi, \psi)$ is the radial coordinate in the physical plane; the upper sign in front of $\mathrm{i}\omega$ corresponds to $\varphi = \varphi_1$ while the lower does to $\varphi = \varphi_2$. A similar form of the condition of radiation was used in computations of noise radiation from ducted fans by Meyer (1996), Özyörük and Long (1995) and Roy and Eversman (1993).

Below we prove that free acoustical oscillations in the subsonic potential flow through a channel are decaying if condition (2.134) holds on the free portions of the boundary. It can be shown that the equation governing axisymmetric oscillations in axisymmetric flow differs from (2.124) only by the factor y^2 in front of u_ψ, where $y(\varphi, \psi)$ is a known function for the given steady flow. That is why, instead of (2.131), we obtain

$$\omega_2 \int_G [(\rho\lambda)_\lambda \rho^{-1}|u_\varphi|^2 + \rho^2 y^2 |u_\psi|^2 + \rho^{1-\gamma}|\omega|^2\lambda^{-2}|u|^2]\,\mathrm{d}G$$
$$+ \mathrm{Im}\int_\varepsilon^{\psi_2} [\bar{\omega}(\rho\lambda)_\lambda \rho^{-1} u_\varphi \bar{u} - \mathrm{i}|\omega|^2\rho^{1-\gamma}|u|^2]\Big|_{\varphi=\varphi_1}^{\varphi=\varphi_2}\,\mathrm{d}\psi$$
$$+ \mathrm{Im}\int_{\varphi_1}^{\varphi_2} [\bar{\omega}\rho^2 y^2 u_\psi \bar{u}]\Big|_{\psi=\varepsilon}^{\psi=\psi_2}\,\mathrm{d}\varphi = 0. \quad (2.135)$$

Assume that $u \in C(\overline{G}_0)$, $u_\varphi \in C(\overline{G}_0)$, and that u_ψ is continuous in $\overline{G}_0$ except, possibly, at the segment $\psi = 0$, $\varphi_1 \leqslant \varphi \leqslant \varphi_2$. The condition of symmetry $u_y = 0$ is to hold on the x-axis, that is why from the relation $u_y = \varphi_y u_\varphi + \psi_y u_\psi = \lambda \sin\theta u_\varphi + y\rho\lambda\cos\theta u_\psi$ we get $yu_\psi \to 0$ as $y \to 0$. (Actually, this is linked with the definition of the stream function for axisymmetric flow $\psi_x = -y\rho\lambda\sin\theta$, $\psi_y = y\rho\lambda\cos\theta$ which yields the relation $\psi = \rho\lambda y^2/2 + o(y^2)$.) Hence, $y^2 u_\psi \to 0$ as $\psi \to 0$. Therefore, passing to the limit as $\varepsilon \to 0$ in (2.135) and using the boundary condition $u_\psi = 0$ at $\psi = \psi_2$, we get

$$\omega_2 \int_{G_0} [(\rho\lambda)_\lambda \rho^{-1}|u_\varphi|^2 + \rho^2 y^2 |u_\psi|^2 + \rho^{1-\gamma}|\omega|^2\lambda^{-2}|u|^2]\,\mathrm{d}G \\ + \operatorname{Im}\int_0^{\psi_2} [\bar{\omega}(\rho\lambda)_\lambda \rho^{-1} u_\varphi \bar{u} - \mathrm{i}|\omega|^2 \rho^{1-\gamma}|u|^2]\Big|_{\varphi=\varphi_1}^{\varphi=\varphi_2}\,\mathrm{d}\psi = 0. \quad (2.136)$$

Due to the condition of radiation (2.134), the second integral can be represented as

$$\operatorname{Im}\int_0^{\psi_2}\left[-\frac{(\rho\lambda)_\lambda}{\rho\lambda}\sqrt{\frac{2}{\gamma+1}} - \rho^{1-\gamma}\right]|\omega|^2|u|^2\Big|^{\varphi=\varphi_2}\,\mathrm{d}\psi \\ + \operatorname{Im}\int_0^{\psi_2}\left[-\frac{(\rho\lambda)_\lambda}{\rho\lambda}\sqrt{\frac{2}{\gamma+1}} + \rho^{1-\gamma}\right]|\omega|^2|u|^2\Big|^{\varphi=\varphi_1}\,\mathrm{d}\psi. \quad (2.137)$$

We assume that

$$-\frac{(\rho\lambda)_\lambda}{\rho\lambda}\sqrt{\frac{2}{\gamma+1}} + \rho^{1-\gamma} < 0 \quad \text{at } \varphi = \varphi_1,\ 0 \leqslant \psi \leqslant \psi_2. \quad (2.138)$$

This can be provided by placing the portion $\varphi = \varphi_1$ of the boundary in the outer media at a large distance from the channel inlet, where $\lambda(\varphi_1, \psi)$ is small enough (Figure 2.24). Due to (2.138), both (2.137) and the second term in (2.136) are non-positive. Consequently, the values of $\omega = \omega_1 + \mathrm{i}\omega_2$ with negative imaginary parts ω_2 are not spectral. Furthermore, if ω is real-valued, i.e. $\omega_2 = \operatorname{Im}\omega = 0$, then from (2.134) and (2.136) we get $u = u_\varphi = 0$ at $\varphi = \varphi_1, \varphi = \varphi_2; 0 \leqslant \psi \leqslant \psi_2$. That is why, owing to the known results on the Cauchy problem for elliptic equations (Lavrent'ev 1957), we conclude that $u \equiv 0$ in G_0. Thus, real values of ω are not spectral as well. Hence, if ω is an eigenfrequency, then $\operatorname{Im}\omega > 0$, which means that free acoustical oscillations are decaying.

Further increase of the accuracy of far-field boundary conditions can be based on a family of partial differential operators B_m of Bayliss and Turkel (1982). They assumed that the solution in the far-field has a wave-like behaviour and then constructed a set of far-field boundary condition operators using asymptotic expansions. The B_1 operator, e.g. is accurate to $O(d^{-5/2})$, where d is the Prandtl–Glauert transformed spherical distance of the boundary point from the reference point which is taken to be the approximate location of the sound source.

Tam and Webb (1993) derived radiation and outflow boundary conditions based on the asymptotic solutions of the linearized Euler equations. Their boundary condition operator and that of Bayliss and Turkel are essentially the same.

2.8 Nonisentropic Flow

A model of steady nonisentropic inviscid flow may be helpful when examining, e.g. transonic flow through the nozzle of a solid propellant engine. In this model, the entropy is constant along the streamlines, however, it may be different at different streamlines. The governing equations are the Crocco one with respect to the vortex, and the continuity equation

$$(\rho\lambda\cos\theta)_x + (\rho\lambda\sin\theta)_y = 0. \tag{2.139}$$

The latter enables one to introduce the stream function by the relations

$$\psi_x = -\rho\lambda\sin\theta, \qquad \psi_y = \rho\lambda\cos\theta, \tag{2.140}$$

where $\rho(x, y)$ is the density related to the critical value of the density at the zero streamline $\hat{\rho}_*$. In the case of nonisentropic flow, the Bernoulli equation written in the form

$$\lambda^2 + \frac{2a^2}{\gamma - 1} = \frac{\gamma + 1}{\gamma - 1}, \tag{2.141}$$

remains true. The velocity magnitude λ and the sound speed a are normalized by the critical sound speed a_* which is constant throughout the flow field in the homo-energetic flow.

From thermodynamics we have

$$a^2 = \gamma p/\rho = \gamma\rho^{\gamma-1} p/\rho^\gamma = \gamma\rho^{\gamma-1}\mathrm{e}^{S/c_\mathrm{v}}, \tag{2.142}$$

where p is the pressure normalized by $\hat{\rho}_* a_*^2$, and c_v is the specific heat under constant volume, $S(\psi) = c_\mathrm{v}\log p/\rho^\gamma$ is the entropy. Therefore, Equation (2.141) takes the form

$$\lambda^2 + \frac{2\rho^{\gamma-1}}{\gamma - 1}\mathrm{e}^{S/c_\mathrm{v}} = \frac{\gamma + 1}{\gamma - 1}.$$

By resolving it with respect to ρ and using the notation

$$\bar{S} = \frac{S}{(\gamma - 1)c_\mathrm{v}} = \frac{1}{\gamma - 1}\log\frac{p}{\rho^\gamma},$$

we find

$$\rho(\lambda, \psi) = \mathrm{e}^{-\bar{S}(\psi)}\hat{\rho}(\lambda), \quad \text{where } \hat{\rho}(\lambda) = \left(1 - \frac{\gamma - 1}{\gamma + 1}\lambda^2\right)^{1/(\gamma-1)}.$$

Substitution of $\rho(\lambda, \psi)$ into (2.140) yields

$$\psi_x = -\sin\theta\,\lambda\hat{\rho}\mathrm{e}^{-\bar{S}(\psi)}, \qquad \psi_y = \cos\theta\,\lambda\hat{\rho}\mathrm{e}^{-\bar{S}(\psi)}. \tag{2.143}$$

The Crocco equation that governs flow vortex (Oswatitsch 1976, Section 4.5) can be represented in the form

$$(\lambda \sin\theta)_x - (\lambda \cos\theta)_y = \frac{\rho a^2}{\gamma}\frac{\mathrm{d}\bar{S}}{\mathrm{d}\psi}. \tag{2.144}$$

In fact, (2.144) follows from the Euler equation $uv_x + vv_y = -p_y/\rho$ which may be rewritten as

$$v_x - u_y = -\frac{1}{u}\left(\frac{1}{2}\frac{\partial\lambda^2}{\partial y} + \frac{1}{\rho}\frac{\partial p}{\partial y}\right). \tag{2.145}$$

From the Bernoulli Equation (2.141) we have

$$\begin{aligned}\frac{1}{2}\frac{\partial\lambda^2}{\partial y} &= -\frac{1}{\gamma-1}\frac{\partial a^2}{\partial y} = -\frac{\gamma}{\gamma-1}\frac{\partial}{\partial y}(\rho^{\gamma-1}\mathrm{e}^{S/c_\mathrm{v}})\\ &= -\gamma\rho^{\gamma-1}\mathrm{e}^{S/c_\mathrm{v}}\frac{\partial\rho}{\partial y} - \frac{\gamma}{\gamma-1}\rho^{\gamma-1}\frac{\partial \mathrm{e}^{S/c_\mathrm{v}}}{\partial y},\end{aligned}$$

while (2.142) yields

$$\frac{1}{\rho}\frac{\partial p}{\partial y} = \frac{1}{\rho}\frac{\partial}{\partial y}\left(\frac{\rho a^2}{\gamma}\right) = \frac{1}{\rho}\frac{\partial}{\partial y}(\rho^\gamma \mathrm{e}^{S/c_\mathrm{v}}) = \gamma\rho^{\gamma-2}\mathrm{e}^{S/c_\mathrm{v}}\frac{\partial\rho}{\partial y} + \rho^{\gamma-1}\frac{\partial \mathrm{e}^{S/c_\mathrm{v}}}{\partial y}.$$

By summing up the latter two equations and substituting the result into the right-hand side of (2.145), we arrive at the equation

$$\begin{aligned}v_x - u_y &= -\frac{1}{u}\left(1 - \frac{\gamma}{\gamma-1}\right)\rho^{\gamma-1}\frac{\partial \mathrm{e}^{S/c_\mathrm{v}}}{\partial y}\\ &= \frac{1}{u}\frac{\rho^{\gamma-1}}{\gamma-1}\frac{\partial \mathrm{e}^{S/c_\mathrm{v}}}{\partial y} = \frac{1}{u}\frac{\rho^{\gamma-1}}{\gamma-1}\rho u\frac{\mathrm{d}\mathrm{e}^{S/c_\mathrm{v}}}{\mathrm{d}\psi},\end{aligned}$$

i.e. Equation (2.144).

Due to (2.145), one can introduce a generalized potential $\varphi(x, y)$ by the relations

$$\varphi_x = \lambda\cos\theta\cdot \mathrm{e}^E, \qquad \varphi_y = \lambda\sin\theta\cdot \mathrm{e}^E, \qquad E = \frac{1}{\gamma}\int_0^\psi \frac{\mathrm{d}\bar{S}}{\mathrm{d}\psi}\frac{\mathrm{d}\psi}{M^2}, \tag{2.146}$$

where $M(\lambda)$ is the Mach number. The existence of this potential immediately follows from the expressions

$$\begin{aligned}\varphi_{xy} &= \left[(\lambda\cos\theta)_y + \lambda\cos\theta\left(\sin\theta\frac{\partial E}{\partial s} + \cos\theta\frac{\partial E}{\partial n}\right)\right]\mathrm{e}^E,\\ \varphi_{yx} &= \left[(\lambda\sin\theta)_x + \lambda\sin\theta\left(\cos\theta\frac{\partial E}{\partial s} - \sin\theta\frac{\partial E}{\partial n}\right)\right]\mathrm{e}^E,\end{aligned}$$

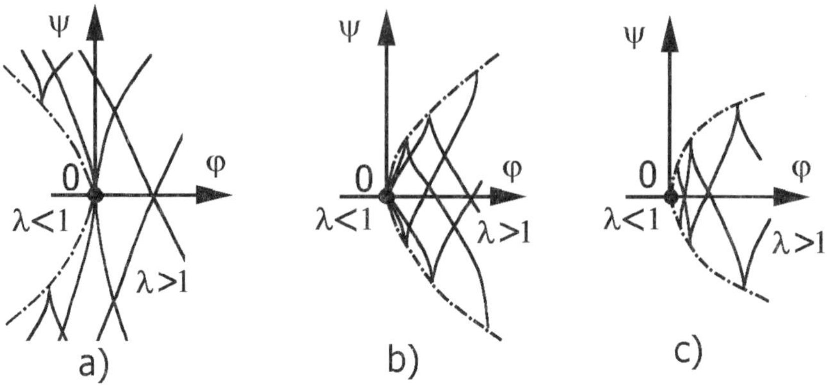

Figure 2.25 Qualitative patterns of the behaviour of the characteristic lines in nonisentropic transonic flow at $\lambda_{02} \neq 0$.

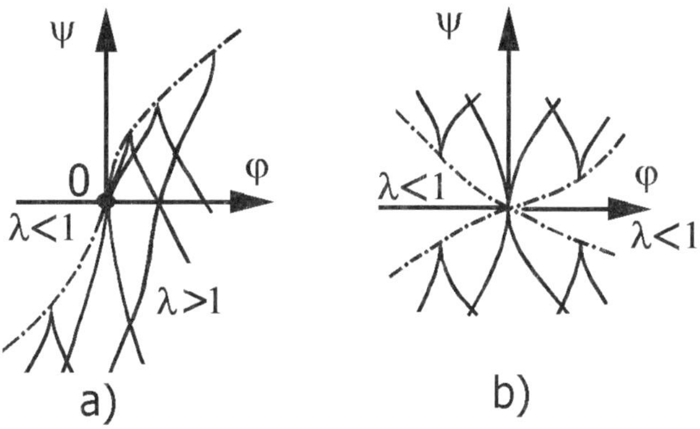

Figure 2.26 Qualitative patterns of the behaviour of the characteristic lines in nonisentropic transonic flow at $\lambda_{02} = 0$.

which prove that

$$\varphi_{xy} - \varphi_{yx} = \left[(\lambda \cos\theta)_y - (\lambda \sin\theta)_x + \lambda \frac{\partial E}{\partial n} \right] \mathrm{e}^E$$
$$= \left[(\lambda \cos\theta)_y - (\lambda \sin\theta)_x + \lambda^2 \frac{\rho}{\gamma M^2} \frac{\mathrm{d}\bar{S}}{\mathrm{d}\psi} \right] \mathrm{e}^E \equiv 0, \qquad (2.147)$$

owing to the vortex Equation (2.144). In these relations the integration is performed from $\psi = 0$ to $\psi(x, y)$ along a path orthogonal to the streamlines (see Figure 2.21).

Equations governing nonisentropic flow in the plane of streamline coordinates (φ, ψ) can be obtained in the same way as Chaplygin's equations in Section 2.1.

Since the potential φ and stream function ψ are now defined by (2.143) and (2.146) instead of (2.4), we obtain the system

$$(M^2-1)\frac{\mathrm{e}^E}{\hat{\rho}}\frac{\log\lambda}{\partial\varphi}=\frac{\partial\theta}{\partial\psi},\qquad \frac{\mathrm{e}^E}{\hat{\rho}}\frac{\partial\theta}{\partial\varphi}=\frac{\partial\log\lambda}{\partial\psi}+\frac{1}{\gamma M^2}\frac{\mathrm{d}\bar{S}}{\mathrm{d}\psi},$$

instead of (2.7) (cf. Verhoff 1999). Let us differentiate the first equation with respect to φ, and the second one with respect to ψ. Then eliminating θ, we get the second-order equation

$$\frac{\partial}{\partial\varphi}\left[(M^2-1)\hat{\rho}^{-1}\mathrm{e}^E\frac{\partial\log\lambda}{\partial\varphi}\right]=\frac{\partial}{\partial\psi}\left[\hat{\rho}\mathrm{e}^{-E}\left(\frac{\partial\log\lambda}{\partial\psi}+\frac{1}{\gamma M^2}\frac{\mathrm{d}\bar{S}}{\mathrm{d}\psi}\right)\right].\tag{2.148}$$

The characteristic lines of Equation (2.148) in the plane (φ,ψ) are defined by the solutions of the ordinary differential equation

$$(M^2-1)\mathrm{e}^{2E}(\mathrm{d}\psi)^2-\hat{\rho}^2(\mathrm{d}\varphi)^2=0,\tag{2.149}$$

where $(M^2-1)=(\gamma+1)[\lambda-1+o(\lambda-1)]$.

Suppose that

$$\left.\begin{aligned}\bar{S}(\psi)&=\gamma[s_0\psi+s_1\psi^2/2+o(\psi^2)],\\ \lambda(\varphi,\psi)&=1+\lambda_{10}\varphi+\lambda_{02}\psi^2+o(|\varphi|+\psi^2),\end{aligned}\right\}\tag{2.150}$$

so that the sonic line $\lambda(\varphi,\psi)-1=0$ is orthogonal to the zero streamline at the origin. Asymptotic analysis of the ordinary differential Equation (2.149) (Shifrin 1966; Kuz'min 1986c) yields the following theorem.

Theorem 2.11. *Assume that $\lambda_{10}>0$ and $\lambda_{02}\neq 0$ in the expansion (2.150) near the origin. Then the qualitative pattern of transonic flow is as follows:*

(i) *in Figure 2.25a if* $s_1-s_0<(\gamma+1)\lambda_{10}^2$;

(ii) *in Figure 2.25b if* $(\gamma+1)\lambda_{10}^2<s_1-s_0\leqslant(\gamma+1)\lambda_{10}^2 9/8$;

(iii) *in Figure 2.25c if* $s_1-s_0>(\gamma+1)\lambda_{10}^2 9/8$.

In the case $\lambda_{10}>0$, $\lambda_{02}=0$, feasible patterns of the nonisentropic flow are presented in Figure 2.26. For $\lambda_{10}<0$ the qualitative patterns are symmetric with respect to the ψ-axis to those shown in Figures 2.25 and 2.26.

We notice that the equations governing nonisentropic inviscid flow admit the occurrence of closed sonic lines in the flow field. Figure 2.27 demonstrates an example of numerical simulation of transonic flow in a channel with the entropy rapidly increasing near the lower wall. It can be seen that the closed sonic line bounds a local supersonic region detached from the wall. The qualitative behaviour of the characteristic lines in the beginning of the supersonic region is expected to be the same as in Figure 2.25c, while in the end of the supersonic region the behaviour of the characteristics is symmetric (with respect to the ψ-axis) to the pattern shown in Figure 2.25b. Details of the numerical method employed are given in Section 4.3.

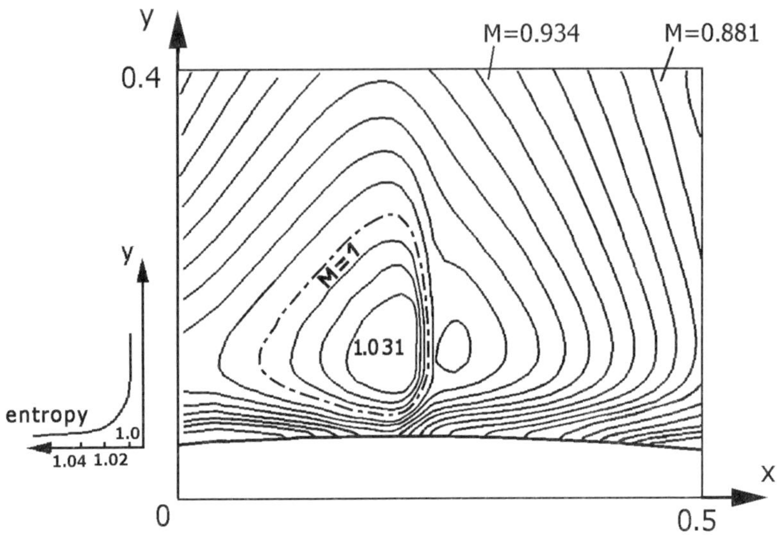

Figure 2.27 IsoMach lines obtained in numerical simulation of nonisentropic transonic flow.

A linearized equation governing the perturbation $u = \varphi'$ of the potential φ in the plane (φ, ψ) in the nearsonic approximation reads

$$(\gamma+1)[(1-\lambda)u_\varphi]_\varphi + \left[\exp\left(-2\bar{S}\frac{\gamma+1}{\gamma}\right)u_\psi\right]_\psi + \frac{d\bar{S}}{d\psi}\exp\left(-2\bar{S}\frac{\gamma+1}{\gamma}\right)u_\psi = 0. \tag{2.151}$$

In order to prove (2.151), we replace in (2.143) ψ by $\psi + \psi'$, θ by $\theta + \theta'$, and λ by $\lambda + \lambda'$, where the prime denotes small perturbation. In this way we obtain the expressions

$$\psi'_x = -\cos\theta \cdot \theta'\lambda\hat{\rho}e^{-\bar{S}} - \sin\theta \cdot (\lambda\hat{\rho})_\lambda \lambda' e^{-\bar{S}} + \sin\theta \cdot \lambda\hat{\rho}\frac{d\bar{S}}{d\psi}\psi' e^{-\bar{S}},$$

$$\psi'_y = -\sin\theta \cdot \theta'\lambda\hat{\rho}e^{-\bar{S}} + \cos\theta \cdot (\lambda\hat{\rho})_\lambda \lambda' e^{-\bar{S}} - \cos\theta \cdot \lambda\hat{\rho}\frac{d\bar{S}}{d\psi}\psi' e^{-\bar{S}}.$$

The derivatives of ψ' in the directions tangent and normal to the streamlines are

$$\frac{\partial\psi'}{\partial s} = \psi'_x \cos\theta + \psi'_y \sin\theta = -\theta'\lambda\hat{\rho}e^{-\bar{S}}, \tag{2.152}$$

$$\frac{\partial\psi'}{\partial n} = \psi'_x \sin\theta + \psi'_y \cos\theta = \lambda'(\lambda\hat{\rho})_\lambda e^{-\bar{S}} - \lambda\hat{\rho}\frac{d\bar{S}}{d\psi}\psi' e^{-\bar{S}}, \tag{2.153}$$

respectively. In the nearsonic approximation, the derivatives (2.146) of the potential φ become

$$\varphi_x = \lambda \cos\theta \exp\left(\frac{\mathrm{d}\bar{S}}{\mathrm{d}\psi}\frac{(\gamma-1)c_{\mathrm{v}}}{\gamma}\right),$$
$$\varphi_y = \lambda \sin\theta \exp\left(\frac{\mathrm{d}\bar{S}}{\mathrm{d}\psi}\frac{(\gamma-1)c_{\mathrm{v}}}{\gamma}\right).$$

A variation of these equations yields

$$\varphi'_x = \left(\lambda' \cos\theta - \lambda \sin\theta\, \theta' + \lambda \cos\theta \frac{\psi'}{\gamma}\frac{\mathrm{d}\bar{S}}{\mathrm{d}\psi}\right)\mathrm{e}^{-\bar{S}}, \tag{2.154}$$
$$\varphi'_y = \left(\lambda' \sin\theta - \lambda \cos\theta\, \beta' + \lambda \sin\theta \frac{\psi'}{\gamma}\frac{\mathrm{d}\bar{S}}{\mathrm{d}\psi}\right)\mathrm{e}^{-\bar{S}}. \tag{2.155}$$

Hence, the derivatives of the potential perturbation φ' in the directions tangent and normal to the streamline are

$$\frac{\partial\varphi'}{\partial s} = \varphi'_x \cos\theta + \varphi'_y \sin\theta = \left(\lambda' + \lambda\frac{\psi'}{\gamma}\frac{\mathrm{d}\bar{S}}{\mathrm{d}\psi}\right)\mathrm{e}^{-\bar{S}},$$
$$\frac{\partial\varphi'}{\partial n} = -\varphi'_x \sin\theta + \varphi'_y \cos\theta = \theta'\lambda\mathrm{e}^{\bar{S}/\gamma}.$$

Now, let us resolve these relations with respect to λ' and θ' and substitute the result into (2.152), (2.153):

$$\frac{\partial\psi'}{\partial s} = -\hat{\rho}\frac{\partial\varphi'}{\partial n}\mathrm{e}^{-\bar{S}(\gamma-1)/\gamma}, \tag{2.156}$$
$$\frac{\partial\psi'}{\partial n} = (\lambda\hat{\rho})_\lambda\left[\frac{\partial\varphi'}{\partial s}\mathrm{e}^{-\bar{S}/\gamma} - \lambda\frac{\psi'}{\gamma}\frac{\mathrm{d}\bar{S}}{\mathrm{d}\psi}\right]\mathrm{e}^{-\bar{S}} - \lambda\hat{\rho}\psi'\frac{\mathrm{d}S}{\mathrm{d}\psi}\mathrm{e}^{-\bar{S}}. \tag{2.157}$$

Using (2.59) and (2.60), one can easily show that

$$\frac{\partial}{\partial s} = \lambda\mathrm{e}^{\bar{S}/\gamma}\frac{\partial}{\partial\varphi}, \qquad \frac{\partial}{\partial n} = \lambda\hat{\rho}\mathrm{e}^{-\bar{S}}\frac{\partial}{\partial\psi}.$$

Therefore, system (2.156), (2.157) becomes

$$\frac{\partial\psi'}{\partial\varphi} = -\hat{\rho}^2 \exp\left(-2\bar{S}\frac{\gamma+1}{\gamma}\right)\frac{\partial\varphi'}{\partial\psi}, \tag{2.158}$$
$$\frac{\partial\psi'}{\partial\psi} = \frac{(\lambda\hat{\rho})_\lambda}{\hat{\rho}}\frac{\partial\varphi'}{\partial\varphi} - \frac{\mathrm{d}\bar{S}}{\mathrm{d}\psi}\psi'. \tag{2.159}$$

Now we differentiate (2.158) with respect to ψ, then (2.159) with respect to φ, and subtract the results,

$$\left[\hat{\rho}^2 \exp\left(-2\bar{S}\frac{\gamma+1}{\gamma}\right)u_\psi\right]_\psi + \left(\frac{(\lambda\hat{\rho})_\lambda}{\hat{\rho}}u_\varphi\right)_\varphi - \frac{\mathrm{d}\bar{S}}{\mathrm{d}\psi}\psi'_\varphi = 0, \tag{2.160}$$

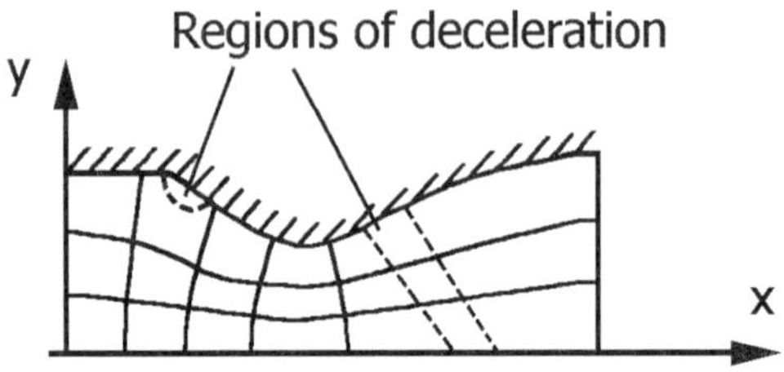

Figure 2.28 Flow in a Laval nozzle with local regions of deceleration.

where $u = \varphi'$ is the perturbation of the generalized potential. Using (2.158) in order to eliminate ψ'_φ, we arrive at the equation

$$\left(\frac{(\lambda\hat{\rho})_\lambda}{\hat{\rho}}u_\varphi\right)_\varphi + \left[\hat{\rho}^2 \exp\left(-2\bar{S}\frac{\gamma+1}{\gamma}\right)u_\psi\right]_\psi + \hat{\rho}^2\frac{\mathrm{d}\bar{S}}{\mathrm{d}\psi}\exp\left(-2\bar{S}\frac{\gamma+1}{\gamma}\right)u_\psi = 0. \tag{2.161}$$

Since we restrict ourselves to the nearsonic approximation, we can use the expansions

$$\lambda\hat{\rho}(\lambda) = 1 + O(\lambda - 1), \qquad (\lambda\hat{\rho})_\lambda = (\gamma+1)(1-\lambda)[1 + O(\lambda - 1)],$$

and omit the higher-order terms in (2.161), thus, we obtain Equation (2.151).

The direct problem for linear perturbations in nonisentropic flow through a Laval nozzle consists of finding a solution of Equation (2.151) that satisfies the usual conditions (2.48) at the nozzle inlet and on the walls. The function $\lambda(\varphi, \psi)$ involved in the coefficients of Equation (2.151) is a solution to system (2.139), (2.144) which describes a given flow in the nozzle.

If $\lambda_\varphi(\varphi, \psi) > 0$ in the domain $\overline{G}$, then the unique solvability of Problem (2.151), (2.48) can be proved in he same way as in Section 2.2. To get rid of the term with the first-order derivative u_ψ in Equation (2.151), one can introduce a new independent variable $\eta(\psi)$ by the relation $\mathrm{d}\eta/\mathrm{d}\psi = \mathrm{e}^{\bar{S}(\psi)}$:

$$(\gamma+1)[(1-\lambda)u_\varphi]_\varphi + [\mathrm{e}^{-2\bar{S}(\eta)/\gamma}u_\eta]_\eta = 0.$$

The coefficient in front of u_η is to satisfy the condition $[\mathrm{e}^{-2\bar{S}/\gamma}]_\varphi \leqslant 0$ by the assumptions of Theorem 1.7. This condition holds since $\bar{S}_\varphi \equiv 0$.

2.9 Flow with a Local Region of Deceleration

The condition of positive acceleration of the given transonic flow played an important part in the previous sections, since it provided the validity of the condition $2\alpha \pm k_x > 0$ over the domain in which a perturbation problem is to be examined. Meanwhile, that condition is important only for the particular technique employed for analysis of the uniqueness of the solution. As shown in Chapter 1, the Fredholm solvability is true under much weaker assumption that $2\alpha \pm k_x > 0$ only at the characteristic points of the type degeneracy set. That is why, using the Fredholm alternative and an advanced technique for analysis of the uniqueness, one can prove the existence of the solution under weaker restrictions on the flow acceleration.

In practice, local regions of deceleration can occur in the subsonic part of the nozzle if the profile of the walls becomes locally steep (Figure 2.28), or in the supersonic part if an unfavourable interplay occurs between the wall curvature and Mach waves approaching the wall. Mathematical examples of flow with variable sign of the acceleration in a channel were given by Ryzhov (1989). Numerical results, which demonstrate examples of transonic flow with local regions of deceleration, will be discussed in Section 4.5.

For simplicity, we shall analyse solvability of the linearized problem that reads as follows: in the domain $G = \{(x, y) \in \mathbb{R}^2;\ 0 < x < l,\ -1 < y < 1\}$, find a solution of the equation

$$(1 - \varphi_x + g)u_{xx} - \varphi_{xx}u_x + \frac{1}{\gamma + 1}u_{yy} = 0, \tag{2.162}$$

where $g(x, y)$ is a given function, endowed by the boundary conditions:

$$\left.\begin{aligned} u &= 0 && \text{at the inlet} && x = 0, \\ u_y &= f^{\pm}(x) && \text{at} && y = \pm 1. \end{aligned}\right\} \tag{2.163}$$

The functions $f^{\pm}$ are perturbations of the angles $\theta(x)^{\pm} = \varphi_y(x, \pm 1)$ which define configuration of the walls.

Theorem 2.12. *Let $\varphi \in W^{2,2}(G)$, $\varphi_x(0, y) < 1 + g$, $\varphi_x(l, y) > 1 + g$, and*

$$\varphi_{xx} + g_x - \mu(1 - \varphi_x + g) > 0 \tag{2.164}$$

for some $\mu = \text{const.}$ in the domain G. Then there exists a unique classical solution $u \in C^1(\overline{G}) \cap C^2(G)$ of Problem (2.162), (2.163).

Condition (2.164) implies that, if $\mu < 0$, then the flow acceleration φ_{xx} may be locally negative in the subsonic region. On the contrary, if (2.164) holds at $\mu > 0$, then the flow acceleration may be locally negative in the supersonic region.

Proof. First we prove the uniqueness of the solution. If there exist two solutions of the problem, then their difference $u(x, y)$ satisfies Equation (2.162) and the boundary conditions (2.163) with vanishing $f^{\pm}(x)$. Let us multiply Equation (2.162) by $u_x e^{-\mu x}$ and integrate the result over the domain G:

$$\int_G \left[(1 - \varphi_x + g)u_{xx} - \varphi_{xx}u_x + \frac{1}{\gamma + 1}u_{yy}\right]u_x e^{-\mu x}\, dG = 0.$$

A simple rearrangement of the integrand yields

$$\int_G \left[(1 - \varphi_x + g)\tfrac{1}{2}(u_x^2)_x e^{-\mu x} - \varphi_{xx}u_x^2 e^{-\mu x} + \frac{1}{\gamma + 1}u_{yy}u_x e^{-\mu x}\right] dG = 0.$$

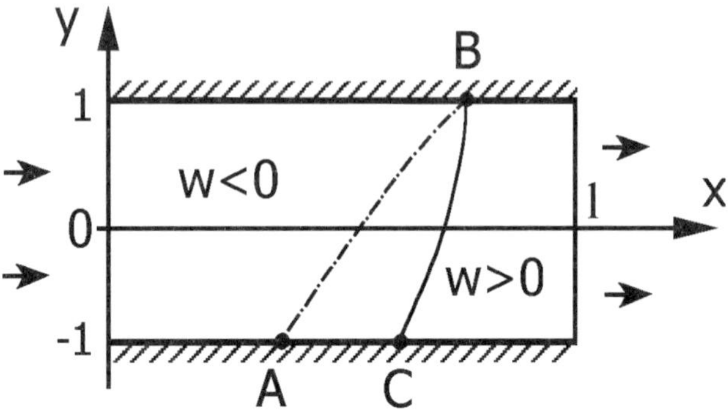

Figure 2.29 A domain and the characteristics line illustrating the proof of Theorem 2.13.

By applying Green's theorem to the left-hand side of the equation, we get

$$\int_G \left[(-\varphi_{xx} - g_x + \mu(1 - \varphi_x + g))\tfrac{1}{2}u_x^2 - \frac{1}{\gamma+1}u_y u_{yx}\right] \mathrm{e}^{-\mu x}\,\mathrm{d}G$$
$$+ \int_\Gamma (1 - \varphi_x + g)\tfrac{1}{2}u_x^2 \mathrm{e}^{-\mu x}\,\mathrm{d}y - \frac{1}{\gamma+1}u_y u_x \mathrm{e}^{-\mu x}\,\mathrm{d}x = 0,$$

$$\int_G (-\varphi_{xx} - g_x + \mu(1 - \varphi_x + g))\tfrac{1}{2}u_x^2 \mathrm{e}^{-\mu x}\,\mathrm{d}G$$
$$+ \int_\Gamma \left[(1 - \varphi_x + g)\tfrac{1}{2}u_x^2\,\mathrm{d}y - \frac{1}{\gamma+1}u_y u_x\,\mathrm{d}x - \frac{1}{\gamma+1}\tfrac{1}{2}u_y^2\,\mathrm{d}y\right]\mathrm{e}^{-\mu x} = 0, \tag{2.165}$$

where the integration over the boundary Γ of G is carried out in the direction that leaves the domain on the left. The integral over G is nonpositive due to the condition (2.164). The second term in the line integral vanishes due to the boundary condition at $y = \pm 1$. Also u_y vanishes at $x = 0$. Therefore, the line integral reduces to the expression

$$\int_\Gamma (1 - \varphi_x + g)\tfrac{1}{2}u_x^2 \mathrm{e}^{-\mu x}\,\mathrm{d}y - \frac{\mathrm{e}^{-\mu l}}{2(\gamma+1)} \int_{-1}^{1} u_y^2(l, y)\,\mathrm{d}y,$$

which is nonpositive due to the inequalities $1 - \varphi_x + g > 0$ at $x = 0$, $1 - \varphi_x + g < 0$ at $x = l$. Therefore, all the terms on the left-hand side of (2.165) are nonpositive, consequently, $u \equiv 0$ in the domain G. Thus, the uniqueness of the solution is proved. The obtained uniqueness and the Fredholm alternative proven in Section 1.4 yield the existence of the classical solution of Problem (2.162), (2.163). □

Another condition, that weakens the requirement of positive acceleration in the subsonic subdomain for Equation (2.162) at $g \equiv 0$, reads (Kuz'min 1992a, p. 199)

$$\varphi_{xx} + 2\mu(1 - \varphi_x) - (\varphi_2 - \varphi_1)\mu^2 \max_{\mathrm{E}_1\mathrm{E}_2} |\varphi_x - 1| > 0.$$

The inequality is supposed to hold in the domain located upstream of the characteristic line E_1E_2 which is tangent to the sonic line at the point E, as shown in Figure 2.7.

Development of the technique for analysis of the uniqueness would make possible a further weakening of the solvability conditions. Below we use a method of auxiliary function in order to prove a uniqueness theorem for flow having no points of contact of the sonic line and the characteristic lines, as sketched in Figure 2.29 and Figure 2.4.

At $g \equiv 0$ Equation (2.162) reduces to

$$-(wu_x)_x + u_{yy} = 0, \tag{2.166}$$

where we used the notation

$$w(x, y) = (\gamma + 1)[\varphi_x(x, y) - 1] = (\gamma + 1)[\lambda(x, y) - 1].$$

The following theorem eliminates the condition of positive acceleration in the subsonic region of the flow field.

Theorem 2.13. *Let the given transonic flow be such that* $w_y < 0$, $w_y^2 - ww_x^2 > 0$ *in the rectangle* $\overline{G}$, *and the sonic line* $w(x, y) = 0$ *be a simple curve with endpoints* A *and* B *on the lower and upper sides of* G, *respectively (Figure 2.29). Suppose that*

$$w_x > 0 \quad \text{at } y = 1,\ 0 \leqslant x \leqslant x_{\mathrm{B}} \quad \text{and} \quad y = -1,\ 0 < x \leqslant x_{\mathrm{C}},$$

where C *is the endpoint of the characteristic line* BC *defined by the equation* $\mathrm{d}x/\mathrm{d}y = \sqrt{w(x, y)}$. *Then Problem (2.166), (2.163) has a unique solution* $u \in C^1(\overline{G}) \cap C^2(G)$.

Proof. The principal effort is again concerned with the proof of the uniqueness. Suppose $u_1(x, y)$ and $u_2(x, y)$ are two different solutions. Then $u = u_1 - u_2$ satisfies Equation (2.166) and boundary conditions (2.163) with $f^{\perp} \equiv 0$. Following Morawetz (1957) and Cook (1978), we introduce the auxiliary function

$$\Psi(x, y) = -\int_{(0,-1)}^{(x,y)} [wbu_x^2 + 2cwu_xu_y + bu_y^2]\,\mathrm{d}y + [wcu_x^2 + 2bu_xu_y + cu_y^2]\,\mathrm{d}x, \tag{2.167}$$

where

$$c = \frac{w_x}{w_y^2 - ww_x^2}, \qquad b = \frac{-w_y}{w_y^2 - ww_x^2} > 0. \tag{2.168}$$

The fact that the line integral (2.167) is path independent is verified below in Lemma 2.14. In the elliptic domain G^+, i.e. at $w < 0$, the function Ψ satisfies the quasilinear elliptic equation $-w\Psi_{xx} + \Psi_{yy} + q_1\Psi_x + q_2\Psi_y = 0$, where the coefficients q_1 and q_2 are bounded in any closed subdomain of G^+ (Cook 1978). Therefore, Ψ satisfies a maximum principle in G^+.

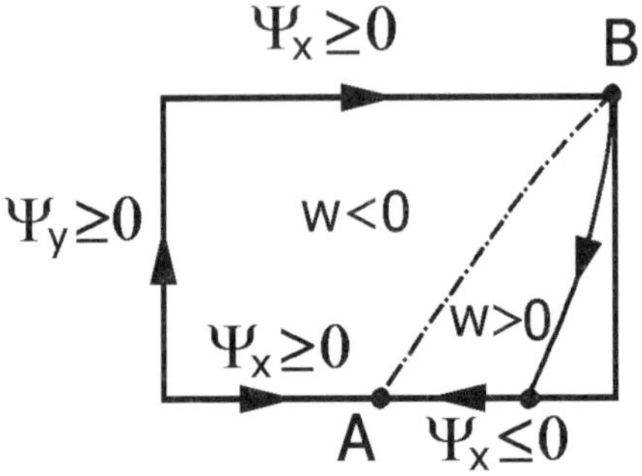

Figure 2.30 The behaviour of the function $\Psi(x, y)$ on the boundary.

In the hyperbolic domain G^-, where $w > 0$, we obtain

$$\begin{aligned}\frac{\mathrm{d}\Psi}{\mathrm{d}y} &= -\left[wbu_x^2 + 2cwu_xu_y + bu_y^2 + (wcu_x^2 + 2bu_xu_y + cu_y^2)\frac{\mathrm{d}x}{\mathrm{d}y}\right] \\ &= -\left[wu_x^2\left(b + c\frac{\mathrm{d}x}{\mathrm{d}y}\right) + 2u_xu_y\left(cw + b\frac{\mathrm{d}x}{\mathrm{d}y}\right) + u_y^2\left(b + c\frac{\mathrm{d}x}{\mathrm{d}y}\right)\right] \\ &= \frac{w_y - w_x\,\mathrm{d}x/\mathrm{d}y}{w_y^2 - ww_x^2}\left(u_x\frac{\mathrm{d}x}{\mathrm{d}y} + u_y\right)^2 \\ &= \frac{1}{w_y + w_x\,\mathrm{d}x/\mathrm{d}y}\left(\frac{\mathrm{d}u}{\mathrm{d}y}\right)^2 = \frac{\mathrm{d}y}{\mathrm{d}w}\left(\frac{\mathrm{d}u}{\mathrm{d}y}\right)^2 \leqslant 0 \qquad (2.169)\end{aligned}$$

along the characteristic lines $\mathrm{d}x/\mathrm{d}y = \pm\sqrt{w}$. Therefore, Ψ attains its maximum on the boundary ∂G of the rectangle G.

Using the conditions (2.163) with $f^{\pm} \equiv 0$, from (2.167) we find

$$\left.\begin{array}{lll}\Psi_y = -wbu_x^2 \geqslant 0 & \text{at the inlet } x = 0, & \\ \Psi_x = -wcu_x^2 \geqslant 0 & \text{at } y = 1,\ 0 \leqslant x \leqslant x_{\mathrm{B}} & \\ & \text{and } y = -1,\ 0 \leqslant x \leqslant x_{\mathrm{A}} & (\text{as } c > 0,\ w < 0), \\ \Psi_x = -wcu_x^2 \leqslant 0 & \text{at } y = -1,\ x_{\mathrm{A}} < x < x_{\mathrm{C}} & (\text{as } c > 0,\ w > 0)\end{array}\right\} \qquad (2.170)$$

(see Figure 2.30 for directions in which Ψ is nondecreasing). These inequalities along with (2.169) prove that Ψ attains a maximum with respect to $\overline{G}$ and $\overline{G}^+$ at point A.

An analysis of the behaviour of Ψ in the vicinity of point A given below in Lemma 2.15 shows that if Ψ attains a maximum with respect to G^+ at point A, then $\Psi \equiv$ const. in G^+. Hence, $\Psi_x \equiv \Psi_y \equiv 0$ in G^+, so that

$$\begin{aligned}-b\Psi_y &= wb^2u_x^2 + 2cwbu_xu_y + b^2u_y^2 \equiv 0, \\ -wc\Psi_x &= (wc)^2u_x^2 + 2cwbu_xu_y + wc^2u_y^2 \equiv 0.\end{aligned}$$

By subtracting the identities, we obtain $wu_x^2 + u_y^2 \equiv 0$ in G^+. Also, $c\Psi_y - b\Psi_x \equiv 0$ proves $u_x u_y \equiv 0$. Consequently, $u_x \equiv u_y \equiv 0$ and since $u = 0$ at the inlet, we get $u \equiv 0$ in $\overline{G}^+$. Due to $\Psi = \text{const.}$ on the sonic line and $\Psi_x \leqslant 0$ on AB, inequality (2.169) yields $\mathrm{d}u/\mathrm{d}y \equiv 0$ and $u \equiv 0$ in G^-. □

Lemma 2.14. *If $u \in C^1(\overline{G}) \cap C^2(G)$ is a solution of Equation (2.166), and the coefficient $w(x, y)$ of the equation satisfies the inequality $w_y^2 - ww_x^2 > 0$ in the domain $\overline{G}$, then the line integral (2.167) is path independent.*

Proof. First, let us show that the functions $c(x, y)$ and $b(x, y)$ defined by (2.168) satisfy the relation $c_y = b_x$. Evidently,

$$
\begin{aligned}
ww_x^2 - w_y^2)^2 c_y &= -w_{xy}(ww_x^2 - w_y^2) + w_x(ww_x^2 - w_y^2)_y \\
&= -w_{xy}(ww_x^2 - w_y^2) + w_x(w_y w_x^2 + 2ww_x w_{xy} - 2w_y w_{yy}) \\
&= w_{xy}(ww_x^2 + w_y^2) + w_x[w_y w_x^2 - 2w_y(ww_x)_x] \\
&= w_{xy}(ww_x^2 + w_y^2) + w_x[-w_y w_x^2 - 2w_y w w_{xx}].
\end{aligned}
$$

On the other hand, we obtain same expression from

$$
\begin{aligned}
(ww_x^2 - w_y^2)^2 b_x &= w_{xy}(ww_x^2 - w_y^2) - w_y(w_x^3 + 2ww_x w_{xx} - 2w_y w_{xy}) \\
&= w_{xy}(ww_x^2 + w_y^2) - w_y(w_x^3 + 2ww_x w_{xx}).
\end{aligned}
$$

Whence, $c_y = b_x$.

Now, let us verify that $b_y = wc_x$. Clearly,

$$
\begin{aligned}
(ww_x^2 - w_y^2)^2 wc_x &= w[-w_{xx}(ww_x^2 - w_y^2) + w_x(ww_x^2 - w_y^2)_x] \\
&= w[-w_{xx}(ww_x^2 - w_y^2) + w_x^4 + 2ww_x^2 w_{xx} - 2w_x w_y w_{xy})] \\
&= w(ww_x^2 w_{xx} + w_y^2 w_{xx} + w_x^4 - 2w_x w_y w_{xy}).
\end{aligned}
$$

On the other hand,

$$
\begin{aligned}
(ww_x^2 - w_y^2)^2 b_y &= w_{yy}(ww_x^2 - w_y^2) - w_y(ww_x^2 - w_y^2)_y \\
&= (ww_x)_x(ww_x^2 - w_y^2) - w_y(w_y w_x^2 + 2ww_x w_{xy} - 2w_y w_{yy}) \\
&= (ww_x)_x(ww_x^2 + w_y^2) - w_x^2 w_y^2 - 2ww_x w_{xy} w_y \\
&= (w_x^2 + ww_{xx})(ww_x^2 + w_y^2) - w_x^2 w_y^2 - 2ww_x w_{xy} w_y \\
&= ww_x^4 + w^2 w_x^2 w_{xx} + ww_y^2 w_{xx} - 2ww_x w_y w_{xy},
\end{aligned}
$$

which coincides with the previous expression.

In order to prove that integral (2.167) is path independent, it is sufficient to show the vanishing of the difference

$$
(\Psi_y)_x - (\Psi_x)_y = -(wbu_x^2 + 2cwu_x u_y + bu_y^2)_x + (wcu_x^2 + 2bu_x u_y + cu_y^2)_y
$$

in the domain G. The terms with u_{xy} on the right-hand side are cancelled:

$$(-2cwu_x - 2bu_y + 2wcu_x + 2bu_y)u_{xy} = 0.$$

Therefore,

$$\begin{aligned}(\Psi_y)_x - (\Psi_x)_y &= -(wbu_x^2)_x - 2u_y(cwu_x)_x - \underline{b_x u_y^2} \\ &\quad + (wc)_y u_x^2 + 2u_x(bu_y)_y + \underline{c_y u_y^2} + 2cu_y u_{yy} \\ &= -(wbu_x^2)_x - 2u_y(cwu_x)_x + (wc)_y u_x^2 + 2u_x(bu_y)_y + 2cu_y u_{yy}.\end{aligned}$$

We differentiate terms in the second and fourth parentheses and then cancel the marked terms using $wc_x = b_y$ and Equation (2.166):

$$\begin{aligned}&= -(wbu_x^2)_x - 2u_y[\underline{c_x wu_x} + \overline{c(wu_x)_x}] \\ &\quad + (wc)_y u_x^2 + 2u_x[\underline{b_y u_y} + bu_{yy}] + \overline{2cu_y u_{yy}} \\ &= -(wbu_x^2)_x + (wc)_y u_x^2 + 2u_x bu_{yy} \\ &= -[(wu_x)_x bu_x + wu_x(bu_x)_x] + (wc)_y u_x^2 + 2u_x bu_{yy} \\ &= u_x[-b(wu_x)_x - w(bu_x)_x + (wc)_y u_x + 2bu_{yy}] \\ &= u_x[-b(wu_x)_x - w(\underline{b_x u_x} + bu_{xx}) + (w_y c + \underline{wc_y})u_x + 2bu_{yy}] \\ &= u_x[-b(wu_x)_x - wbu_{xx} + w_y cu_x + 2bu_{yy}] \\ &= u_x[-b(wu_x)_x - wbu_{xx} - w_x bu_x + 2bu_{yy}] \\ &= u_x 2b[-(wu_x)_x + u_{yy}] = 0.\end{aligned}$$

The obtained identity $(\Psi_y)_x - (\Psi_x)_y \equiv 0$ proves Lemma 2.14. □

Two lemmas represented below follow Morawetz (1966) and Cook (1978).

Lemma 2.15. *Under the assumptions of Theorem 2.13, if the function Ψ defined by (2.167) attains a maximum with respect to G^+ at point* A, *the intersection of $w = 0$ and $y = 0$, then $\Psi \equiv$* const. *in G^+.*

Proof. Suppose that Ψ does have a maximum at point A, and $\Psi(\mathrm{A}) = M$. If P is a point located at x-axis, $x_\mathrm{P} < x_\mathrm{A}$, and $\Psi(\mathrm{P}) = M$, then $\Psi(x, 0) \equiv M$ at $x_\mathrm{P} \leqslant x \leqslant x_\mathrm{A}$. Therefore, by the continuation theorem for the solution of the elliptic equation (Bers 1954), we have $\Psi \equiv M$ in G^+ and, consequently, $\Psi \equiv M$ in G^-.

If Q is a point located on the sonic line in the vicinity of A and $\Psi(\mathrm{Q}) = M$, then $\Psi = M$ and $u =$ const. along the characteristic line $\mathrm{QQ_1}$ and the segment $\mathrm{Q_1A}$. Taking into consideration the boundary condition $u_y(x, -1) = 0$, we get $u \equiv$ const. above $\mathrm{QQ_1A}$ up to the sonic line. Moreover, the Cauchy data along the sonic line with respect to Ψ yield $\Psi =$ const. in G^+.

The last case we have to analyse is the one when $\Psi(\mathrm{P}) < M$ and $\Psi(\mathrm{Q}) < M$ for any points P and Q located in the small vicinity of point A on the x-axis and on the sonic line, respectively (Figure 2.31). In this case, by Lemma 2.16 and the inequality

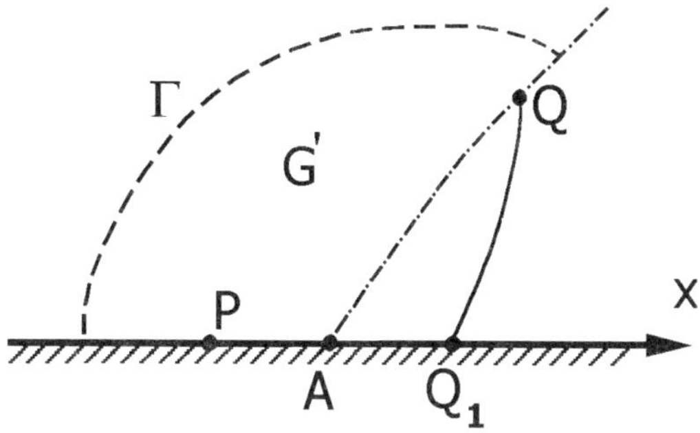

Figure 2.31 A sketch illustrating an analysis of the behaviour of Ψ in the vicinity of point A.

$\Psi_x(x, 0) \geqslant 0$ at $ww_x < 0$, there is an arc Γ connecting $w = 0$ to $y = 0$ along which $\Psi < M - \delta$. Consider the subregion G' of G^+ enclosed by $w = 0$, $y = 0$, and Γ. Choose ε small enough that $\Psi - \varepsilon w < M$ along Γ. Then $\Psi - \varepsilon w$ satisfies equation $-w\Psi_{xx} + \Psi_{yy} + q_1\Psi_x + q_2\Psi_y = 0$ and so must attain its maximum with respect to G' along $\partial G'$. Since $\Psi - \varepsilon w < M$ along Γ, $\Psi - \varepsilon w \leqslant M$ along $w = 0$, the expression $\Psi - \varepsilon w$ must attain its maximum along $y = 0$. That is, there is a point $P = (x, 0)$, $x < x_A$ at which $(\Psi - \varepsilon w)|_P = M$ and $(\Psi - \varepsilon w)_y|_P \leqslant 0$. However, $w_y < 0$ over $\overline{G}$ by the conditions of Theorem 2.13, therefore $\Psi_y|_P < 0$. Meanwhile, $\Psi_y(x, 0) \geqslant 0$ at $ww_y \geqslant 0$ due to the boundary condition. So, we have a contradiction, which proves Lemma 2.15. □

Lemma 2.16. *Under the assumptions of Theorem 2.13, the function Ψ attains its maximum and minimum in any subdomain of G^+ along the boundary of that subdomain. If it attains its maximum at a point* P $\in \partial G^+$, *and if there is a tangent to ∂G^+ at* P, *then $\partial\Psi/\partial n|_P < 0$, where n is any direction pointing into G^+.*

Proof. The function Ψ defined by (2.167) is the solution of the elliptic differential equation $-w\Psi_{xx} + \Psi_{yy} + q_1\Psi_x + q_2\Psi_y = 0$ in G^+, hence, it satisfies both a maximum and a minimum principle within G^+. Thus, if Ψ has a maximum at a point P on ∂G^+, and there is a tangent to ∂G^+ at P, then the derivatives of Ψ at P in a direction pointing into G^+ must be strictly negative (see, for example, Walter 1970; Smirnov 1978). This holds even on $w = 0$ as long as that line is not vertical, which is clearly not the case, since w_y is always negative by the assumptions of Theorem 2.13. □

3

Flow with a Local Supersonic Region

This chapter is mainly concerned with two-dimensional transonic flow which accelerates from subsonic to supersonic velocity and then decelerates smoothly or through shock waves back to the subsonic velocity. For analysis of the solvability of direct problems, we again use the perturbation approach: we assume that a smooth solution $\varphi(x, y)$ describing flow with a local supersonic region is known, and study a small perturbation $u(x, y)$ of $\varphi(x, y)$. This makes it possible to fix the elliptic type of the governing equation at the ends of a channel or at a large distance from the airfoil where an artificial boundary is introduced and far-field conditions are applied.

A shock-free flow with a local supersonic region and a correct boundary-value problem in $W^{2,2}(G)$ can be obtained if one either exempts a portion of the airfoil from boundary condition (Frankl's problem) or imposes a damping condition on a portion of the airfoil in the end of the local supersonic region.

On the other hand, we face the nonexistence of a smooth solution to the boundary-value problem if the slip condition or a velocity distribution is prescribed over the full airfoil. This is the case when shock waves, possibly weak ones, arise in the flow.

3.1 Particular Smooth Solutions and Shockless Airfoils

Practical interest in supercritical airfoil design was enhanced in the 1970s by the need to boost the commercial efficiency of aircraft. This need has forced the cruise conditions into the high subsonic speed regime. At high subsonic freestream Mach numbers, however, local supersonic regions arise over the airfoil similarly to those near the nozzle throat shown in Figure 2.1b.

Flow with a local supersonic region typically contains shock waves as discussed in detail in Section 3.3. Particular smooth solutions, that describe *shock-free* flows with a local supersonic region, are therefore *unstable* to small perturbation of boundary conditions. However, construction of shock-free flows and corresponding shockless airfoils is of practical importance for the wing design because the shock waves, that

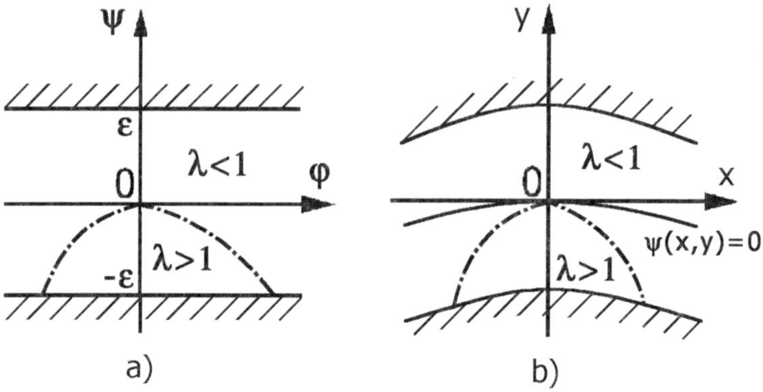

Figure 3.1 Construction of an example of smooth flow with a local supersonic region by imposing appropriate initial data on the zero streamline $\psi(x, y) = 0$.

result from small perturbations of the freestream Mach number or the angle of attack, are typically weak and the flow is nearly shock-free.

Existence of exact solutions with a local supersonic region over a solid wall was first established by Taylor (1930b), who used power series expansions in the vicinity of the origin in the physical plane. The existence actually follows from a general theorem of PDE theory, as pointed out by Frankl (1933). His arguments look most simple if one considers the problem in the (φ, ψ)-plane as follows.

Let us prescribe a subsonic velocity distribution $\lambda(\varphi, 0)$ along the zero streamline $\psi = 0$ with the sonic maximum $\lambda(0, 0) = 1$ at the origin, e.g. we can take $\lambda(\varphi, 0) = 1/(1+\varphi^2)$. In addition, we set the angle $\theta(\varphi, 0)$ of the velocity inclination to decrease from positive values at $\varphi < 0$ to negative ones at $\varphi > 0$, e.g. $\theta(\varphi, 0) = -\varphi$. Due to the Cauchy–Kowalewsky Theorem, in an ε-vicinity of the line $\psi = 0$ there exists a smooth solution to the Chaplygin equations (2.7) written in the natural coordinates,

$$\theta_\psi + \lambda_\varphi \frac{(\lambda\rho)_\lambda}{\lambda\rho^2} = 0, \qquad \rho\lambda_\psi - \lambda\theta_\varphi = 0. \tag{3.1}$$

and endowed with the prescribed initial data $\lambda(\varphi, 0)$, $\theta(\varphi, 0)$. From the second equation of the system (3.1), we find $\lambda_\psi = \lambda\theta_\varphi/\rho$ on the zero streamline. Since $\theta_\varphi(\varphi, 0) < 0$, we obtain $\lambda_\psi(\varphi, 0) < 0$. Therefore, the sonic line is located in the lower half-plane and contacts the axis $z = 0$ at the origin (Figure 3.1a). Due to the one-to-one correspondence of the (φ, ψ)- and (x, y)-planes, we obtain the pattern shown in Figure 3.1b in the physical plane. Then accepting two streamlines $\psi = \pm\varepsilon$ for the walls, one gets an example of shock-free flow with a local supersonic region.

A slightly different way for constructing examples of shock-free flows with a local supersonic region would be the prescription of either the Dirichlet or Neumann condition on the sonic line for Equation (2.72) with respect to the stream function in the hodograph plane. In this case, solving a boundary-value problem in the subsonic region under appropriate conditions on the rest of the boundary, one would find the solution in the subsonic region. The solvability of this problem follows from classical results for degenerate elliptic equations (Smirnov 1966). Then one can continue the solution across the sonic line into the hyperbolic region as a solution of the Cauchy problem with data on the sonic line.

An exact solution to the Chaplygin equations, which describes shock-free flow with a local supersonic region over a smooth wall, was obtained by Ringleb (1940). This classical solution is represented in many books on aerodynamics (see, for example, Oswatitsch 1976; Cole and Cook 1986). A simple solution describing flow with supersonic regions over a wavy wall was given by von Kármán and Fabri (1950). Other methods for constructing shock-free flows were proposed by Goldstein *et al.* (1948), Cherry (1949) and Takami (1956).

A first practical tool for the design of shockless airfoils under given freestream conditions was suggested by Garabedian and Korn (1971). The method is based on a numerical solution of the governing equations in the hodograph plane. To obtain a solution in the subsonic region under the given far-field conditions, Garabedian and Korn extended the hodograph equations in their characteristic form into the complex domain and used a complex four-dimensional space. Thus in the subsonic part of the flow field the solution is governed by the hyperbolic equations. This enabled the authors to develop a relatively fast algorithm for computation of the solution. Further developments of their method were made by Bauer *et al.* (1977) and Chen and Garabedian (1998).

To construct shockless airfoils, Niewland and Spee (1973) superposed a suitable set of particular solutions in the hodograph plane. Boerstoel and Huizing (1974) developed this approach and supplied a variety of thick shock-free airfoils.

Experiments by Pearcey (1962) showed the possibility of realizing almost shock-free transonic flow in practice. Then an extensive experimental investigation was performed on a series of symmetric airfoils by Spee (1969). The results indicated that the designed shock-free flow can in principle be approached arbitrarily closely.

> Consequently, shock-free transonic flows cannot be called "unstable" in any physically useful sense. However, there is a smaller or larger degree of sensitivity of the flow variations in Mach number or incidence that can be expressed in terms of change in drag...
>
> (Niewland and Spee 1973)

In the early development of transonic flow theory, the hodograph technique was used extensively. However, the complicated nature of the hodograph image of the real flow kept this technique from becoming a popular tool for designers. In particular, its utility was restricted by the difficulties in imposing boundary conditions for an airfoil of specified shape. Also, some difficulties arose because of a feasible multi-valued mapping from the supersonic region of the physical plane into the hodograph plane.

Sobieczky (1971) suggested a so-called rheograph plane in order to solve successively the problem in the subsonic region and find the flow deflection angle on the sonic line, and after that to integrate into the supersonic region. Later, Sobieczky

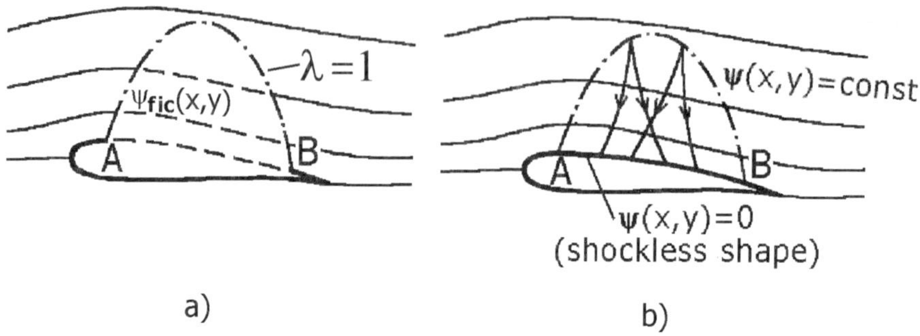

Figure 3.2 A scheme of the shock-free airfoil design by the fictitious gas method.

(1978) proposed a *fictitious gas* method which simplifies construction of a solution in the subsonic region. In the fictitious gas method, one specifies the fore and aft parts of the airfoil as well as an initial guess for the middle part of the airfoil in the (x, y)-plane. At the first step of computations, the Bernoulli law (2.2) in the supersonic region is replaced by the fictitious law,

$$\rho = \left(1 + \frac{\lambda - 1}{n}\right)^{-n} \quad \text{if } \lambda > 1, \tag{3.2}$$

where as usual λ is the flow velocity normalized by the critical speed of sound. Employment of the relation (3.2) with $n \simeq 0.6$–0.8 makes the type of the full-potential equation elliptic in the region where $\lambda > 1$. This enables us to use available codes for solving elliptic type equations in order to find the potential $\varphi_{\text{fic}}(x, y)$ of fictitious flow over the specified airfoil under the Neumann condition $\partial\varphi_{\text{fic}}/\partial n = 0$ on the airfoil. The obtained solution $\varphi_{\text{fic}}(x, y)$ yields a smooth sonic line $\lambda = 1$ and a distribution of the flow deflection angle θ along it (see Figure 3.2a). The fictitious potential φ_{fic} and stream function ψ_{fic} satisfy the relation $\partial\varphi_{\text{fic}}/\partial n = \partial\psi_{\text{fic}}/\partial s$, where s and n are the directions tangent and normal to the initial airfoil, respectively. Hence, $\psi_{\text{fic}}(\text{A}) = \psi_{\text{fic}}(\text{B})$ at the ends of the sonic line.

At the second step of the fictitious gas method, using the obtained angle θ on the sonic line, one can integrate the full-potential Equation (2.38) into the supersonic region, e.g. with the method of characteristics, and obtain the actual flow there (Figure 3.2b). Let $\varphi(x, y)$ and $\psi(x, y)$ be the velocity potential and stream function of the obtained flow. Then the equation $\psi(x, y) = 0$, i.e. $\mathrm{d}y/\mathrm{d}x = -\psi_x/\psi_y$, determines a shockless shape of the airfoil $y_0(x)$. Thus, if the airfoil streamline can be continued in the supersonic region downstream of the sonic point A without intersecting the limit lines, then it represents a shockless airfoil. The streamline $\psi(x, y) = 0$ will meet the given aft part of the airfoil at the sonic point B ensuring the closure of the airfoil. This follows from the observation that the problem in the supersonic region can in general be solved in the hodograph plane (λ, θ) under the given distribution of $\psi = \psi_{\text{fic}}$ along the sonic line, such that $\psi = \psi_{\text{fic}} = 0$ at points A and B. The flow quantities are not changed at the sonic line. Therefore, the streamwise momentum and normal pressure gradient at points A and B are unchanged. Consequently, the local flow curvature must be the same (Sobieczky *et al.* 1979).

On the other hand, if the zero streamline, having been continued downstream of the point A, met a limit line in the supersonic region, this would mean that configurations of the fore and aft parts of the airfoil or the value of the parameter n in (3.2) had been inappropriately chosen for the given freestream Mach number.

The fictitious gas method is thus very close to the direct prescription of the location of the sonic line in the physical plane. If it is prescribed, then the full boundary of the subsonic region is fixed, and a condition is posed on each portion of the boundary. In particular, the Mach number $M = 1$ is given on the portion of the boundary assigned for the sonic line. Therefore, one can compute successively the solution in the subsonic and supersonic regions. However, in this case, the supersonic region is completely ignored at the first step of computations in distinction to the fictitious gas method. That is why one must eventually provide the closure of the airfoil contour (i.e. the condition that the streamline $\psi(x, y) = 0$ meets the given aft part of the airfoil at the sonic point B) iteratively by choosing appropriate positions of the points A and B on the airfoil.

Since the full-potential equation is equivalent to the system of the continuity and momentum equations in the case of irrotational shock-free flow, the above methods virtually provide a tool for constructing shock-free inviscid transonic flows governed by the Euler equations. Effects of viscosity were incorporated into the fictitious gas method by Nebeck *et al.* (1980).

Methods of airfoil design can be based on approaches different from the concept of shock-free flow. The approaches may depend on the specific goals of the design, e.g. the need to achieve the maximum efficiency or a specific velocity distribution that would provide appropriate control of boundary-layer behaviour. The relation between the shockless airfoils and the ones of maximum efficiency was clearly characterized by Sobieczky and Seebass (1984) as follows.

> While the shock wave has its own inherent losses, its predominant adverse effect derives from the boundary-layer separation that inevitably occurs as the shock wave becomes stronger and moves farther aft on the wing surface... Airfoils and wings that operate supercritically and that do so without shock waves obviously avoid the adverse effects usually associated with supercritical flight Mach numbers. Generally, the best efficiency will be obtained when such airfoils are operated at a Mach number or a lift coefficient slightly higher than that for which shock-free flow exists. This slight increase in the flight Mach number provides an increase in the aircraft's efficiency proportional to the increment Mach number. The losses associated with the shock wave that immediately appears have been determined experimentally to increase with a rather high power of the increment in Mach number; thus, small increases in lift coefficient or a freestream Mach number above those for which shock-free flow

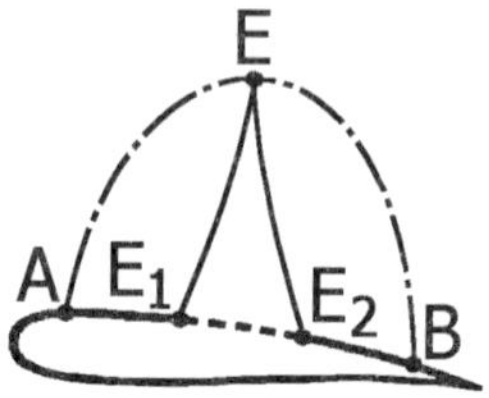

Figure 3.3 Finding airfoil's arc E_1E_2 which must be free of boundary conditions for the correctness of the problem in the class of smooth solutions.

occurs lead to an improved efficiency. Eventually, however, these losses overtake the linear gains, and this trade-off determines the most efficient flight Mach number for a given lift coefficient. Because the operating conditions for maximum efficiency involve a weak shock wave, one can argue that it is not essential to find shock-free flows to obtain efficient designs. This is no doubt true. But the relative ease with which shock-free flows can be found in both two and three dimensions provides a direct path to efficient airfoil and wing designs. Because of the past success of airfoils that were designed to be shock-free at specific operating conditions and the ease with which such airfoils, and indeed wings, can be found, we adopt the view that at supercritical Mach numbers shock-free configurations are of paramount importance in achieving the maximum gains possible through careful aerodynamic design.

3.2 A Modification of the Frankl Problem

Frankl (1947) analysed transonic flow with local supersonic regions over a symmetric airfoil and suggested physical arguments in favour of the correctness of a problem in which a part of the airfoil is free of boundary conditions. Let us consider a smooth flow with a supersonic region (Figure 3.3) and construct two characteristic lines EE_1 and EE_2 emanating from a point E of the sonic line AB and intersecting the airfoil at points E_1 and E_2. Frankl showed that, at a given freestream velocity, the imposing of the slip condition on the airfoil outside the arc E_1E_2 would be sufficient for the correctness of the boundary-value problem in the class of continuous flows. He set both the nonlinear and linearized problems in the physical and hodograph planes and formulated a solvability theorem.

A rigorous proof of the uniqueness of the solution to the linearized problem was given by Morawetz (1956, 1957) in the hodograph plane. She employed an auxiliary function which was monotonous along the characteristic lines in the supersonic region and satisfied an equation of elliptic type in the subsonic region. Cook (1978) proved a similar uniqueness theorem directly in the plane (x, y). Two important consequences of the uniqueness were derived:

(∗) If a smooth solution to the Frankl problem exists, then it is uniquely determined in the domain outside the airfoil including the triangular subdomain EE_1E_2. The obtained normal component of the velocity on the arc E_1E_2 thus depends on the freestream Mach number and the airfoil shape outside E_1E_2. Since the normal component of the velocity yields a shockless shape of the arc E_1E_2, the Frankl problem is virtually associated with airfoil adaptation to changing flow conditions.

(∗∗) If a boundary condition were posed on the arc E_1E_2 (in addition to the slip condition on E_1ABE_2), then the problem would be overdetermined in the class of smooth solutions. Therefore, the solution of this problem is expected to contain singularities or shock waves.

We study below a modification of the Frankl problem in which a velocity distribution instead of the slip condition is prescribed on the arcs AE_1 and BE_2 or on some of their portions PE_1 and E_2Q. We demonstrate that the above-mentioned result on the uniqueness remains true in this case.

Thus, we turn from the problem of airfoil adaptation to a problem of the shockless airfoil design in which a target velocity is prescribed on PE_1 and E_2Q while the arc E_1E_2 is free of boundary conditions. The study is performed for the linearized nearsonic equations in the same framework as that of Morawetz and Cook. We consider an infinite domain and use asymptotic far-field expansions for a boundary condition at infinity.

In the nearsonic approximation, steady two-dimensional flow of inviscid gas is governed by the von Kármán Equation (2.10),

$$(\gamma+1)(1-\varphi_x)\varphi_{xx}+\varphi_{yy}=0, \tag{3.3}$$

where $\varphi(x,y)$ is the nondimensional velocity potential and $\gamma>1$ is the ratio of specific heats. As discussed in Chapter 2, the slip condition, which implies the vanishing of the normal velocity, $\varphi_n=(\nabla\varphi,\boldsymbol{n})=0$, can be replaced in case of a thin airfoil by the condition $\varphi_y=\theta^{\pm}(x)$ on the x-axis, where $\theta^{\pm}$ are the angles made by the tangents to the upper and lower surfaces of the airfoil with the x-axis. The asymptotic expansion for the solution of the small disturbance equation at $r^2=x^2+y^2K_{\mathrm{sim}}\to\infty$ reads (Cole and Cook 1986)

$$\begin{aligned}\phi(x,y)=&-\frac{\Gamma\vartheta}{2\pi}+\frac{\gamma+1}{K_{\mathrm{sim}}}\left(\frac{\Gamma}{4\pi}\right)^2\frac{\log r}{r}\cos\vartheta\\&-\frac{\gamma+1}{K_{\mathrm{sim}}}\left(\frac{\Gamma}{8\pi}\right)^2\frac{\cos 3\vartheta}{r}+\frac{D_1\cos\vartheta}{2\pi r}+\frac{D_2\sin\vartheta}{2\pi r}+O\left(\frac{1}{r^2}\right),\end{aligned} \tag{3.4}$$

where Γ is the circulation of the velocity field over the airfoil,

$$K_{\mathrm{sim}}=(1-M_\infty^2)\delta^{-2/3}$$

is the transonic similarity parameter, δ is the thickness of the airfoil,

$$\vartheta=\arctan(yK_{\mathrm{sim}}^{1/2}/x).$$

For simplicity, we consider a symmetric airfoil at zero incidence, hence, $\Gamma = 0$. We assume that both the bow and trailing edges of the airfoil are cusped in order to avoid a singularity of $\varphi(x, y)$ which arises at the nose of the blunt airfoil in the framework of the small disturbance theory (Rusak 1993).

The constant D_1 in the asymptotic expansion (3.4) is associated with the wave drag caused by shock waves (Cole and Cook 1986, Section 4.3.1). In the shock-free flow, the wave drag vanishes, therefore, $D_1 = 0$. In addition, $D_2 = 0$ due to the symmetry of the flow with respect to the x-axis. That is why the asymptotic expansion for the velocity potential $\varphi(x, y)$ in the case at hand reads $\varphi(x, y) = \lambda_\infty x + O(r^{-2})$ as $r \to \infty$.

The nonlinear problem in question is formulated as follows. Find a solution $\varphi(x, y)$ of Equation (3.3) in the half-plane $G = \{(x, y) \in \mathbb{R}^2;\ y > 0,\ -\infty < y < \infty\}$ such that $\varphi_x > 1$ in a subdomain adjoining the x-axis (Figure 3.4), and the following boundary conditions are true:

$$\left.\begin{aligned} &\varphi_y = \theta(x) && \text{at } y = 0,\ -\infty < x < x_{\mathrm{P}} \text{ and } x_{\mathrm{Q}} < x < \infty,\\ &\varphi_x = \lambda(x) && \text{at } y = 0,\ x_{\mathrm{P}} < x < x_{\mathrm{E}_1} \text{ and } x_{\mathrm{E}_2} < x < x_{\mathrm{Q}},\\ &\varphi(x, y) = \lambda_\infty x + O\left(\frac{1}{r^2}\right) && \text{at } r^2 = x^2 + y^2 K_{\mathrm{sim}} \to \infty. \end{aligned}\right\} \tag{3.5}$$

Here $\lambda(x)$ and $\theta(x)$ are given functions; $\theta(x) \equiv 0$ at $x < 0$ and $x > l$ due to the flow symmetry with respect to the x-axis.

Let $\varphi(x, y)$ be a smooth solution to (3.3), (3.5) that describes shock-free flow over a shockless airfoil. The perturbation $u = \varphi - \hat{\varphi}$ of the solution $\varphi(x, y)$ is governed by Equation (2.20),

$$(1 - \varphi_x + u_x)u_{xx} - \varphi_{xx}u_x + \frac{1}{\gamma + 1}u_{yy} = 0, \tag{3.6}$$

and satisfies the boundary conditions which follow from (3.5):

$$\left.\begin{aligned} &u_y = \theta'(x) && \text{at } y = 0,\ -\infty < x < x_{\mathrm{P}} \text{ and } x_{\mathrm{Q}} < x < \infty,\\ &u_x = \lambda'(x) && \text{at } y = 0,\ x_{\mathrm{P}} < x < x_{\mathrm{E}_1} \text{ and } x_{\mathrm{E}_2} < x < x_{\mathrm{Q}},\\ &u = O\left(\frac{1}{x^2 + y^2}\right) && \text{as } x^2 + y^2 \to \infty. \end{aligned}\right\} \tag{3.7}$$

The linearization of (3.6) yields

$$(1 - \varphi_x)u_{xx} - \varphi_{xx}u_x + \frac{1}{\gamma + 1}u_{yy} = 0. \tag{3.8}$$

By introducing the function $w = (\gamma + 1)(\varphi_x - 1)$, which is the deviation of the nondimensional velocity from the critical speed of sound, we represent Equation (3.8) in the form

$$(-wu_x)_x + u_{yy} = 0. \tag{3.9}$$

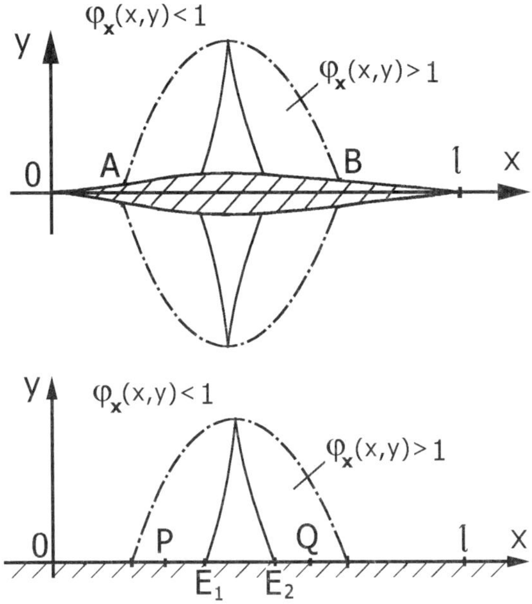

Figure 3.4 The domain in which problem (3.3), (3.5) is formulated.

This equation obviously reduces to (1.1) with the notation $-w = k$. However, we shall use the form (3.9) following Cook (1978).

Theorem 3.1. *Let the given transonic flow be such that* $w \in C^3(\overline{G})$, $w_y < 0$, $w_y^2 - ww_x^2 > 0$ *at* $y \geqslant 0$, $-\infty < x < \infty$, *and the sonic line* $w(x, y) = 0$ *be a simple curve with endpoints* A *and* B *on the* x*-axis. Suppose that*

$$w_x > 0 \quad \text{at} \quad y = 0,\ x_{C_1} < x < x_C \quad \text{and} \quad x_{C_2} < x < \infty,$$

$$w_x < 0 \quad \text{at} \quad y = 0,\ x_C < x < x_{C_2} \quad \text{and} \quad -\infty < x < x_{C_1},$$

and the flow acceleration vanishes, i.e. $w_x = 0$, *at the points* C, C_1, C_2 *(see Figure 3.5).*

Then Problem (3.7), (3.9) with respect to the perturbation $u(x, y)$ *of the velocity potential can have at most one solution* $u \in C^1(\overline{G}) \cap C^2(G)$.

Proof. Suppose $u_1(x, y)$ and $u_2(x, y)$ are two different solutions. Then $u = u_1 - u_2$ satisfies Equation (3.9) and the boundary conditions (3.7) with $\theta' = \lambda' \equiv 0$. As in Section 2.9, we introduce the auxiliary function

$$\Psi(x, y) = -\int_{(0,0)}^{(x,y)} [wbu_x^2 + 2cwu_xu_y + bu_y^2]\,dy + [wcu_x^2 + 2bu_xu_y + cu_y^2]\,dx, \tag{3.10}$$

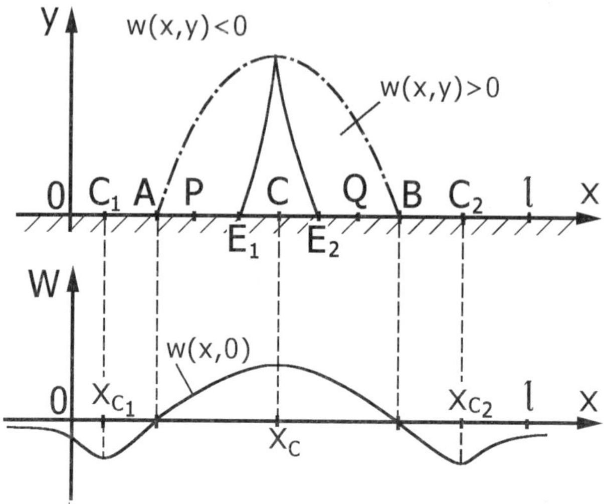

Figure 3.5 A behaviour of the given function $w(x, y)$ in the upper half-plane and on the x-axis. The acceleration $w_x(x, 0)$ vanishes at the points C, C_1, C_2.

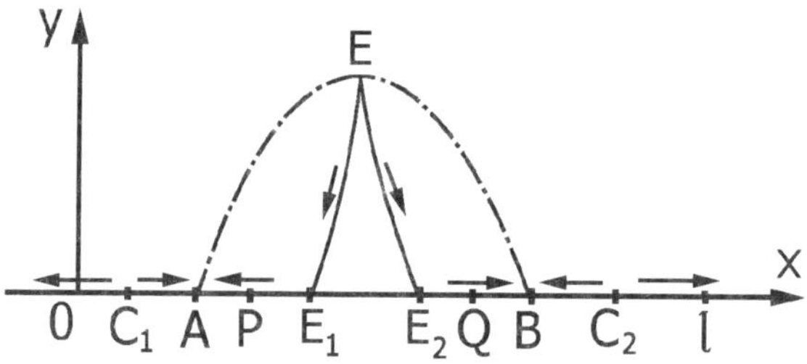

Figure 3.6 The directions in which the auxiliary function Ψ is nondecreasing along the characteristic lines and the x-axis.

where

$$c = \frac{w_x}{w_y^2 - ww_x^2}, \qquad b = \frac{-w_y}{w_y^2 - ww_x^2} > 0.$$

In the elliptic domain $G^+ = G \cap (w < 0)$, the function Ψ satisfies a quasilinear elliptic equation, hence, Ψ satisfies a maximum principle in G^+. In the hyperbolic domain $G^- = G \cap (w > 0)$, the derivative of Ψ along the characteristic lines $\mathrm{d}x/\mathrm{d}y = \pm w^{1/2}$ is nonpositive due to the expression (2.169):

$$\frac{\mathrm{d}\Psi}{\mathrm{d}y} = \frac{\mathrm{d}y}{\mathrm{d}w}\left(\frac{\mathrm{d}u}{\mathrm{d}y}\right)^2 \leqslant 0. \tag{3.11}$$

Therefore, Ψ attains its maximum on the boundary $y = 0, -\infty < x < \infty$, of the half-plane $y > 0$.

It follows from (3.10) that

$$\Psi_x(x, y) = -[wcu_x^2 + 2bu_x u_y + cu_y^2]. \tag{3.12}$$

Using the boundary conditions (3.7) with $\theta' = \lambda' \equiv 0$, from (3.12) we find (see Figure 3.6)

$$\left.\begin{aligned}
\Psi_x &= -wcu_x^2 \leqslant 0 \quad \text{at} \quad y = 0,\ x_A \leqslant x \leqslant x_P \quad (\text{as } c > 0,\ w > 0),\\
\Psi_x &= -cu_y^2 \leqslant 0 \quad \text{at} \quad y = 0,\ x_P \leqslant x \leqslant x_{E_1} \quad (\text{as } c > 0),\\
\Psi_x &= -cu_y^2 \geqslant 0 \quad \text{at} \quad y = 0,\ x_{E_2} \leqslant x \leqslant x_Q \quad (\text{as } c < 0),\\
\Psi_x &= -wcu_x^2 \geqslant 0 \quad \text{at} \quad y = 0,\ x_Q \leqslant x \leqslant x_B \quad (\text{as } c < 0,\ w > 0).
\end{aligned}\right\} \tag{3.13}$$

Due to the condition $u_y(x, 0) = 0$ outside the segment AB, from (3.12) we obtain

$$\left.\begin{aligned}
\Psi_x &= -wcu_x^2 \leqslant 0\\
&\qquad \text{at } y = 0,\ -\infty < x \leqslant x_{C_1} \text{ and } y = 0,\ x_B \leqslant x < x_{C_2},\\
\Psi_x &= -wcu_x^2 \geqslant 0\\
&\qquad \text{at } y = 0,\ x_{C_1} < x \leqslant x_A \text{ and } y = 0,\ x_{C_2} \leqslant x < \infty.
\end{aligned}\right\} \tag{3.14}$$

Inequalities (3.13), (3.14) show that Ψ attains a maximum with respect to $\overline{G}$ and $\overline{G}^+$ at point A, or point B, or at infinity.

Since the asymptotic expansions for w and u read

$$u = O(r^{-2}), \qquad w = (\gamma + 1)[\lambda_\infty - 1 + O(r^{-3})],$$

we have $u_x, u_y = O(r^{-3})$, $w_x = O(r^{-4})$,

$$\Psi_x, \Psi_y \approx \frac{u_x^2}{w_x} = \frac{O(r^{-6})}{O(r^{-4})} = O(r^{-2}) \quad \text{at } r \to \infty.$$

Hence, Ψ is continuous at infinity and cannot attain a maximum there due to the well-known properties of elliptic equations.

The analysis of the behaviour of $\Psi(x, y)$ in the vicinity of point A made in Lemma 2.15 shows that Ψ can have a maximum with respect to G^+ at point A only if $\Psi(x, y) \equiv$ const. in G^+. The same refers to point B. Hence, $\Psi_x \equiv \Psi_y \equiv 0$ in G^+, so that

$$-b\Psi_y = wb^2u_x^2 + 2cwbu_x u_y + b^2u_y^2 \equiv 0,$$
$$-wc\Psi_x = (wc)^2u_x^2 + 2cwbu_x u_y + wc^2u_y^2 \equiv 0.$$

By subtracting the identities, we obtain $wu_x^2 + u_y^2 \equiv 0$ in G^+. In addition, $u_x u_x = 0$ due to $c\Psi_y - b\Psi_x \equiv 0$. Consequently, $u_x \equiv u_y \equiv 0$ and since $u = 0$ at infinity, we have $u \equiv 0$ in $\overline{G}^+$. Due to $\Psi =$ const. on the sonic line and $\Psi_x \leqslant 0$ on AB, inequality (3.11) yields $\mathrm{d}u/\mathrm{d}y \equiv 0$ and $u \equiv 0$ in G^-. □

For incompressible flow, Lighthill (1945) demonstrated that a solution to the problem with specified velocity distribution on an airfoil generally does not exist unless the prescribed distribution satisfies a certain integral constraint arising from the requirement that the freestream velocity be equal to a specified value. The additional requirements that the airfoil be at a given incidence and its contour be closed were shown by Lighthill to lead to two additional constraints on the prescribed velocity distribution. Thus, in order to ensure that a solution to the problem exists, some freedom must be permitted in the prescribed velocity (or the pressure which is related to the velocity by the Bernoulli law) to allow the three constraints to be satisfied. This was rigorously proved only for incompressible and subsonic flows (see Woods 1955; Daripa 1991). For transonic flow, Volpe and Melnik (1984) suggested a numerical method in which the target pressure distribution contained a free parameter in order to provide a solvability condition which is analogous to Lighthill's first integral constraint. The method was based on solving the full-potential equation in the polar coordinates. Numerical experiments confirmed that the target pressure distribution must satisfy a constraint if the freestream velocity is specified.

The result formulated in Theorem 3.1 shows that the numerical methods, in which a finite number of free parameters are introduced into the velocity distribution over the supersonic arc AB of the airfoil, cannot guarantee the absence of shocks in the flow. The imposing of the longitudinal or normal velocity is a 'rigid' boundary condition which, being prescribed along the full supersonic portion of the airfoil, overdetermines the problem in the class of smooth solutions. The introduction of a few integral constraints into the prescribed velocity along the supersonic arc does not suffice for the correctness of the problem.

In practice of airfoil design, the shock waves appearing in the flow under the smooth target velocity distribution may be very weak and often smeared out by the numerical viscosity. At the same time, some numerical studies did demonstrate that the imposing of a smooth velocity/pressure along the airfoil leads to the arising of internal shock waves of notable strength even though the target pressure is smooth. Giles and Drela (1987) worked out a design method, based on the simultaneous solutions for multiple streamlines, coupled through the position of, and pressure at, the streamline interfaces. This allowed the specification of either the airfoil shape or the surface pressure distribution. The system of Euler equations was formulated in a conservative manner and solved by a fast Newton method. The method was used to design an airfoil with the lift coefficient $C_L = 1.00$ at $M_\infty = 0.70$. The starting point for the design procedure was the NACA 65-412 airfoil. Variation of the angle of attack made it possible to obtain the specified lift coefficient C_L. The solution revealed a strong shock wave on the upper surface of the airfoil, leading to a high wave drag. An attempt to design an improved airfoil allowed the geometry to vary along almost the entire length of the upper surface and specified a shock-free surface pressure distribution with its minimum near the leading edge. Figure 3.7 shows the obtained Mach number isolines with a weak shock and an indentation in the sonic line configuration. The authors mentioned that the shock is displaced off the airfoil. A more plausible interpretation of the obtained pattern is that the shock wave

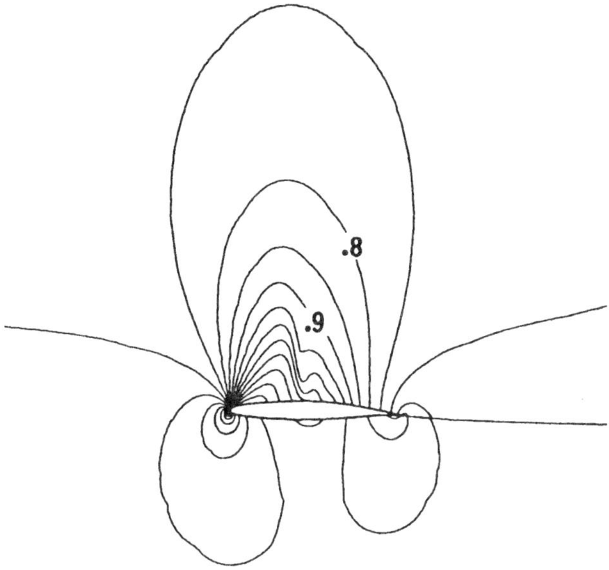

Figure 3.7 Mach number contours over an airfoil designed by prescribing a smooth target pressure distribution. Computations of Giles and Drela (1987). Copyright © 1987 AIAA. Reproduced with permission.

actually meets the airfoil and is reflected from it as an expansion wave due to the prescribed smooth pressure distribution along the airfoil. The reflection point must be an expansion corner, which was evidently smoothed out in the computations because of the coarse mesh of 132 × 32 employed and, consequently, considerable numerical viscosity. Such reflection is similar to a shock-wave reflection from a free boundary separating supersonic flow from a media at rest (see Oswatitsch 1976, Section 8.7).

The imposing of a smooth velocity/pressure distribution along the airfoil can also lead to the appearance of a concave portion of the designed airfoil, as pointed out by Sobieczky (1989).

At present, there are a great deal of numerical methods of airfoil design which are based on solving the direct problem and the 'trial-and-error' approach to transonic flow for a given supercritical freestream Mach number. In these methods, direct solutions are sought with an airfoil shape modified iteratively to minimize the difference between the computed pressures and a prescribed target pressure distribution (see Hirose *et al.* 1987; Obayashi 1995; Ilinca *et al.* 1998). We would mention again that the imposing of a target pressure or velocity distribution over the supersonic arc of the airfoil is correct if the aim of the design is different from obtaining a shock-free flow. However, this necessarily leads to the arising of shock waves, possibly weak ones. Consequently, the methods of computational fluid dynamics (CFD) in which the airfoil design is carried out by imposing a smooth pressure or velocity distribution over the supersonic arc do not provide strictly shock-free flow. This notice is conso-

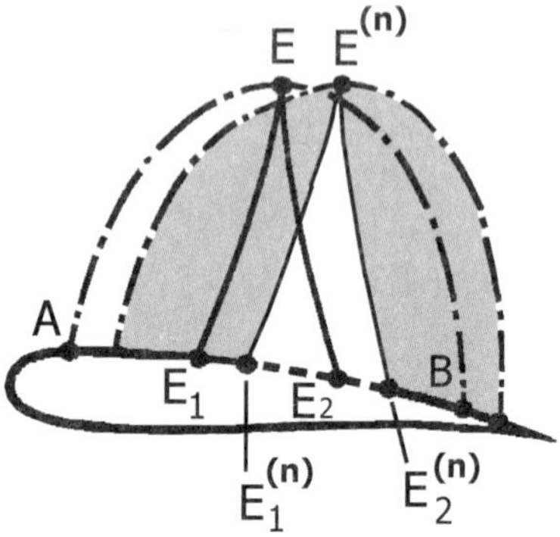

Figure 3.8 A scheme of shifts of the airfoil arc E_1E_2 in the course of iterations aimed at shape adaptation.

nant with the approach of Bartelheimer (1995), who developed a method involving a smoothing algorithm for the design of transonic airfoils/wings, and mentioned that 'not every prescribed target pressure will have a corresponding geometry… Therefore, some experience is necessary to avoid ill-posed problems and to achieve an improved design.'

The above-mentioned result on the uniqueness of the Frankl problem leads to an algorithm for airfoil shape adaptation as follows. Consider a smooth flow over a shockless airfoil designed for a given freestream Mach number M_∞. Let us construct two characteristic lines EE_1 and EE_2 emanating from a point E of the sonic line as shown in Figure 3.8. If M_∞ is slightly changed, then we get the Frankl problem with respect to the perturbation $u(x, y)$ of the velocity potential in the domain located above the curve AE_1EE_2B under the specified perturbation of M_∞ and the slip condition on the supersonic portions AE_1, E_2B and the subsonic arc BA of the airfoil. This perturbation problem can be solved by a simple modification of existing solvers for the full-potential or Euler equations in the transonic range. In the course of iterations, the sonic line changes its location following the perturbation obtained at each iteration step. Consequently, the positions of the point E on the sonic line and the points E_1, E_2 on the airfoil change.

Some effort may be needed in order to provide the closure of the airfoil contour. Indeed we can fix, e.g. the point E_1, so that in the course of iterations we obtain a sequence of points $E_2, E_2^{(1)}, E_2^{(2)}, \ldots, E_2^{(n)}$. It goes without saying that, having obtained the solution of the problem in the transonic region above the curve $AE_1E^{(n)}E_2^{(n)}B$, one can use values of that solution on the characteristic lines $E^{(n)}E_1$ and $E^{(n)}E_2^{(n)}$ in order to solve the Goursat problem in the triangular domain $E^{(n)}E_1E_2^{(n)}$ and to find values of u_n on the arc $E_1E_2^{(n)}$. These values establish a new angle of velocity inclination on the airfoil and, consequently, the shockless shape of the airfoil for the disturbed flow. However, one cannot guarantee that the airfoil streamline continued downstream of the point E_1 will meet point $E_2^{(n)}$ and provide the closure of the airfoil. In order to ensure the closure, one can, for instance, adjust appropriately the locations of points $E_1, E_1^{(1)}, E_1^{(2)}, \dots, E_1^{(n)}$ which would finally provide the closure. Alternatively, depending on the specified perturbation of the freestream velocity, one may adjust appropriately the sequence $E_2, E_2^{(1)}, E_2^{(2)}, \dots, E_2^{(n)}$ in order to provide the closure. The advantage of this method is that just a short arc of the airfoil is supposed to be varied in the beginning or in the end of the supersonic region for accomplishing the necessary adaptation.

3.3 Concepts of the Arising of Shock Waves

The inevitability of the arising of shock waves in a local supersonic region can be elucidated by a simple analogy with a linear equation of hyperbolic type in a domain with vanishing cross-sectional area at $x \to l$, as sketched in Figure 1.23. In this case, there exists a singularity of the solution at the exit point $x = l$ of the domain due to the accumulative phenomenon in the vicinity of that point.

The situation is much more complicated for nonlinear equations, which give rise to a variety of feasible wavy structures of inviscid flow in the supersonic region. In this section, we assume the existence of a solution to the full-potential or Euler equations, which represents flow with a local supersonic region, and analyse *a posteriori* properties of that solution. The flow is irrotational as long as shock waves intervene, hence, it can be described by the potential $\varphi(x, y)$, which obviously determines the nondimensional velocity $\lambda(x, y) = |\nabla\varphi|$, the flow deflection angle $\theta = \arctan(\varphi_y/\varphi_x)$, and the Mach number $M(x, y)$ linked with $\lambda(x, y)$ by the relation (2.3).

3.3.1 Small local curvature of the airfoil

A basic idea that addresses the nature of formation of shock waves in a local supersonic region was suggested by Nikolsky and Taganov (1946). They proved the arising of a shock in the case when the curvature of an airfoil portion is too small.

In this case, a shock wave is formed by the coalescence of compression waves emanating from the airfoil's portion with the small curvature. The arising of the compression waves can be accounted for by isentropic relations as follows. The Prandtl–Meyer function

$$\zeta(M) := \left(\frac{\gamma+1}{\gamma-1}\right)^{1/2} \arctan\left[\frac{\gamma-1}{\gamma+1}(M^2-1)\right]^{1/2} - \arctan(M^2-1)^{1/2}$$

is an increasing function of the Mach number $M \geqslant 1$. If $\varphi(x, y)$ is the velocity potential of a supersonic flow, then the difference $\theta - \zeta(M)$ is constant along the characteristic lines of the first family defined by the ordinary differential equation $dy/dx = \tan[\theta + \arcsin(M^{-1})]$ (left-running characteristics). Therefore, the following relation is true,

$$\theta_a(x) - \zeta(M_a(x)) = \theta_+(x), \tag{3.15}$$

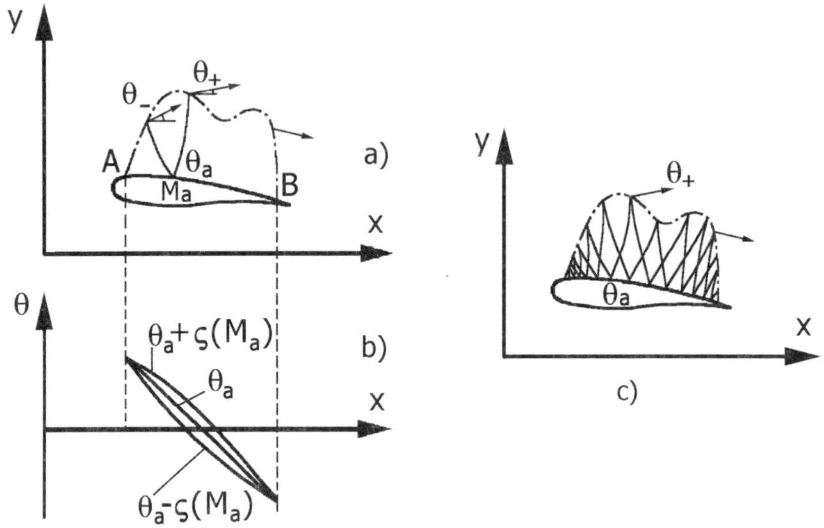

Figure 3.9 Dependence of $\theta_a \pm \zeta(M_a)$ on x in a shock-free flow.

where the subscript 'a' refers to a point with coordinates $(x, y_a(x))$ on the airfoil, while '+' denotes the point of the sonic line which is connected to the point $(x, y_a(x))$ by a left-running characteristic (see Figure 3.9a).

Similarly, the sum $\theta + \zeta(M)$ is constant along the characteristic lines of the second family $dy/dx = \tan[\theta + \arcsin(M^{-1})]$ (right-running characteristics), that is why

$$\theta_a(x) + \zeta(M_a(x)) = \theta_-(x), \tag{3.16}$$

where '$-$' corresponds to the point which is located on the sonic line and connected to the point $(x, y_a(x))$ by a right-running characteristic.

As proved by Nikolsky and Taganov (1946), in the shock-free flow the angles $\theta_\pm$ *decrease* as the observation point moves along the sonic line in the direction that leaves the subsonic region on the left. This actually follows from the Chaplygin equations (3.1) which reduce to $\theta_\psi = 0$, $\theta_\varphi = \rho\lambda_\psi/\lambda$ on the sonic line. Therefore, $d\theta = (\rho/\lambda)\lambda_\psi d\varphi$, and since $\lambda_\psi = -\lambda_\varphi\, d\varphi/d\psi$ on the sonic line, we obtain

$$d\theta = -\frac{\rho}{\lambda}\left(\frac{d\varphi}{d\psi}\right)^2 \lambda_\varphi\, d\psi \leqslant 0. \tag{3.17}$$

The equality $d\theta = 0$ occurs only at the points where $\lambda_\varphi = 0$, i.e. the sonic line contacts streamlines.

Thus,

$$\frac{d\theta_\pm(x)}{dx} \leqslant 0 \quad \text{along the sonic line,}$$

and the left-hand sides of (3.15) and (3.16) are decreasing functions of x:

$$\frac{\mathrm{d}[\theta_{\mathrm{a}}(x) \pm \zeta(M_{\mathrm{a}}(x))]}{\mathrm{d}x} \leqslant 0 \quad \text{along the airfoil,}$$

where $\theta_{\mathrm{a}}(x)$ is determined by the shape of the airfoil and decreases with increasing x for the convex airfoil (Figure 3.9b).

If the local curvature of the airfoil curvature is small, so that the angle $\theta_{\mathrm{a}}(x)$ turns out to decrease more slowly than $\theta_-(x)$ does along the sonic line

$$\left|\frac{\mathrm{d}\theta_{\mathrm{a}}(x)}{\mathrm{d}x}\right| < \left|\frac{\mathrm{d}\theta_-(x)}{\mathrm{d}x}\right|, \tag{3.18}$$

then (3.16) yields

$$\frac{\mathrm{d}\zeta(M_{\mathrm{a}})}{\mathrm{d}x} = \frac{\mathrm{d}[\theta_-(x) - \theta_{\mathrm{a}}(x)]}{\mathrm{d}x} < 0. \tag{3.19}$$

Consequently, $\mathrm{d}M_{\mathrm{a}}/\mathrm{d}x < 0$, i.e. the flow decelerates along that portion of the airfoil. Therefore, there exists a compression wave emanating from the airfoil. This compression wave may occur in both shock-free flow and flow with shock waves. However, its strength depends on the deceleration $|\mathrm{d}M_{\mathrm{a}}/\mathrm{d}x|$ of the flow along the airfoil.

(i) If the deceleration $|\mathrm{d}M_{\mathrm{a}}/\mathrm{d}x|$ is not strong, then the difference $\theta_{\mathrm{a}} - \zeta(M_{\mathrm{a}})$ on the left-hand side of (3.15) is a decreasing function of x due to the decreasing θ_{a}. Such a situation takes place in the supersonic region over a shockless airfoil at on-design conditions (Figure 3.9b,c).

(ii) However, if the local curvature is too small, then the deceleration $|\mathrm{d}M_{\mathrm{a}}/\mathrm{d}x|$ becomes strong. Consequently, the difference $\theta_{\mathrm{a}} - \zeta(M_{\mathrm{a}})$ *increases* along a portion $x_* < x < x^*$ of the airfoil contour (Figure 3.10), and due to (3.15) the inequality $\mathrm{d}\theta_+/\mathrm{d}x \leqslant 0$ is violated. The violation is impossible in the shock-free transonic flow. Hence, one can conclude that there exists a shock wave formed by the coalescence of left-going characteristic lines emanating from the airfoil.

In practice, if the scheme (ii) of shock formation occurs, it typically takes place near the nose of the airfoil where the curvature $\mathrm{d}\theta_{\mathrm{a}}/\mathrm{d}s$ drops rapidly. One thus can say that the shock is formed because of the slow decrease of $\theta_{\mathrm{a}}(x)$, i.e. slow rotation of the tangent to the airfoil, with increasing x. The term 'slow' implies here the rate of the decrease of θ_{a} as compared with that of the angle θ_- on the sonic line, see (3.18). In other words, it is the fast decrease of θ_-, i.e. fast rotation of the velocity vector on the sonic line in the clockwise direction, that causes the above pattern of shock wave formation.

With increasing freestream Mach number M_∞, the rate of the velocity vector rotation on the sonic line decreases as theoretical examples and numerical simulations demonstrate. Then due to (3.19) the flow deceleration near the same portion of the airfoil with a small local curvature becomes weak or may turn into an acceleration. Therefore, with the increasing M_∞ the origin of the shock wave moves away from

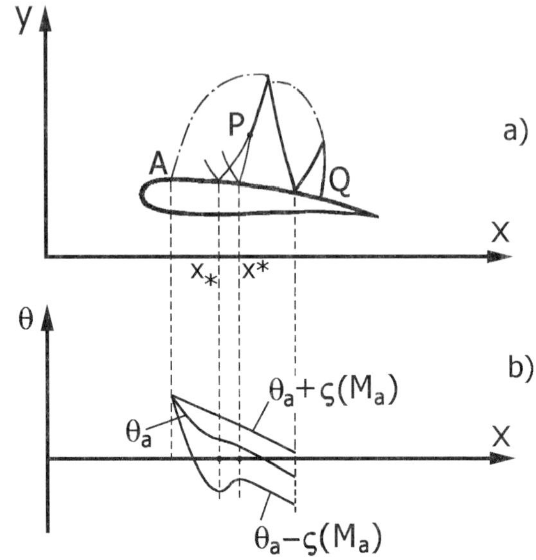

Figure 3.10 Direct formation of a shock wave in the supersonic region due to the small local curvature of the airfoil.

the airfoil towards the sonic line, and the scheme shown in Figure 3.10 may fail to exist.

We notice that the left-running characteristics originating at the airfoil used to be interpreted as expansion waves, while the right-running characteristics originating on the sonic line as compression waves (Guderley 1957, Section 6.1; Ferrari and Tricomi 1968, p. 298; Moulden 1984, p. 165). This interpretation needs to be clarified. First, the above assertion is true only for transonic flows with monotonously decreasing difference $\theta_a - \zeta(M_a)$ along the airfoil. At the same time, if $\theta_a - \zeta(M_a)$ happens to increase locally (see item (ii) above), then the left-running characteristics $\xi = \text{const.}$ form a compression wave, which makes the flow velocity *decrease* along the right-running characteristics $\eta = \text{const.}$ (Figure 3.11a). This follows from the relations $\theta_- = \theta + \zeta(M)$, $\theta_a - \zeta(M_a) = \theta - \zeta(M)$, which yield

$$2\zeta(M) = \theta_- - [\theta_a - \zeta(M_a)]. \tag{3.20}$$

Hence, the local increase of the expression in brackets results in a decrease of $\zeta(M)$ along $\eta = \text{const.}$

Furthermore, in the typical case when the difference $\theta_a - \zeta(M_a)$ is decreasing along the airfoil up to the terminating shock wave, the left-running characteristics indeed force the flow velocity to increase along the characteristics $\eta = \text{const.}$ However, if there is a deceleration $dM_a/dx < 0$ along the wall (see item (i) above), it results in a compression wave which tends to slow down the flow *along the streamlines*. The occurrence of either acceleration or deceleration along the streamlines depends on the

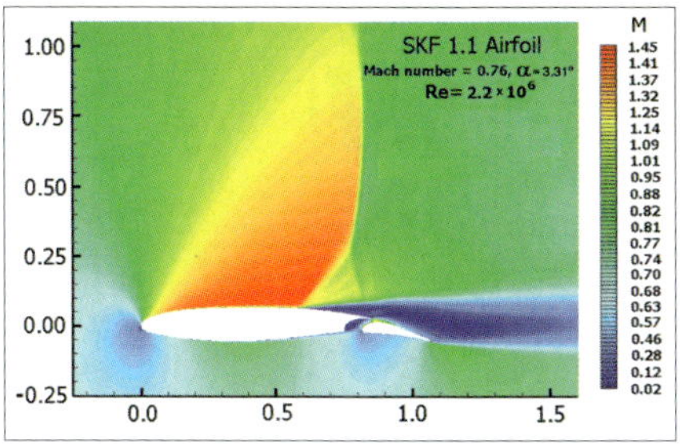

Plate 1 Transonic flow separation. Mach number isolines over SKF 1.1 airfoil obtained in numerical simulation of Zingg, Nelson and Godin (2001). Reproduced by permission of the authors.

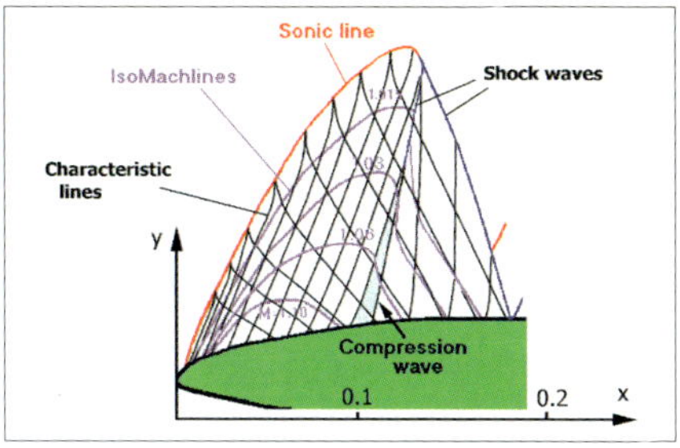

Plate 2 Recomputation of a fragment of the local supersonic region over the X63T18S airfoil at $M_\infty = 0.766$.

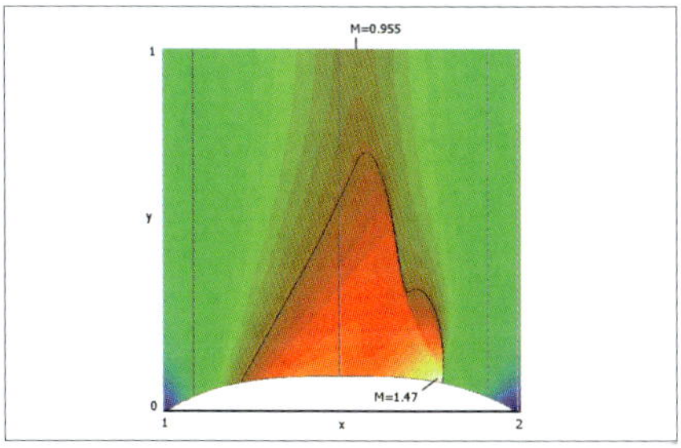

Plate 3 Flow pattern over the airfoil (4.4) with the vanishing curvature at the midpoint ($y_{max} = 0.1$, $n = 3$, $M'_{out} = 0.675$).

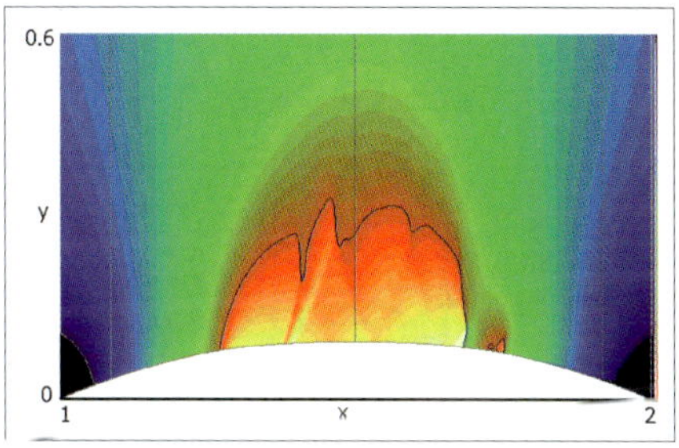

Plate 4 Recomputation of the flow presented in Plate 9 on a fine mesh of 801 x 341 ($\Delta x = \Delta y = 0.002$ in the supersonic region).

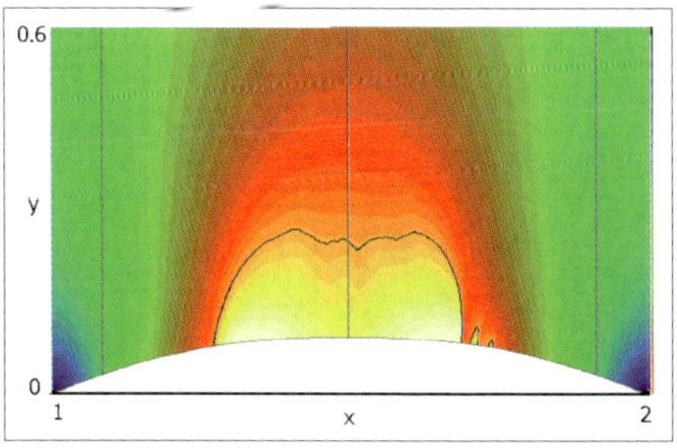

Plate 5 Flow over the shockless airfoil $y_0(x)$ at $M'_{out} = 0.673$.

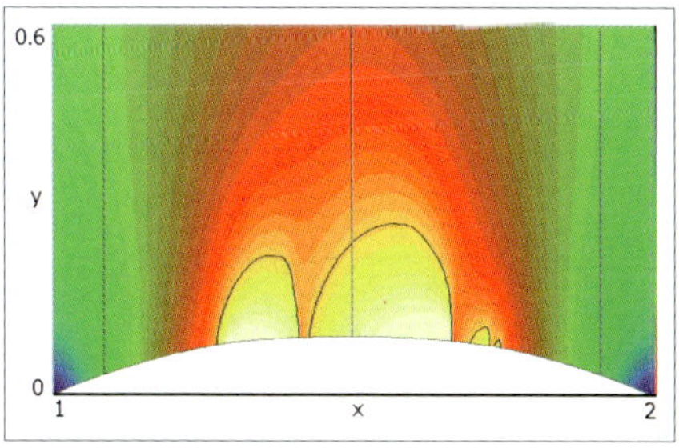

Plate 6 Flow over the shockless airfoil $y_0(x)$ at $M'_{out} = 0.671$

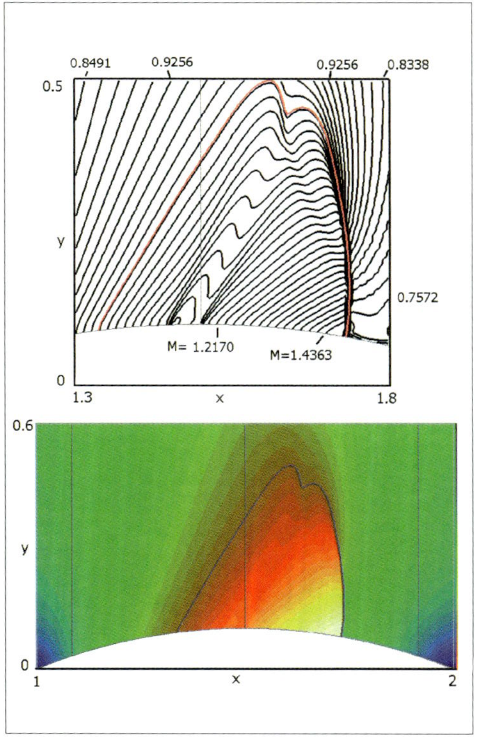

Plate 7 Flow over the circular arc airfoil (4.1) modified by a straight segment at $1.45 < x < 1.50$ ($M'_{out} = 0.675$, $\Delta x = \Delta y = 0.004$).

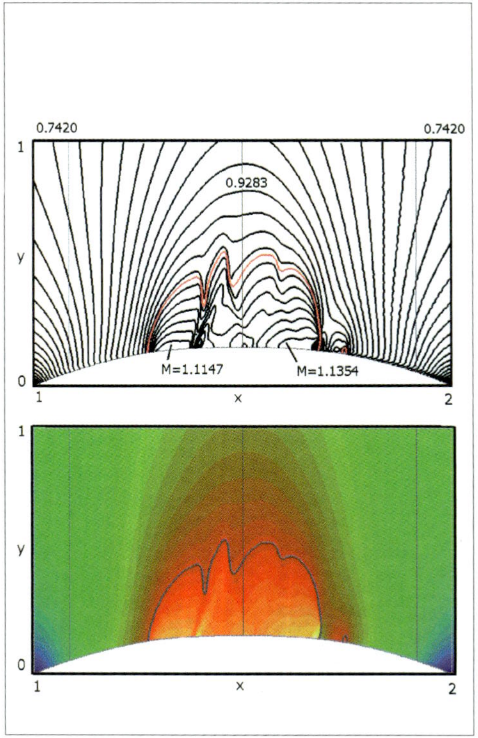

Plate 8 Mach number isolines over the shockless airfoil with perturbation (4.5) of height $h = 2 \cdot 10^{-4}$ imposed at $1.37 < x < 1.40$ ($\Delta x = \Delta y = 0.004$ in the supersonic region).

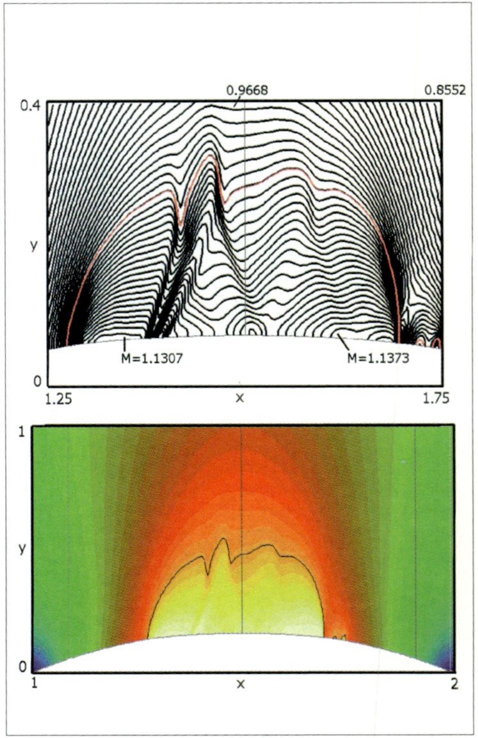

Plate 9 Mach number isolines over the shockless airfoil with perturbation (4.5) of height $h = 1.5 \cdot 10^{-4}$ imposed at $1.37 < x < 1.40$ ($\Delta x = \Delta y = 0.004$ in the supersonic region).

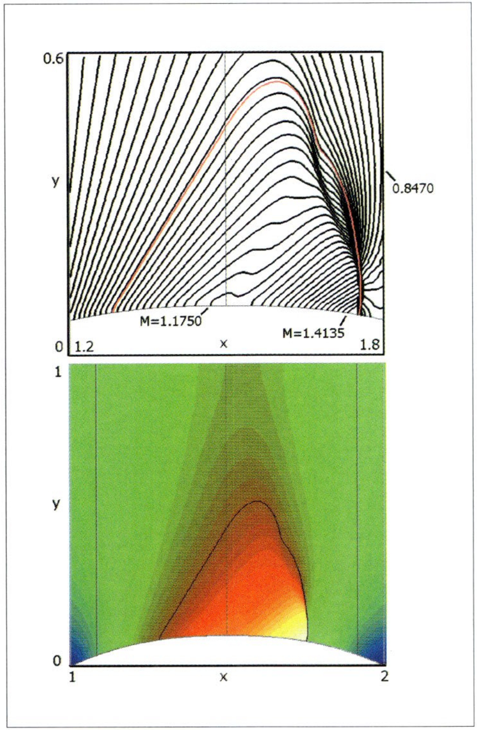

Plate 10 Weaker shock, which is originated by the profile (4.4) with $n = 2.5$, $y_{max} = 0.1$, interacts with the sonic line yielding an inflection point of the sonic line.

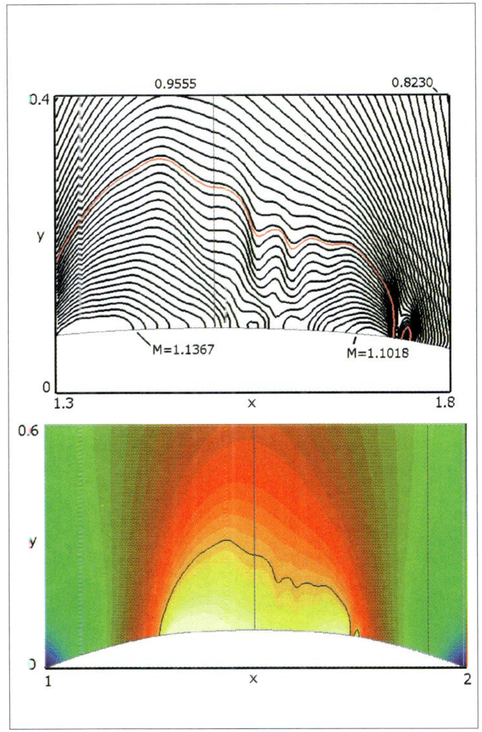

Plate 11 Transonic flow over the airfoil (4.6) at $p = -0.10$ and $M'_{out} = 0.675$ ($\Delta x = \Delta y = 0.004$).

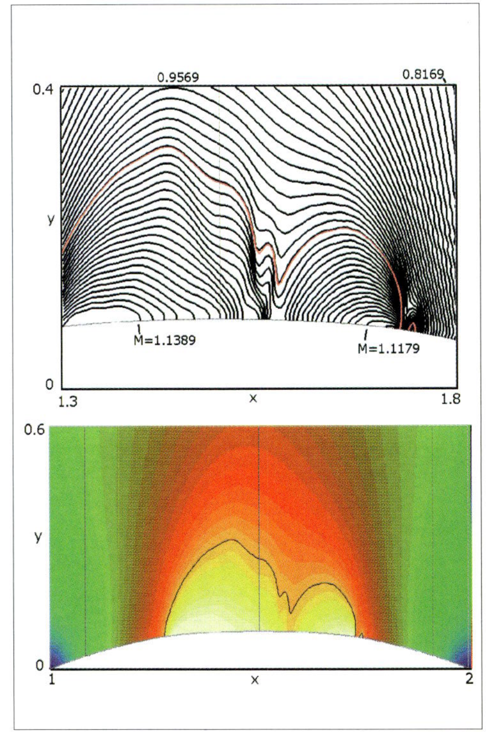

Plate 12 Transonic flow over the airfoil (4.6) at $p = -0.12$ and $M'_{out} = 0.675$ ($\Delta x = \Delta y = 0.004$).

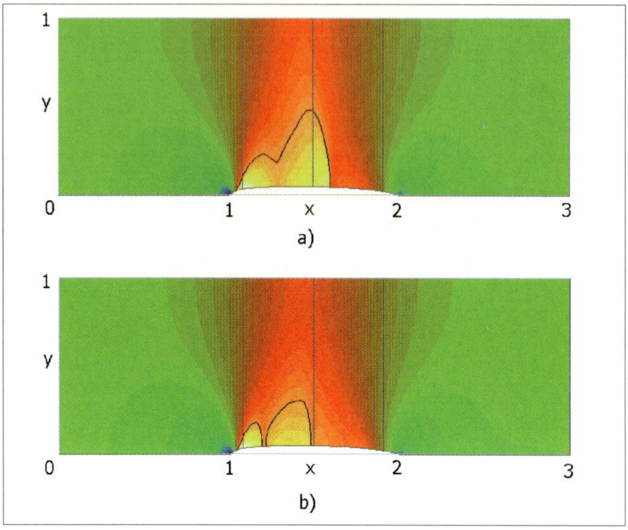

Plate 13 Flow over a K–2 type airfoil. a) $M'_{\text{out}} = 0.772$, b) $M'_{\text{out}} = 0.770$.

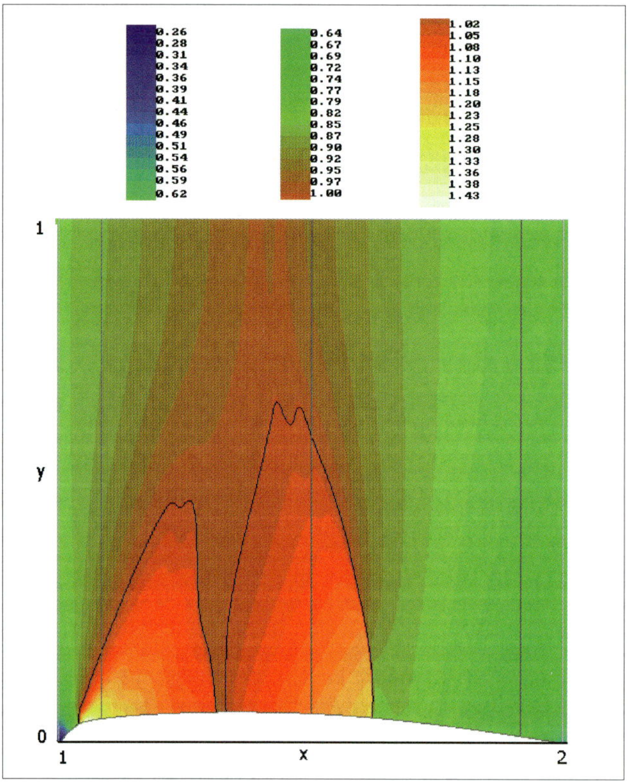

Plate 14 A RAE type airfoil in the channel. An instantaneous pattern of transonic flow due to a change of M'_{out} from 0.76 to 0.75.

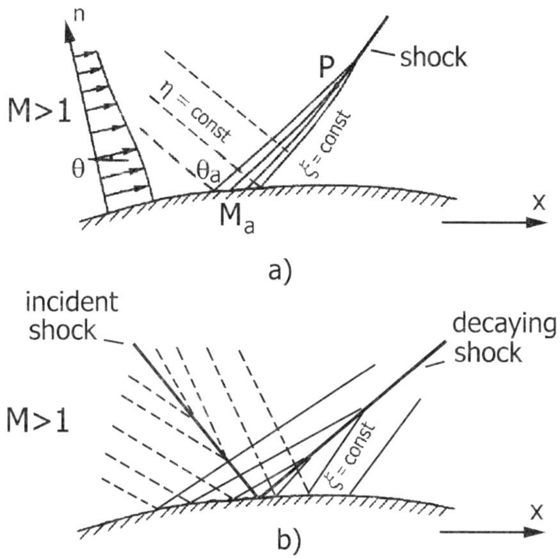

Figure 3.11 Nonuniform supersonic flow over a convex wall: (a) formation of a shock wave by compression waves emanating from the wall; (b) the weakening of the reflected shock by expansion waves emanating from the wall.

rate of decrease of $\theta_a - \zeta(M_a)$ in (3.20) as compared with that of θ_-. If the quantity $\theta_a - \zeta(M_a)$ decreases more slowly than θ_- does (the local curvature of the airfoil is small), then one ends up with the deceleration.

Therefore, the left-going characteristics can in general result in either an expansion wave (and they typically do so in the beginning of a local supersonic region) or a compression wave. The latter is quite similar to the occurrence of the compression wave in a purely supersonic nonuniform flow over a convex wall if an appropriate velocity distribution is prescribed at an inlet section (Figure 3.11a). In this case, one observes the merging of left-running characteristics into a shock wave at a point P. At the same time, if the curvature of the wall is large enough, then the left-running characteristics yield expansion waves for any velocity distribution at the inlet section as well as in the case of an oblique shock reflection from the wall (Figure 3.11b).

3.3.2 Coalescence of compression waves emanated from the rear portion of the sonic line

Another concept of the arising of shocks was suggested in a well-known book by Guderley (1957). It establishes the formation of a shock wave due to the coalescence of compression waves emanating from the rear portion of the sonic line. In this case, the subsonic region plays the role of a permeable obstacle which declines streamlines towards the airfoil and causes compression waves in the same way as a concave wall does in the classical problem of supersonic flow over a smooth rigid wall. The shock

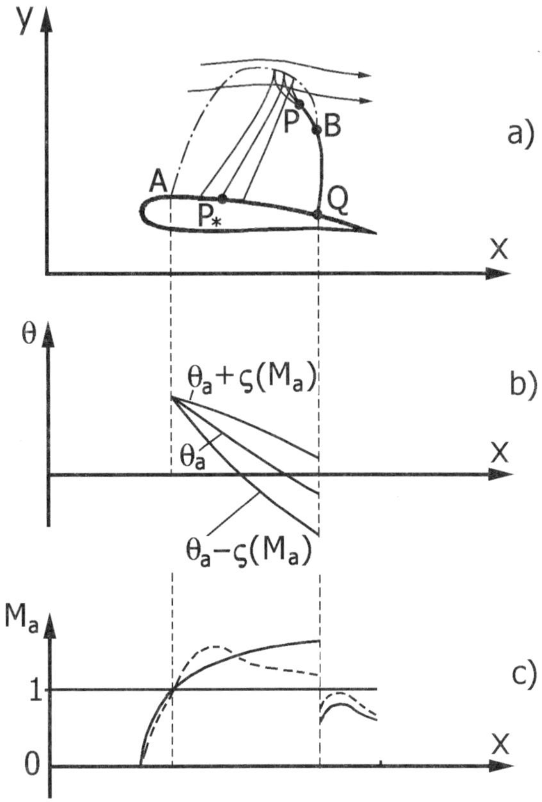

Figure 3.12 Formation of a shock wave due to the strong deceleration in a vicinity of the rear part of the sonic line.

origin P is embedded in the supersonic region, the flow velocity behind the portion PB of the shock remains supersonic (Figure 3.12a). The sonic line terminates on the shock wave at the point B. Details of the flow field were analysed by Sobieczky and Stanewsky (1976) using the hodograph variables.

The difference $\theta_a - \zeta(M_a)$ in this case decreases along the airfoil with increasing x (Figure 3.12b), otherwise one would obtain the direct shock formation shown in Figure 3.10. The flow velocity may either increase along the airfoil up to the point Q (the continuous curve in Figure 3.12c) or have a maximum (the dashed curve in Figure 3.12c). The latter situation takes place if $\theta_a - \zeta(M_a)$ happens to change slowly along a portion of the airfoil as discussed in the previous section. If so, then the compression waves emanating from the vicinity of the point P_* (Figure 3.12a) contribute to the stronger deceleration of the flow near the sonic line. This results in a significant drop of the quantity $\theta_+ = \theta_a - \zeta(M_a)$ on a short arc of the sonic line, and generates stronger compression waves emanating from the sonic line.

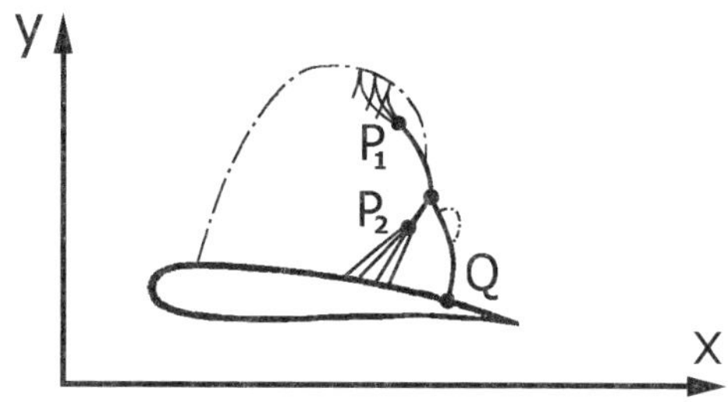

Figure 3.13 A scheme of shock wave formation due to both flow deceleration near the sonic line and the small curvature of the airfoil.

Also, a combined scheme is feasible in which the upper end of a complex shock configuration forms due to the flow deceleration near the sonic line, while the lower end of the shock forms due to the deceleration caused by the small local curvature of the airfoil (Figure 3.13).

If the local supersonic region terminates by a shock wave, then the shock must meet the convex airfoil surface normally and bend upstream of its foot (point Q) in the framework of the inviscid flow model. The curvature of the shock wave at the foot is infinite, and there exists a logarithmic singularity in the pressure and the velocity gradients with the distance along the airfoil behind the shock wave,

$$\left.\frac{\mathrm{d}M}{\mathrm{d}x}\right|_{x_Q} \approx |\log(x - x_Q)|(M_a - 1)^{-3/2}\left(\tfrac{1}{2}(\gamma + 1)\right)^{3/4}\left|\left.\frac{\mathrm{d}\theta_a}{\mathrm{d}x}\right|_{x_Q}\right., \tag{3.21}$$

where M_a is the Mach number in front of the shock foot Q. The existence of the singularity was established with asymptotic expansions by Gadd (1960) and Oswatitsch and Zierep (1960). A more general analysis of the flow field in the vicinity of point Q was given by Cole and Cook (1986).

The singularity (3.21) gives a flavour of difficulties which are to be overcome in rigorous mathematical studies of the solvability of boundary-value problems for transonic flow with the shock wave terminating the local supersonic region. The smoothness of the coefficients of the equation governing a small perturbation was an essential point for the technique developed in Chapters 1 and 2. At the same time, the singularity (3.21) causes the nonsmoothness of the coefficients at the point Q even in the case when one introduces a cut along the shock wave PQ (Figure 3.12) in order to avoid the velocity jump in the domain where the solution is to be found.

We notice that the sonic line makes an acute angle with the flow direction at point A, as $\theta_\varphi < 0$ for the convex airfoil and $\lambda_\psi < 0$ due to the second equation in the system (3.1).

In the nose area of a blunt airfoil, there is obviously a stagnation point. Similarly, the tip of an airfoil with a finite leading-edge angle is a stagnation point. This follows from the condition that the velocity must be continuous in the subsonic domain including the corner point of the boundary. In the streamline coordinates (φ, ψ), the stagnation point yields a singularity in the Chaplygin Equations (3.1) and in the mapping $(\varphi, \psi) \leftrightarrow (x, y)$. This singularity was resolved and incorporated into a numerical scheme by Dokolin *et al.* (1992).

At the trailing edge of the airfoil, physical considerations yield the well-known Kutta–Joukowski condition: in the attached flow, two streamlines which correspond to the upper and lower surfaces of the airfoil must leave the airfoil at the trailing edge; moreover, they must have the same direction immediately behind the trailing edge, and the pressures on both streamlines are to be identical. At the same time, the velocity potential must be admitted to have a jump across the trailing-edge streamline if it is treated as a single valued function.

In inviscid steady flow with a nonzero trailing-edge angle, the Kutta condition reduces to the condition that the trailing-edge pressure be the same as the total pressure, and the local velocity vanishes.

A rigorous analysis of a boundary-value problem with Kutta–Joukowski condition at the trailing edge was carried out by Ciavaldini *et al.* (1986). The authors investigated steady plane subsonic flow past an airfoil with a sharp trailing edge. The problem was formulated in a finite domain bounded by a far-field boundary on which conditions were prescribed by the asymptotic expansion of the flow at infinity. Using variational methods in Sobolev spaces, the authors proved that there exists a unique solution of the problem satisfying the Kutta–Joukowski condition. An approximating sequence of linear problems was introduced and *a priori* estimates were derived which guarantee the convergence of the algorithm. Also, the regularity of the solution was studied.

3.3.3 Accumulative phenomenon

In a fundamental paper, Busemann (1949) discussed the propagation of a short disturbance in the flow with a local supersonic region as follows:

> A small depression of symmetrical shape may be applied at the axis of symmetry where the highest supersonic Mach number occurs. When this depression is created suddenly, the nonsteady flow within the supersonic region also has a restricted disturbance area, and the main disturbance travels through the outgoing alley of Mach waves. This alley ends on the sonic line and gives access through the subsonic region to every other point in the flow...
>
> The subsonic flow spreads around the end of the alley a disturbance distribution with decreasing amplitudes, but a more concentrated reaction must be expected to travel down the supersonic alley that starts at this end of the first alley (Figure 3.14). The second alley gets contact to a third alley by the reflection on the solid walls of the body, and a zig-zag disturbance in the supersonic region will be the result. It may be an interesting problem to determine the conditions near the sonic line where the disturbance reverses, but the only thing that really matters

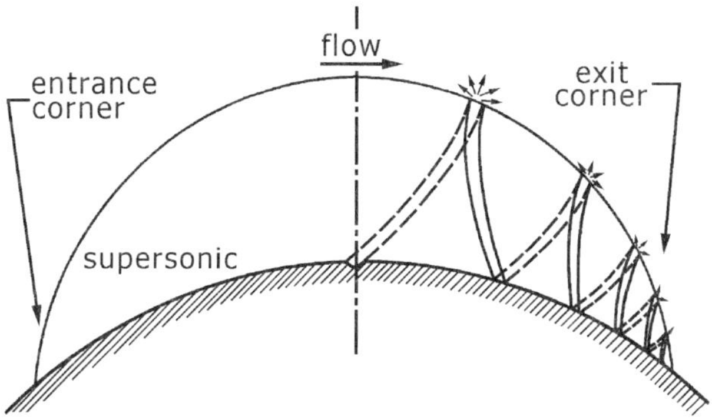

Figure 3.14 The nonsymmetric disturbance pattern. Reproduced from Busemann (1949).

> is the alternative, whether the disturbance on this zig-zag decreases or increases.
>
> The first guess may be that the intensity will decrease because of the contributions to the subsonic region. If this guess holds true, the original depression can be made small enough to avoid shock formation before the disturbance dies completely out. In this case a potential flow is guaranteed... The opposite guess – that the disturbance is increasing in such a manner that, finally, a shock wave must occur – has a completely different aspect. Even the smallest depression can only lead to a flow that is regular everywhere except at the point where the sonic line reaches the body contour. This is the point where the zig-zag repetitions are infinitely crowded...

In the sequel, arguments were found in favour of the second scheme of propagation of the small disturbance. Guderley (1953) analysed perturbation of the transonic potential vortex, i.e. the flow whose potential in the physical plane is given by the simple expression $\varphi(x, y) = \arctan(y/x)$. Since

$$\varphi_x = \frac{y}{x^2 + y^2}, \qquad \varphi_y = \frac{-x}{x^2 + y^2},$$

the velocity magnitude is constant along the concentric circles which are the streamlines: $\lambda = |\nabla\varphi| = (x^2 + y^2)^{-1/2}$. Using the Bernoulli law (2.2), one can easily verify that the continuity equation $(\rho\varphi_x)_x + (\rho\varphi_y)_y = 0$ is satisfied. At $x^2 + y^2 > 1$ the velocity is subsonic, while at

$$\frac{\gamma - 1}{\gamma + 1} < x^2 + y^2 < 1$$

it is supersonic. Two streamlines may be chosen to represent the walls; one of them can lie at a subsonic, the other at a supersonic velocity. An interferogram of the transonic vortex is given, for example, in Oswatitsch (1976, Section 6.9). The interferogram demonstrates Mach waves which emanate from the inner wall due to its considerable roughness, propagate to the circular sonic line, and reflect back to the inner wall.

If a small bump is introduced on the inner wall, so that the wall shape is changed in some interval $\theta_1 < \theta < \theta_2$ of the θ-coordinate, then we obtain a small perturbation problem. Guderley discussed particular solutions to that problem as well as a general solution obtained in the form of a series in the hodograph plane. An analysis of the solutions showed that, in the transonic potential vortex, a perturbation of the wall (either in the subsonic or in the supersonic region) will create a wavy flow pattern downstream of that perturbation. The pattern will have its largest amplitudes in the supersonic region, although it is felt in the subsonic region as well. In the vicinity of the bump, the waves have the magnitude of the original perturbation. If the perturbation lies in the subsonic region, they will even attenuate with distance from the perturbed spot. However, if the local Mach number at the wall in the supersonic region approaches one, the amplitudes of perturbation downstream of the bump will increase.

Another question is what happens to the wavy flow pattern, i.e. streamline and velocity oscillations, when the sonic line of the initial flow terminates on the wall at a point B, as in the shock-free flow with a local supersonic region past a shockless airfoil. Following Busemann (1949), Guderley (1953) discussed the accumulation of singularities propagating in the local supersonic region and multiply reflected from the airfoil and the sonic line. A *weak singularity* was supposed to arise on the wall/airfoil due to the nonsmoothness in the wall curvature. A pattern, in which a singular Mach wave is reflected from the wall and the sonic line an infinite number of times as the observation point approaches the endpoint B of the sonic line, was described. With some general arguments related to the uniqueness reasons and a simpler boundary-value problem, Guderley inferred that the accumulation of singular waves at the point B is incompatible with the potential flow. This requires introduction of shock waves into the scheme of flow.

For a smooth perturbation of the wall, Guderley and Busemann showed that in the hodograph plane the pattern can be treated in the same way as in the case of a potential vortex. The occurrence of oscillatory particular solutions, whose amplitude increases toward the exit corner B of the supersonic region, has been exposed by an analysis of particular solutions for the perturbation potential in the hodograph plane, since the solutions cause physical difficulties close to the point B. In addition, Guderley and Busemann gave an example for which the treatment is particularly simple. The solution contained the term $\theta^{-1/4}\sin(\theta^{-1/2}2(\tau^*)^{3/2})$, where $\tau^* = \text{const.}$, with the vanishing flow deflection angle $\theta \to 0$ as one approaches the exit point B of the local supersonic region. Some boundary-value problems were discussed in which wavy flow patterns do occur. An analysis indicated that the amplitude of these waves increases as one approaches the end of the supersonic region.

In Section 4.1, we shall present a numerical confirmation of the accumulative phenomenon at flow conditions which are very close to shock-free ones. At such conditions, it takes a few reflections of compression waves from the sonic line and

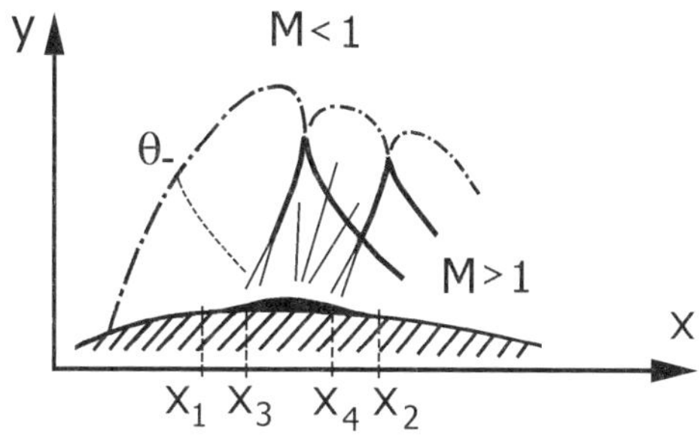

Figure 3.15 A scheme of the direct formation of shock waves by a short bump on the airfoil. The resulting profile may remain convex or become concave near the points x_3 and x_4.

the airfoil to get steep velocity gradients. This eventually results in the formation of a shock wave according to either the scheme pointed out in Figure 3.10 or the one of Figure 3.12.

3.3.4 Short perturbation of the airfoil shape

As discussed above, in shock-free flow with a local supersonic region the arc AB of the airfoil is necessarily convex and its local curvature cannot be too small. At the same time, of practical interest is an analysis of the generation of shock waves by a short smooth bump that reduces the curvature of the airfoil contour or makes it locally concave. In this case, the scheme of the arising of shock waves caused by the bow and tail portions of the bump is virtually the same as shown in Figure 3.10. In addition, we can obtain asymptotic estimates for the height of the bump that is responsible for the direct formation of shock waves.

We consider the bump configuration

$$\Delta y = f(x) = h \sin^4\left(\pi \frac{x - x_1}{x_2 - x_1}\right) \quad \text{at } x_1 \leqslant x \leqslant x_2, \tag{3.22}$$

which is concave near its bow and tail and convex in the vicinity of the middle point. Depending on h, the resulting airfoil with superimposed perturbation (3.22) may remain convex or become locally concave as sketched in Figure 3.15. The fourth power of the sine is chosen to avoid discontinuities of the first-order velocity derivatives at the endpoints $x = x_1$, $x = x_2$. The discontinuities would propagate along the characteristics of the first family emanating from the endpoints. Having been reflected from the sonic line, such weak discontinuities would yield weak shock waves, as demonstrated by examples of self-similar solutions (Landau and Lifshits 1987).

We denote by $\Delta\theta$ the angle made by the tangent to the bump (3.22) with the x-axis. Assuming that h is small enough, we obtain

$$\Delta\theta \approx \frac{\mathrm{d}f}{\mathrm{d}x} = \frac{4\pi h}{x_2 - x_1} \sin^3\left(\pi \frac{x - x_1}{x_2 - x_1}\right) \cos\left(\pi \frac{x - x_1}{x_2 - x_1}\right).$$

The second differentiation and a simple analysis of maxima yield

$$\left.\frac{\mathrm{d}\Delta\theta}{\mathrm{d}x}\right|_{\max} \approx \frac{9}{4}\frac{\pi^2 h}{(x_2 - x_1)^2} \tag{3.23}$$

at the points

$$x_3 = x_1 + \frac{x_2 - x_1}{\pi}\arcsin(\tfrac{3}{8})^{1/2},$$
$$x_4 = x_2 - \frac{x_2 - x_1}{\pi}\arcsin(\tfrac{3}{8})^{1/2}.$$

If the maximum curvature of the perturbation (3.22) is greater than the local curvature of the basic airfoil,

$$\left.\frac{\mathrm{d}\Delta\theta}{\mathrm{d}x}\right|_{\max} > \left|\frac{\mathrm{d}\theta_{\mathrm{a}}}{\mathrm{d}x}\right|_{x=x_3 \text{ or } x=x_4},$$

then the resulting profile contains concave arcs. Therefore, using (3.23) we arrive at the estimate for the amplitude h of the bump (3.22) which yields the direct formation of shock waves:

$$h > \frac{4(x_2 - x_1)^2}{9\pi^2}\left|\frac{\mathrm{d}\theta_{\mathrm{a}}}{\mathrm{d}x}\right|_{x=x_3 \text{ or } x=x_4}. \tag{3.24}$$

Now we turn to another estimate, which refines (3.24). However, it requires the knowledge of the Mach number M_{a} on the airfoil at the x-coordinate where the perturbation (3.22) is superimposed. The sharper estimate is

$$h > \frac{2(x_2 - x_1)^2}{9\pi^2}\left|\frac{\mathrm{d}[\theta_{\mathrm{a}} - \zeta(M_{\mathrm{a}})]}{\mathrm{d}x}\right|_{x=x_3 \text{ or } x=x_4}. \tag{3.25}$$

In order to derive (3.25), we notice that the existence of a concave arc of the wall/airfoil is a sufficient but not necessary condition for shock formation. That is why, we suppose that the airfoil with the imposed perturbation (3.22) remains convex along the whole supersonic portion AB of the airfoil including the vicinities of the points $x = x_2$ and $x = x_4$. Due to (3.17), the angle θ_+ must decrease in the shock-free flow, and the same applies to the angle $\theta_+ + \Delta\theta_+$ in the perturbed flow with the bump (3.22). Therefore, if the perturbed flow is shock free, then $\mathrm{d}(\theta_+ + \Delta\theta_+)/\mathrm{d}x \leqslant 0$. Consequently, the condition of shock formation can be written as $\mathrm{d}(\theta_+ + \Delta\theta_+)/\mathrm{d}x > 0$, or due to (3.15) as

$$\frac{\mathrm{d}[\theta_{\mathrm{a}} + \Delta\theta - \zeta(M_{\mathrm{a}} + \Delta M)]}{\mathrm{d}x} > 0. \tag{3.26}$$

Assume that the perturbation (3.22) of the airfoil shape leaves the angle θ_- almost unchanged, in other words, that the perturbation $\Delta\theta_-$ of the angle θ_- on the sonic line (see Figure 3.15) is a small quantity of higher order as compared with $\Delta\theta$: $\Delta\theta_- = o(\Delta\theta)$. This assumption is justified by the downstream propagation of flow perturbations in the supersonic region and by their small influence on the value of θ_- through the subsonic region after penetrating it at the end of the outgoing alley. Then

$$\theta_- = \theta_{\mathrm{a}} + \zeta(M_{\mathrm{a}}) \simeq \theta_{\mathrm{a}} + \Delta\theta + \zeta(M_{\mathrm{a}} + \Delta M),$$

whence

$$\zeta(M_a + \Delta M) = \zeta(M_a) - \Delta\theta.$$

Inserting this relation into (3.26), we obtain a condition of the arising of shock waves,

$$\frac{d[\theta_a + \Delta\theta - \zeta(M_a) + \Delta\theta]}{dx} > 0,$$

or

$$2\frac{d\Delta\theta}{dx}\bigg|_{max} > \frac{d[\zeta(M_a) - \theta_a]}{dx},$$

that is,

$$2\frac{d\Delta\theta}{dx}\bigg|_{max} > \left|\frac{d[\theta_a - \zeta(M_a)]}{dx}\right|.$$

Using (3.23), we arrive at the inequality (3.25). If this inequality holds at $x = x_3$ and $x = x_4$, then both the bow and tail of the perturbation (3.22) yield shock waves.

The shock waves generated by the bump (3.22) propagate toward the sonic line and then reflect back from it to the airfoil. We postpone an analysis of feasible schemes of shock-wave reflection from the sonic line until Section 4.2.

3.4 Flow with a Local Supersonic Region over a Porous Boundary

In this section, we study a direct boundary-value problem with respect to small perturbation of a given flow. As shown in Section 3.2, if the slip condition is prescribed over the full supersonic arc of the airfoil, or the slip condition is replaced by a velocity distribution on a portion of that arc, then the problem is overdetermined in the class of smooth solutions. The nonexistence of a smooth solution demonstrates that shock-free flow with a local supersonic region is unstable to small steady perturbation of the Neumann or Dirichlet boundary data with respect to the potential. Many attempts to tackle this problem in a class of functions admitting shock waves faced formidable mathematical difficulties.

At small deviations of the airfoil's shape from shock-free conditions, the arising of shock waves occurs in the very end of the local supersonic region due to the accumulation of perturbations multiply reflected from the sonic line and the airfoil, as discussed in the previous section. On the other hand, recent numerical and experimental studies showed that employment of appropriate control through a porous or perforated portion of the airfoil weakens shock waves in the supersonic region. The perforated surface plays the role of a damping boundary condition, which may prevent the accumulative phenomenon and, thus, provide smoothness of the flow. Therefore, we employ below a boundary condition which models the perforation of the airfoil and exerts a damping effect on the flow perturbation. That approach makes it possible to avoid shock waves and singularities and to concentrate on difficulties caused by the mixed type of the governing equations.

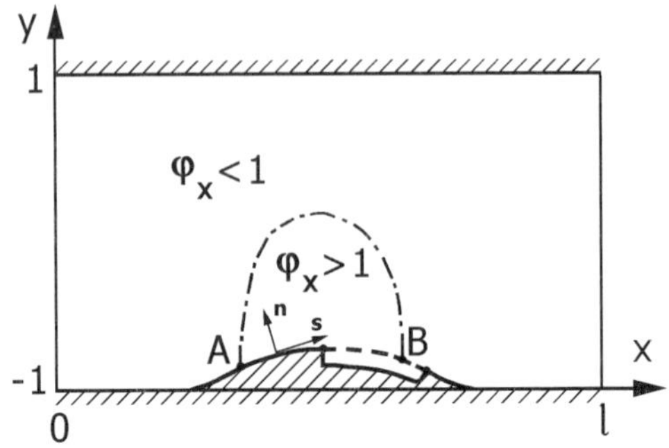

Figure 3.16 Transonic flow with a local supersonic region over a permeable airfoil. Reproduced from Kuz'min (2001) by permission of Birkhäuser.

3.4.1 Formulation of the problem. Uniqueness of the solution

Let $\varphi(x, y)$ be a smooth solution to the von Kármán Equation (2.10):

$$(\gamma + 1)(1 - \varphi_x)\varphi_{xx} + \varphi_{yy} = 0. \tag{3.27}$$

Suppose that $\varphi(x, y)$ describes shock-free flow with a local supersonic region over an impermeable long bump modelling an airfoil and located on the lower wall $y = -1$ of a channel.

The slip condition $\varphi_n = 0$ on the thin bump, where n is the inner normal direction, can be replaced as usual by the condition $\varphi_y = \theta^-(x)$ on a straight segment $y = -1$. The same flow field governed by the velocity potential $\varphi(x, y)$ develops over a perforated airfoil if an active control is introduced in such a way that the pressure $p_c(x)$ in the cavity beneath the perforation is maintained equal to the flow pressure on the perforation (Figure 3.16). However, if the airfoil shape or the pressure in the cavity are slightly changed, then the flow field undergoes a small perturbation. The potential $\hat{\varphi}(x, y)$ of the perturbed flow satisfies Equation (3.27) and the relation on the perforated surface

$$\hat{\varphi}_n - \varphi_n = R^- \cdot (\hat{\varphi}_s - \varphi_s), \tag{3.28}$$

where s is the tangential direction, and the factor $R^-(x)$ depends on the porosity of the surface. The relation (3.28) is a linearized version of Darcy's law (see, for example, Chen *et al.* 1989),

$$v_n = -\frac{\sigma}{\rho_\infty v_\infty}(p - p_c), \tag{3.29}$$

for the transpiration velocity v_n on a perforated or porous surface, where σ is the porosity factor, ρ_∞ and v_∞ are the freestream density and velocity, p is the flow pressure on the perforated surface, p_c is the pressure in the cavity beneath the perforation. In order to obtain (3.28) from (3.29), we invoke the classical theory of small perturbations in compressible flow, in which the perturbation $p(x) - p_1(x)$ of the inviscid flow pressure $p_1(x)$ on a rigid wall is related to the perturbation $v_s(x)$ of the tangential velocity $v_1(x)$ by the expression $p - p_1 = -\rho_1 v_1 v_s$. In case of the active

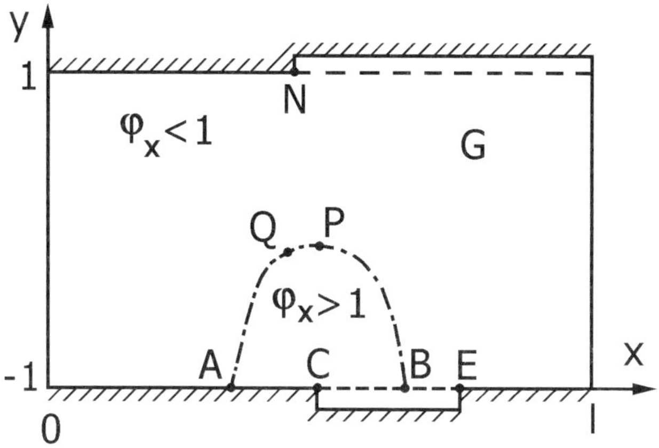

Figure 3.17 Domain in question in the plane (x, y). Reproduced from Kuz'min (2001) by permission of Birkhäuser.

flow control, p_c may be imposed equal to the pressure p_1 of the unperturbed flow. Then the replacement of $p - p_c$ by $-\rho_1 v_1 v_s$ on the right-hand side of (3.29) yields the relation

$$v_n = \frac{\sigma}{\rho_\infty v_\infty} \rho_1 v_1 v_s.$$

By substituting $v_n = v_1(\theta - \theta_1)$, $v_s = v - v_1$ into this relation, one arrives at the condition $\theta - R_* v = \theta_1 - R_* v_1$, where

$$R_*(x) = \frac{\sigma(x)\rho_1}{\rho_\infty v_\infty}.$$

Finally, using the expressions $\theta = \varphi_y$, $v/a_* = \varphi_x$, we obtain the required oblique derivative condition (3.28) with $R^- = R_* a_*$. A similar linearization of the Darcy law (3.29) was used, for example, by Kraft and Lo (1977).

Using the approximate relation $\hat{\varphi}_n = \hat{\varphi}_y - \theta^-(x) + f^-(x)$, where $f^-(x)$ is the perturbation of the flow deflection angle $\theta^-(x)$, we replace (3.28) by the relation on a segment of the line $y = -1$:

$$\hat{\varphi}_y - \varphi_y + f^- = R^- \cdot (\hat{\varphi}_x - \varphi_x) \quad \text{at } y = -1.$$

Therefore, for the perturbation $u = \varphi - \hat{\varphi}$ of the solution $\varphi(x, y)$, we obtain the condition

$$u_y - R^-(x) \cdot u_x = f^-(x) \quad \text{at } y = -1.$$

In order to simplify the proof of the uniqueness theorem, we employ the oblique derivative condition $u_y - R^+(x)u_x = f^+(x)$ on a portion of the upper wall $y = +1$ of the channel as well. The latter resembles the perforation used in practice for attenuating reflections from the walls of a wind tunnel (Velichko *et al.* 1995; Lifshitz *et al.* 1996).

Thus, the perturbation $u = \varphi - \hat{\varphi}$ of the solution $\varphi(x, y)$ is governed by the equation

$$(1 - \varphi_x + u_x)u_{xx} - \varphi_{xx}u_x + \frac{1}{\gamma + 1}u_{yy} = 0, \tag{3.30}$$

and satisfies the boundary conditions

$$\left.\begin{aligned} u &= 0 && \text{at the inlet} && x = 0, \\ u_x &= 0 && \text{at the outlet} && x = l, \\ u_y - R^{\pm}(x)u_x &= f^{\pm}(x) && \text{at} && y = \pm 1, \end{aligned}\right\} \tag{3.31}$$

where $R^{\pm}(x) = 0$ corresponds to the impermeable portions of the walls, while $R^{\pm}(x) \neq 0$ does to the perforated ones (Figure 3.17). Since the flow is shock free, no Rankine–Hugoniot conditions are required in the domain under consideration $G = \{(x, y) \in \mathbb{R}^2;\ 0 < x < l,\ -1 < y < 1\}$.

Along with (3.30), let us consider the linear equation

$$L_g u := (1 - \varphi_x + g)u_{xx} - \varphi_{xx}u_x + \frac{1}{\gamma + 1}u_{yy} = 0, \tag{3.32}$$

where $g(x, y)$ is a given function. In the case $g \equiv 0$, Equation (3.32) reduces to (3.9), i.e.

$$(-wu_x)_x + u_{yy} = 0, \tag{3.33}$$

where $w = (\gamma + 1)(\varphi_x - 1)$.

For simplicity we suppose that there is only one point C on the segment $y = 0$, $0 \leqslant x \leqslant l$, at which $w_x(x, 0) = 0$. This is true, e.g. for Ringleb's flow. If there are two more points C_1 and C_2 where $w_x(x, 0)$ vanishes due to the velocity minima at subsonic portions of the airfoil, as in Figure 3.5, then the analysis can be performed as in Section 3.2.

Theorem 3.2. *Let the given transonic flow be such that $w \in C^3(\overline{G})$, $w_y < 0$, $w_y^2 - ww_x^2 > 0$ in the domain $\overline{G}$, and the sonic line $w = 0$ be a simple curve with endpoints* A *and* B *on the lower wall (Figure 3.17). Suppose that*

$$\begin{aligned} w_x > 0 \quad &\text{at } y = -1,\ 0 \leqslant x < x_C, \text{ and } y = 1,\ 0 \leqslant x < x_N, \\ w_x < 0 \quad &\text{at } y = -1,\ x_C < x \leqslant l, \text{ and } y = 1,\ x_N < x \leqslant l, \end{aligned}$$

so that the flow acceleration vanishes, i.e. $w_x = 0$, at points N *and* C *located on the upper and lower walls, respectively.*

If the porosity coefficient $R^{\pm} \in C^2[0, l]$ satisfies the following conditions.

On the upper wall:

$$\left.\begin{aligned} R^{+}(x) &= 0 && \text{at} && 0 \leqslant x \leqslant x_N, \\ R^{+}(x) &< -w\tilde{w}_x && \text{at} && x_N < x \leqslant l. \end{aligned}\right\}$$

On the lower wall:

$$\left.\begin{array}{lll} R^-(x) = 0 & at & 0 \leqslant x \leqslant x_C \text{ and } x_E \leqslant x \leqslant l, \\ |w\tilde{w}_x| < R^-(x) < 1/|\tilde{w}_x| & at & x_C < x < x_B, \\ 0 < R^-(x) < 1/|\tilde{w}_x| & at & x_B \leqslant x < x_E, \end{array}\right\} \tag{3.34}$$

where

$$\tilde{w}_x = \frac{w_x}{|w_y| + \sqrt{w_y^2 - ww_x^2}}.$$

Then the linear Problem (3.31), (3.33) with respect to the perturbation $u(x, y)$ can have at most one solution $u \in C^1(\overline{G}) \cap C^2(G)$.

Proof. Suppose $u_1(x, y)$ and $u_2(x, y)$ are two different solutions. Then $u = u_1 - u_2$ satisfies Equation (3.33) and the boundary conditions (3.31) with $f^{\pm} \equiv 0$. As in Sections 2.9 and 3.2, we use the auxiliary function,

$$\Psi(x, y) = -\int_{(0,-1)}^{(x,y)} [wbu_x^2 + 2cwu_x u_y + bu_y^2]\,\mathrm{d}y + [wcu_x^2 + 2bu_x u_y + cu_y^2]\,\mathrm{d}x, \tag{3.35}$$

where

$$c = \frac{w_x}{w_y^2 - ww_x^2}, \qquad b = \frac{-w_y}{w_y^2 - ww_x^2} > 0.$$

In the elliptic domain $G^+ = G \cap (w < 0)$, the function Ψ satisfies a maximum principle. In the hyperbolic domain $G^- = G \cap (w > 0)$, the derivative of Ψ along the characteristic lines $\mathrm{d}x/\mathrm{d}y = \pm w^{1/2}$ is nonpositive due to the expression (2.168):

$$\frac{\mathrm{d}\Psi}{\mathrm{d}y} = \frac{\mathrm{d}y}{\mathrm{d}w}\left(\frac{\mathrm{d}u}{\mathrm{d}y}\right)^2 \leqslant 0. \tag{3.36}$$

Therefore, Ψ attains its maximum on the boundary ∂G of the rectangle G.

Using the boundary conditions (3.31) with $f^{\pm} \equiv 0$, from (3.35) we find

$$\left.\begin{array}{ll} \Psi_y = -wbu_x^2 \geqslant 0 & \text{at the inlet } x = 0, \\ \Psi_y = -bu_y^2 \leqslant 0 & \text{at the outlet } x = l, \\ \Psi_x = -wcu_x^2 \geqslant 0 & \text{at } y = 1,\ 0 \leqslant x \leqslant x_N \\ & \text{and } y = -1,\ 0 \leqslant x \leqslant x_A \quad (\text{as } c > 0,\ w < 0), \\ \Psi_x = -wcu_x^2 \leqslant 0, & \text{at } y = -1,\ x_E \leqslant x \leqslant l \quad (\text{as } c < 0,\ w < 0) \\ & \text{and } y = -1,\ x_A < x < x_C \quad (\text{as } c > 0,\ w > 0) \end{array}\right\} \tag{3.37}$$

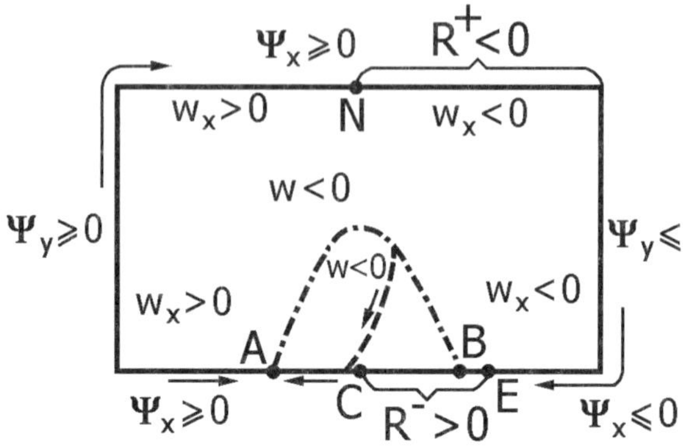

Figure 3.18 The behaviour of the function Ψ on the boundary of the domain G. Reproduced from Kuz'min (2001) by permission of Birkhäuser.

(see Figure 3.18 for directions in which Ψ is nondecreasing). On the portions of the boundary ∂G where $R^{\pm} \neq 0$, we replace u_y by $R^{\pm}u_x$ and represent Ψ_x in the form

$$\begin{aligned}\Psi_x &= -(wcu_x^2 + 2bu_xu_y + cu_y^2)\\ &= [-ww_x + 2w_yR^{\pm} - w_x(R^{\pm})^2]\frac{u_x^2}{w_y^2 - ww_x^2}\\ &= -w_x(R^{\pm} - r_1)(R^{\pm} - r_2)\frac{u_x^2}{|w_y^2 - ww_x^2|},\end{aligned} \tag{3.38}$$

in which the expressions

$$r_1 = \frac{w_y + \sqrt{w_y^2 - ww_x^2}}{w_x}, \qquad r_2 = \frac{w_y - \sqrt{w_y^2 - ww_x^2}}{w_x},$$

are the coordinates of the intersections of the parabola $\Psi_x(R^{\pm})$ with the R-axis (Figure 3.19). Due to conditions (3.34), the right-hand side of (3.38) yields

$$\left.\begin{array}{lll}\Psi_x \geqslant 0 & \text{at } y = 1,\ x_N < x \leqslant l, & \\ & \text{as } R^+ < r_1 < r_2,\ w_x < 0 & \text{(Figure 3.19a)},\\ \Psi_x \leqslant 0 & \text{at } y = -1,\ x_C < x < x_B, & \\ & \text{as } r_1 < R^- < r_2,\ w_x < 0 & \text{(Figure 3.19b)},\\ \Psi_x \leqslant 0 & \text{at } y = -1,\ x_B < x \leqslant x_E, & \\ & \text{as } 0 < R^- < r_2,\ w_x < 0 & \text{(Figure 3.19c)}.\end{array}\right\} \tag{3.39}$$

These inequalities along with (3.36), (3.37) prove that Ψ attains a maximum with respect to $\overline{G}$ and $\overline{G}^+$ at point A.

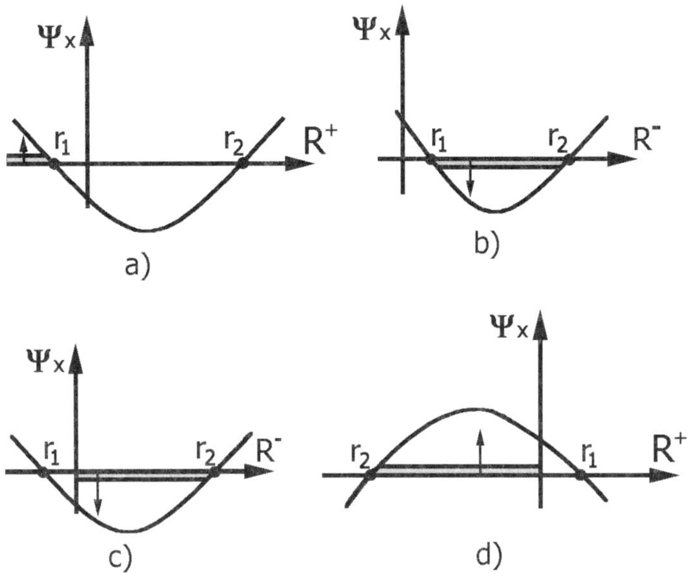

Figure 3.19 Dependence of Ψ_x on the porosity coefficient $R^{\pm}$. Reproduced from Kuz'min (2001) by permission of Birkhäuser.

An analysis of the behaviour of $\Psi(x, y)$ in the vicinity of point A (Lemma 2.15) shows that Ψ can have a maximum with respect to G^+ at point A only if $\Psi(x, y) \equiv$ const. in G^+. As in Sections 2.9 and 3.2, this proves $u(x, y) \equiv 0$ in G^-. □

Remark 3.3. Theorem 3.2 remains true if the oblique derivative condition in (3.31) is extended to the left of point N on the segment $y = 1$, and $-1 < R^+(x) \cdot \tilde{w}_x \leqslant 0$ holds in (3.33) at $0 \leqslant x \leqslant x_N$ instead of $R^+(x) = 0$. In this case (3.38) yields $\Psi_x \geqslant 0$ at $y = 1, 0 \leqslant x \leqslant x_N$, owing to $r_2 < R^+ \leqslant 0 < r_1$, $w_x > 0$ (Figure 3.19d).

In addition, Theorem 3.2 is true if the oblique derivative condition is extended on $y = -1$ to the left of point C : $R^-(x) \geqslant 0$ at $x_A \leqslant x \leqslant x_C$. Now $\Psi_x \leqslant 0$ at $y = -1$, $x_A \leqslant x \leqslant x_C$, due to $r_2 < r_1 < 0 \leqslant R^-$, $w_x > 0$.

3.4.2 Solvability of the nonlinear problem

Theorem 3.4. *Let $\varphi(x, y) \in W^{6,2}(G)$ be a smooth solution of Equation (3.27), such that the sonic line and points* A, B, C, E, N *are located as pointed out in Theorem 3.2.*

Suppose that $w = (\gamma+1)(\varphi_x - 1)$ satisfies the inequalities $w_y < 0$, $w_y^2 - ww_x^2 > 0$, in $\overline{G}$, and the acceleration w_x is positive $w_x > 0$ on a portion AP *of the sonic line, while $w_x < 0$ on* PB, *so that w_x vanishes at the point* P. *Moreover, assume that w_x/w_y is an increasing function of x along the sonic line* APB.

Let the porosity coefficient $R^{\pm}(x)$ satisfy conditions (3.34), except for the left half-vicinities of points C *and* N, *where the inequalities $R^-(x) > 0$ and $-1 <$*

$R^+(x) \cdot \tilde{w}_x < 0$ hold, respectively, as mentioned in Remark 3.3. Also, let $R^-(x_B) < 2w_y/(7w_x)$ at point B.

Then for any perturbation $f^\pm$ of $\theta^\pm$ vanishing in the small vicinities of $x = 0$, $x = l$, and having sufficiently small norm $\|f^\pm\|_{W^{3,2}(0,l)}$, there exists a unique solution $u \in W^{4,2}(G)$ of the nonlinear perturbation Problem (3.30), (3.31).

Proof. The linear Problem (3.31), (3.32) can be reduced to the problem for the nonhomogeneous equation

$$L_g u := (1 - \varphi_x + g)u_{xx} - \varphi_{xx}u_x + \frac{1}{\gamma + 1}u_{yy} = f_g \in W^{3,2}(G), \tag{3.40}$$

endowed with the homogeneous boundary conditions

$$\left.\begin{aligned} u &= 0 \quad \text{at the inlet } x = 0, \\ u_x &= 0 \quad \text{at the outlet } x = l, \\ u_y - R^\pm u_x &= 0 \quad \text{at } y = \pm 1. \end{aligned}\right\} \tag{3.41}$$

This can be accomplished by substituting $u + \hat{u}$ for u in (3.31), (3.32), where $\hat{u}$ is a function which satisfies the boundary conditions (3.31).

From (3.40), (3.41), with the change of variables $x = t$, $y = -\tilde{x}$, we obtain Problem (A.1), (A.2) of the appendix in the rectangle D of the auxiliary plane $(\tilde{x}, t)$ with $k(\tilde{x}, t) = 1 - \varphi_x + g$, $\alpha(\tilde{x}, t) = \varphi_{xx}$, $a(\tilde{x}, t) = 1/(\gamma + 1)$, $\lambda = 0$, $R_1(t) = R^+(x)$, $R_2(t) = R^-(x)$, $T = l$.

Let us verify that all the conditions required in Lemma A.1 are true. Evidently,

$$\left.\begin{aligned} 2\alpha + k_t &= \varphi_{xx} + g_x, & 2\alpha - k_t &= 3\varphi_{xx} - g_x, \\ 2\alpha - 3k_t &= 5\varphi_{xx} - 3g_x, & 2\alpha - 5k_t &= 7\varphi_{xx} - 5g_x. \end{aligned}\right\} \tag{3.42}$$

Since $\varphi_{xx} = w_x/(\gamma + 1)$ is strictly positive on the arc AQ of the sonic line, where Q is a point located on AP near point P (see Figure 3.17), the expressions (3.42) are positive on AQ at sufficiently small g. Inequality (A.3) reduces to

$$R^-(x_B) < 2\min_{QB}\left|\frac{\varphi_{xy} - g_y}{3\varphi_{xx} + g_x}\right| = 2\min_{QB}\left|\frac{w_y - (\gamma + 1)g_y}{3w_x + (\gamma + 1)g_x}\right|. \tag{3.43}$$

The minimum on the right-hand side of (3.43) is attained at point B, as the function w_x/w_y increases along QB. Therefore, (3.43) is true at sufficiently small g by the conditions of Theorem 3.4. Inequality (A.5) becomes

$$R^-(x_B) < 2\left|\frac{w_y - (\gamma + 1)g_y}{7w_x + (\gamma + 1)g_x}\right|_{x_B},$$

which is also true.

The condition $\varphi \in W^{6,2}(G)$ yields $\varphi_x \in C^3(\overline{G})$ due to the embedding theorems for the two-dimensional domain (Adams 1975). If $g \equiv 0$, then all the coefficients of

(3.40) are smooth enough. Therefore, Lemma A.1 ensures the Fredholm solvability of the linear Problem (3.40), (3.41), as well as the validity of *a priori* estimates,

$$\|u\|_{W^{n,2}(G)} \leqslant m\|L_g u - \lambda u\|_{W^{n-1,2}(G)}, \qquad n = 1, 2, 3, 4, \quad m > 0, \tag{3.44}$$

for noneigenvalues of λ. A solution $u \in W^{4,2}(G)$ belongs to $C^2(\overline{G})$, hence, Theorem 3.2 guarantees the uniqueness of the solution. The uniqueness along with the Fredholm alternative prove the existence of the solution $u \in W^{4,2}(G)$ to Problem (3.40), (3.41) at $g \equiv 0$.

The solvability of the linear problem remains valid if $g \neq 0$ and $\|g\|_{W^{3,2}(G)}$ is small enough. To prove this, it is sufficient to verify the validity of the estimates (3.44). Let us show, for instance, that (3.44) is true at $n = 4$ if it is so at $n = 1, 2, 3$. By differentiating (3.40) twice with respect to x, we obtain the relation

$$\begin{aligned}(1 - \varphi_x + g)(u_{xx})_{xx} - 3\varphi_{xx}(u_{xx})_x + (u_{xx})_{yy}/(\gamma + 1)\\ = (f_g)_{xx} - g_{xx}u_{xx} - 2g_x u_{xxx} + 3\varphi_{xxx}u_{xx} + \varphi_{xxxx}u_x.\end{aligned} \tag{3.45}$$

Thus, (u_{xx}) can be considered as a solution of the second-order Equation (3.45), which is similar to (3.40). Consequently, one can use (3.44) with $n = 2$ in order to estimate $\|u_{xx}\|_{W^{2,2}(G)}$ by the norm of the right-hand side of (3.45),

$$\|u_{xx}\|_2 \leqslant m\|(f_g)_{xx} - g_{xx}u_{xx} - 2g_x u_{xxx} + 3\varphi_{xxx}u_{xx} + \varphi_{xxxx}u_x\|_1, \tag{3.46}$$

where $\|\cdot\|_n$ denotes $\|\cdot\|_{W^{n,2}(G)}$ as usual. Hereafter the constants $m > 0$ are independent of u and, in general, different in different formulae. Due to the embedding theorems, we find

$$\|g_{xx}u_{xx}\|_1 \leqslant \|g_{xx}\|_1\|u_{xx}\|_{C(\overline{G})} \leqslant m\|g\|_3\|u_{xx}\|_2$$

and

$$\|g_x u_{xxx}\|_1 \leqslant \|g_x\|_{C(\overline{G})}\|u_{xxx}\|_1 \leqslant m\|g_x\|_2\|u_{xx}\|_2 \leqslant m\|g\|_3\|u_{xx}\|_2.$$

The fourth and fifth terms in (3.46) are estimated by (3.44) with $n = 3$ and $n = 2$. Therefore, using the simple inequality

$$\|g\|_3\|u_{xx}\|_2 \leqslant \tfrac{1}{2}\varepsilon\|u_{xx}\|_2 + \frac{1}{2\varepsilon}\|g\|_3,$$

under sufficiently small $\|g\|_3$, from (3.46) we obtain the estimate $\|u_{xx}\|_2 \leqslant m\|f_g\|_3$. Then Equation (3.40) enables one to estimate $\|u_{yy}\|_2$ by $\|u_{xx}\|_2$, $\|f\|_2$, $\|u\|_3$; as a consequence, we arrive at (3.44) with $n = 4$.

Let us define the operator Tg from $W^{3,2}(G)$ into $W^{3,2}(G)$ as follows. For any $g \in W^{3,2}(G)$, we assign $Tg = u_x$ where u is the solution of the equation $L_g u = 0$ endowed with the boundary conditions (3.31). Then the nonlinear Problem (3.30), (3.31) can be rewritten in the form $Tg = g$. Provided that the norm $\|g\|_3 \leqslant r$ is sufficiently small, the operator Tg is a contracting one in the norm of $W^{1,2}(G)$, since

$$\begin{aligned}\|Tg_1 - Tg_2\|_1 &= \|u_{1x} - u_{2x}\|_1 \leqslant \|u_1 - u_2\|_2 \leqslant m\|L_{g_1}(u_1 - u_2)\|_1 \\ &= m\|(g_1 - g_2)u_{2xx}\|_1 \leqslant m\|g_1 - g_2\|_1 \cdot \|u_{2xx}\|_2 \\ &\leqslant mr\|g_1 - g_2\|_1,\end{aligned}$$

where we have used the relation

$$L_{g_1}u_1 - L_{g_2}u_2 = L_{g_1}(u_1 - u_2) + (g_1 - g_2)u_{2xx} = 0.$$

In addition, under sufficiently small $f^{\pm}$, the operator Tg maps a ball $B_r \subset W^{3,2}(G)$ of radius r into B_r, since from (3.44) one finds

$$\|u_x\|_3 \leqslant mr\||f^+| + |f^-|\|_{W^{3,2}(0,l)} < r.$$

Hence, the principle of contracting mappings ensures that successive approximations $g^{(i+1)} = Tg^{(i)}$, $i = 1, 2, \ldots$, converge to a function $g \in W^{1,2}(G)$. On the other hand, owing to the weak compactness of B_r, there exists a subsequence which weakly converges to $g^* \in W^{3,2}(G)$. Due to the uniqueness, one gets $g = g^* \in W^{3,2}(G)$. Then $u = \int_0^x g \, dx$ is the solution to Problem (3.30), (3.31). □

We notice that in the case of the impermeable bump ($R^- \equiv 0$ at $0 \leqslant x \leqslant l$), the proof of Theorem 3.4 fails because of the arguments pointed out in Remark A.2. This is the case of the nonexistence of a smooth solution to the problem with the slip condition on the full supersonic arc AB as discussed in the beginning of this section and in Section 3.2.

3.5 Flow that Decelerates from Supersonic to Subsonic Velocities

3.5.1 Continuous problem

Theorem 3.4 showed that, though transonic flow with a local supersonic region is typically accompanied by the arising of shock waves, the employment of an appropriate control through a porous portion of the wall/airfoil can prevent the formation of shocks and provide smoothness of the flow.

A similar situation occurs in the flow decelerating from supersonic to subsonic velocities in a curved channel. Formulation of the problem for inviscid flow (which may substantially deviate from the sonic conditions) in the plane of the natural coordinates is as follows. In the rectangle $G = \{(\varphi, \psi) \in \mathbb{R}^2;\ \varphi_1 < \varphi < \varphi_2,\ -1 < \psi < 1\}$, find a solution $\Phi(\varphi, \psi)$ of Equation (2.42),

$$k(\Phi_\varphi)\Phi_{\varphi\varphi} + \Phi_{\psi\psi} = 0, \tag{3.47}$$

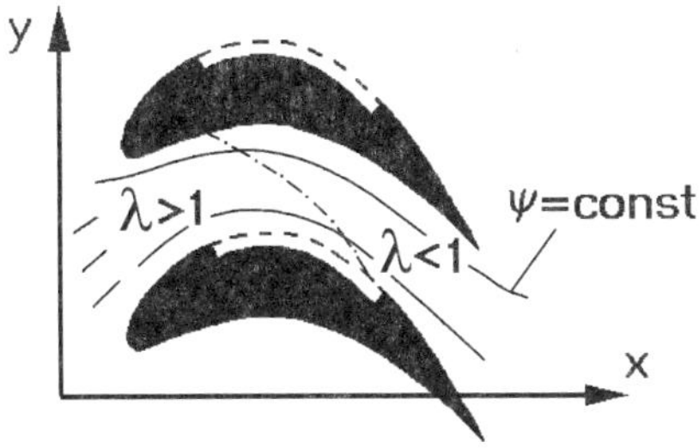

Figure 3.20 A scheme of flow that decelerates from supersonic to subsonic velocities.

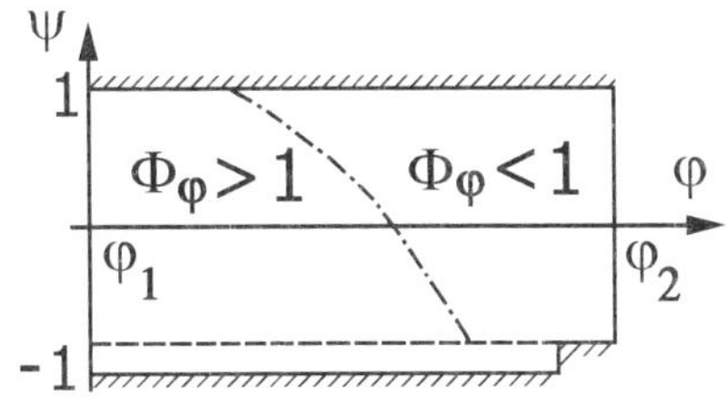

Figure 3.21 Domain in question in the plane (φ, ψ).

where

$$k(\Phi_\varphi) = \tfrac{1}{2}(\gamma + 1)(1 - \lambda^2)\rho^{-(\gamma+1)}, \qquad \Phi_\varphi = 1 + \int_1^\lambda \frac{\rho(\lambda)}{\lambda}\, \mathrm{d}\lambda,$$

which satisfies the boundary conditions

$$\left.\begin{array}{lr} \Phi = h_1(\psi),\ \Phi_\varphi = \lambda_{\text{in}}(\psi) > 1 & \text{at the inlet } \varphi = \varphi_1, \\ \Phi_\psi - h_2(\psi),\ \Phi_\varphi < 1 & \text{at the outlet } \varphi = \varphi_2, \\ \Phi = g^+(\varphi) & \text{on thc upper wall } \psi = 1, \\ \Phi_\psi - R(\varphi)\Phi_\varphi = g^-(\varphi) & \text{on the lower wall } \psi = -1, \end{array}\right\} \tag{3.48}$$

where $\lambda_{\text{in}}(\psi)$, $g^\pm(\varphi)$, $h_{1,2}(\psi)$ are given functions, and $R(\varphi) \geqslant 0$. The conditions (3.48) require that the velocity be supersonic at the inlet, and subsonic (but unprescribed) at the outlet of the channel (see Figures 3.20 and 3.21). We assume $R(\varphi)$ to be a smooth function such that $R \equiv R_0$ at $\varphi_1 \leqslant \varphi \leqslant \varphi_2 - \varepsilon$, and $R \equiv 0$ in the small vicinity of $\varphi = \varphi_2$. The Dirichlet condition prescribed on the upper wall may be associated with a design problem in which the flow velocity along the wall is specified, while the shape of the wall is to be found.

Due to (3.48), the type of Equation (3.47) is hyperbolic at $\varphi = \varphi_1$. Therefore, a necessary condition for the existence of a smooth solution to Problem (3.47), (3.48) is the consistency of initial and boundary data $\lambda_{\text{in}}(\psi)$, $h_1(\psi)$ and $\theta^\pm(\varphi)$ at the corner points $\varphi = \varphi_1$, $\psi = \pm 1$ of the domain G.

Theorem 3.5. *Let $\Phi \in W^{6,2}(G)$ be a smooth solution of Problem (3.47), (3.48) such that the flow acceleration is negative: $\Phi_{\varphi\varphi}(\varphi, \psi) \leqslant \delta < 0$ in the domain G, and the sonic line be a simple curve with the endpoints on the lower and upper walls. Suppose that $\Phi_{\varphi\psi} < 0$ on the sonic line (i.e. the velocity decreases across the sonic line with increasing ψ), and $R_0 > 0$ is small enough.*

Then for any perturbation $f^{\pm}$ of $g^{\pm}$ vanishing in the small vicinities of $\varphi = \varphi_1$, $\varphi = \varphi_2$, and having sufficiently small norm $\|f^{\pm}\|_{W^{3,2}(\varphi_1\varphi_2)}$ there exists a unique solution $\hat{\Phi} \in W^{4,2}(G)$ of Problem (3.47), (3.48) with $f^{\pm}(\varphi) + g^{\pm}(\varphi)$ instead of $\theta^{\pm}(\varphi)$ on the walls.

Proof. We sketch below a proof of Theorem 3.5 under the assumption of nearsonic flow conditions when (3.47) reduces to the von Kármán equation in the plane (φ, ψ):

$$(\gamma + 1)(1 - \Phi_{\varphi})\Phi_{\varphi\varphi} + \Phi_{\psi\psi} = 0. \tag{3.49}$$

It can be seen from Equation (3.49) and conditions (3.48) that the perturbation $u = \Phi - \hat{\Phi}$ is governed by the equation

$$[(1 - \Phi_{\varphi} + \tfrac{1}{2}u_{\varphi})u_{\varphi}]_{\varphi} + \frac{1}{\gamma + 1}u_{\psi\psi} = 0, \tag{3.50}$$

and satisfies the boundary conditions

$$\left.\begin{array}{rll} u = 0,\ u_{\varphi} = 0 & & \text{at the inlet } \varphi = \varphi_1, \\ u = 0 & & \text{at the outlet } \varphi = \varphi_2, \\ u = f^{+}(\varphi) & & \text{on the upper wall } \psi = 1, \\ u_{\psi} - Ru_{\varphi} = f^{-}(\varphi) & & \text{on the lower wall } \psi = -1. \end{array}\right\} \tag{3.51}$$

Along with (3.50), consider the linear equation

$$L_g u := (1 - \Phi_{\varphi} + \tfrac{1}{2}g)u_{\varphi\varphi} - (\Phi_{\varphi\varphi} - \tfrac{1}{2}g_{\varphi})u_{\varphi} + \frac{1}{\gamma + 1}u_{\psi\psi} = 0, \tag{3.52}$$

where $g(\varphi, \psi)$ is a given function. The linear Problem (3.51), (3.52) easily reduces to the problem for the nonhomogeneous equation endowed with the homogeneous boundary conditions.

Since the coefficient $(\Phi_{\varphi\varphi} - g_{\varphi}/2)$ in (3.52) is negative in the case of the decelerating transonic flow, the standard solvability conditions do not hold (see (3.42)). However, the employment of the oblique derivative condition on the lower wall makes it possible to prove the unique solvability of the linear Problem (3.51), (3.52) in $W^{4,2}(G)$ under certain assumptions with respect to the coefficient $R(\varphi)$ (Kuz'min 2000b). The solvability of the linear problem enables one to define the operator $Tg = u_{\varphi}$ where u is the solution of the equation $L_g u = 0$ endowed with conditions (3.51). Then the solvability of the equation $Tg = g$ equivalent to the nonlinear Problem (3.50), (3.51) can be proved in a usual way with the principle of contracting mappings. □

For the impermeable lower wall, $R(\varphi) \equiv 0$, we get the *nonexistence* of a smooth solution to Problem (3.47), (3.48). This virtually follows from Theorem 2.6, since replacing φ by $\varphi_2 + \varphi_1 - \varphi$ in (3.47), (3.48), we obtain a problem similar to (2.42), (2.46) with the positive acceleration of the flow and the prescribed extra Φ and Φ_φ at the outlet, which overdetermines the problem. The fact that we use the Dirichlet condition instead of the Neumann one at $\psi = 1$ does not influence the validity of the uniqueness established in Theorem 2.6.

3.5.2 Finite-element approximations

Denote by $\mathcal{T}_h$ a decomposition of $\overline{G}$ into closed triangular subsets, where h is the discretization parameter, and assume that the triangulations $\mathcal{T}_h$ are regular and quasi-uniform. In order to treat finite-element approximations to transonic flow in a channel, we need a weak formulation of the problem. In order to manage the oblique derivative condition at $\psi = -1$, we add and subtract the term $ru_{\varphi\psi}$, where

$$r(\varphi, \psi) = \tfrac{1}{2}(1 - \psi)R(\varphi),$$

in Equation (3.50). Then multiplying the equation by $\chi \in W^{1,2}(G)$, integrating over G, and using Green's formula, we obtain

$$\int_G [(\gamma + 1)(1 - \Phi_\varphi + \tfrac{1}{2}u_\varphi)u_\varphi \chi_\varphi + u_\psi \chi_\psi + u_\psi (r\chi)_\varphi - u_\varphi (r\chi)_\psi] \, \mathrm{d}G$$
$$= \int_\Gamma [(\gamma + 1)(1 - \Phi_\varphi + \tfrac{1}{2}u_\varphi)u_\varphi + u_\psi r]\chi \, \mathrm{d}\psi + (u_\varphi r - u_\psi)\chi \, \mathrm{d}\varphi, \quad (3.53)$$

where the integration over the boundary Γ is carried out in the counterclockwise direction. Due to the boundary conditions (3.51), the integral over Γ reduces to

$$-\int_{\varphi_1}^{\varphi_2} f^- \chi|_{\psi=-1} \, \mathrm{d}\varphi + \int_{\varphi_1}^{\varphi_2} u_\psi \chi|_{\psi=1} \, \mathrm{d}\varphi$$
$$+ \int_{-1}^{1} (\gamma + 1)(1 - \Phi_\varphi + \tfrac{1}{2}u_\varphi)u_\varphi \chi|_{\varphi=\varphi_2} \, \mathrm{d}\psi.$$

The second and third terms vanish if we require $\chi \in \hat{V}$, where $\hat{V}$ is the subspace of functions which belong to $W^{1,2}(G)$ and vanish at the outlet and on the upper wall of the channel:

$$\hat{V} = \{\chi \in W^{1,2}(G);\ \chi(\varphi_2, \psi) = 0;\ \chi(\varphi, 1) = 0\}.$$

Let V be the subspace of functions which belong to $W^{1,2}(G)$ and vanish at both the inlet and the outlet,

$$V = \{u \in W^{1,2}(G);\ u(\varphi_1, \psi) = u(\varphi_2, \psi) = 0\},$$

and $V^{\mathrm{ad}} = \{u \in V;\ u(\varphi, 1) = f^+\}$ be the class of admissible functions. Theorem 3.5 yields the existence of a solution $u \in V^{\mathrm{ad}}$ to Problem (3.53).

Formulation of a discretized problem corresponding to (3.53) is as follows. Find $u_h \in V_h^{\text{ad}}$ such that

$$a(u_h, \chi_h) = l(\chi_h), \quad \text{for all } \chi_h \in \hat{V}_h, \tag{3.54}$$

where

$$a(u_h, \chi_h) = \int_G [(\gamma + 1)(1 - \Phi_\varphi + \tfrac{1}{2}u_{h\varphi})u_{h\varphi}\chi_{h\varphi} + u_{h\psi}\chi_{h\psi} + u_{h\psi}(r\chi_h)_\varphi - u_{h\varphi}(r\chi_h)_\psi]\,\mathrm{d}G,$$

$$l(\chi_h) = -\int_{\varphi_1}^{\varphi_2} f^- \chi_h|_{\psi=-1}\,\mathrm{d}\varphi,$$

and $\hat{V}_h \subset \hat{V}$, $V_h^{\text{ad}} \subset V$ are the classes of continuous functions which are linear on every triangle $\tau_i \in \mathcal{T}_h$; elements of V_h^{ad} are equal to f^+ at the vertices located on $\psi = 1$. Problem (3.54) can be reformulated as the minimization problem (Glowinski and Pironneau 1978)

$$\mathcal{J}_h(u_h) = \min_{\phi_h \in V_h^{\text{ad}}} \mathcal{J}_h(\phi_h) \tag{3.55}$$

for the functional

$$\mathcal{J}_h(\phi_h) = \tfrac{1}{2}\|\nabla\xi_h(\phi_h)\|^2_{L^2(G)}, \tag{3.56}$$

where the function $\xi_h(\phi_h) \in \hat{V}_h$ is the solution to the discrete weak Poisson problem,

$$\int_G \nabla\xi_h \nabla\chi_h\,\mathrm{d}G = a(\phi_h, \chi_h) - l(\chi_h), \qquad \forall\chi_h \in \hat{V}_h. \tag{3.57}$$

Theorem 3.6. *Under the assumptions of Theorem 3.5, a sequence $\{u_h\}$ of solutions of the discrete minimization Problem (3.55) for the functional defined by (3.56), (3.57) converges to a solution of the continuous problem in the space V as $h \to 0$.*

Proof.

Step 1. By the finite-element interpolation theory (Ciarlet 1978, p. 70), for every $\Phi \in W^{2,2}(G)$ there exists an interpolant $\Phi_h^I \in V_h$ of Φ satisfying the inequality $\|\Phi - \Phi_h^I\|_{W^{1,2}(G)} \leqslant ch\|\Phi\|_{W^{2,2}(G)}$, where c is a constant independent of Φ and h. In particular, such an interpolant $\hat{\Phi}_h \in V_h \cap V^{\text{ad}}$ exists for the solution $\hat{\Phi} \in W^{6,2}(G) \cap V^{\text{ad}}$ of the continuous Problem (3.53).

If Φ_h is a solution of the minimization Problem (3.55), then it satisfies the inequality

$$0 \leqslant \mathcal{J}_h(\Phi_h) = \min_{\phi_h \in V_h \cap V^{\text{ad}}} \mathcal{J}_h(\phi_h) \leqslant \mathcal{J}_h(\hat{\Phi}_h).$$

Due to definition (3.56), (3.57) we get

$$\begin{aligned}
2\mathcal{J}_h(\hat{\Phi}_h) &= \|\nabla\xi_h(\hat{\Phi}_h)\|^2_{L^2}\\
&= a(\hat{\Phi}_h, \xi_h) - l(\xi_h)\\
&= a(\hat{\Phi}_h, \xi_h) - a(\hat{\Phi}, \xi_h)\\
&= \int_G [(\tilde{K}(\hat{\Phi}_{h\varphi}) - \tilde{K}(\hat{\Phi}_{\varphi}))\xi_{h\varphi} + (\hat{\Phi}_{h\psi} - \hat{\Phi}_{\psi})\xi_{h\psi}]\,\mathrm{d}G,\\
&\leqslant c\|\hat{\Phi} - \hat{\Phi}_h\|_{W^{1,2}}\|\xi_h\|_{W^{1,2}} \to 0 \quad \text{as } h \to 0,
\end{aligned}$$

therefore,

$$\mathcal{J}_h(\Phi_h) \to 0 \quad \text{as } h \to 0. \tag{3.58}$$

Step 2. The sequence $\{\Phi_h\}_{h\in(0,h_0)}$ is bounded in $V \cap V^{\mathrm{ad}}$, therefore, the weak compactness yields the existence of a subsequence $\{\Phi_h\}$ and a function $\Phi^* \in V \cap V^{\mathrm{ad}}$ such that $\Phi_h \rightharpoonup \Phi^*$ weakly in $W^{1,p}(G)$ for every $p \in (1, \infty)$ (Berger 1989). Then the strong convergence

$$\|\Phi_h - \Phi^*\|_{W^{1,2}(G)} \to 0$$

can be shown using (3.58) and the auxiliary functionals

$$\mathcal{I}_h(\eta) := -\int_G \nabla\Phi_h \cdot \nabla\Pi_h\eta\,\mathrm{d}G,$$

where Π_h is the interpolation operator.

Step 3. The function Φ^* obtained is a solution of the continuous Problem (3.53). In order to prove this, we find due to (3.58)

$$\begin{aligned}
&|a(\Phi^*, \chi) - l(\chi)|\\
&\quad= \lim_{h\to 0} |a(\Phi_h, \chi) - l(\chi)|\\
&\quad= \lim_{h\to 0} |a(\Phi_h, \chi - R_h\chi) - l(\chi - R_h\chi) + a(\Phi_h, R_h\chi) - l(R_h\chi)|
\end{aligned} \tag{3.59}$$

for an arbitrary but fixed $\chi \in \hat{V}$, where $R_h\chi \in \hat{V}_h$ is the Ritz projection of χ defined by

$$\int_G \nabla R_h\chi \cdot \nabla\chi_h\,\mathrm{d}G = \int_G \nabla\chi \cdot \nabla\chi_h \mathrm{d}G \quad \text{for all } \chi_h \in \hat{V}_h.$$

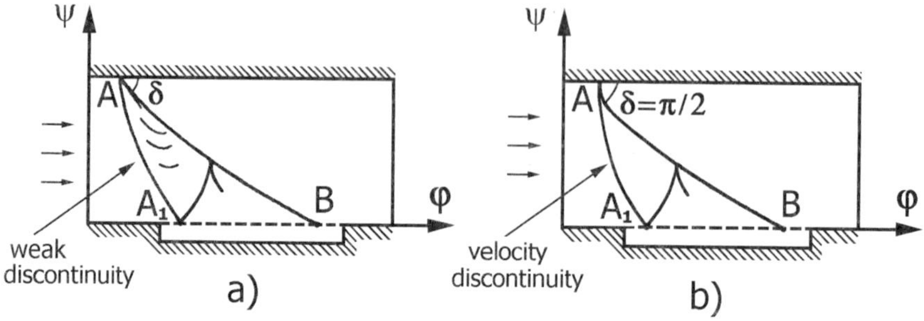

Figure 3.22 Dependence of the bow perturbation on the angle δ of the sonic line inclination to the streamline at point A.

Now, we replace the last two terms in (3.59) by the integral of $\nabla \xi_h \cdot \nabla R_h \chi$ according to (3.57):

$$
\begin{aligned}
&= \lim_{h \to 0} \left| a(\Phi_h, \chi - R_h\chi) - l(\chi - R_h\chi) + \int_G \nabla\xi_h(\Phi_h) \cdot \nabla R_h \chi \, \mathrm{d}G \right| \\
&\leqslant \lim_{h \to 0} \Big(c_1 \|\Phi_h\|_V \cdot \|\chi - R_h\chi\|_V + c_2 \|f^{\pm} + \theta^{\pm}\|_{L^\infty(G)} \cdot \|\chi - R_h\chi\|_V \\
&\qquad + \left| \int_G \nabla\xi_h(\Phi_h) \cdot \nabla R_h \chi \, \mathrm{d}G \right| \Big) \\
&\leqslant \lim_{h \to 0} c \left(\|\chi - R_h\chi\|_V + \left| \int_G \nabla\xi_h(\Phi_h) \cdot \nabla R_h \chi \, \mathrm{d}G \right| \right) \\
&\leqslant \lim_{h \to 0} c (\|\chi - R_h\chi\|_V + \|\nabla\xi_h(\Phi_h)\|_{L^2(G)} \cdot \|\nabla R_h\chi\|_{L_2(G)}) = 0,
\end{aligned}
$$

$\forall \chi \in \hat{V}$, since the first term tends to zero due to the properties of the Ritz projection, and the second term does due to (3.56), (3.58). Hence, Φ^* is a solution of Problem (3.53). □

Remark 3.7. A similar analysis of the convergence of finite-element approximations can be performed for the shock-free flow with a local supersonic region over a porous airfoil treated in Section 3.4.

3.5.3 Asymptotics in the vicinity of the beginning of the sonic line

Consider a decelerating transonic flow with a sonic line AB which is nonorthogonal to the streamlines at each point (Figure 3.22a). Theorem 3.5 establishes conditions of the existence of a smooth solution to the perturbation problem for the decelerating flow. If the conditions are violated, or the boundary conditions are different from (3.48), then a shock wave may arise in the supersonic region. In addition, a weak singularity may originate at the beginning A of the sonic line and propagate along the characteristic line AA_1. One may associate the weak singularity with discontinuities of the first-order derivatives of the flow velocity $\lambda(\varphi, \psi)$.

A study of the behaviour of the solution in the vicinity of point A in the framework of the linearized model can be done with Equation (2.56),

$$\left[\frac{(\rho\lambda)_\lambda}{\rho} u_\varphi\right]_\varphi + (\rho^2 u_\psi)_\psi = 0,$$

where $u(\varphi, \psi)$ is the perturbation of the velocity potential, $\rho(\lambda)$ is related to the given velocity $\lambda(\varphi, \psi)$ by the Bernoulli law (2.2).

The initial conditions at the inlet of the channel make it possible to find the solution downstream of the inlet up to the characteristic line AA_1. Suppose that the obtained distribution of $u(\varphi, \psi)$ on AA_1 can be represented in the form

$$u|_{AA_1} = r_0 \cdot (1-\psi)^r + o((1-\psi)^r), \qquad r,\ r_0 = \text{const.}, \quad r \geqslant \tfrac{3}{2}. \tag{3.60}$$

An asymptotic analysis of a Tricomi-like problem in the vicinity of point A under the slip condition on the wall (Kuz'min 1992b) yields the expansion for the derivative $u_\varphi(\varphi, \psi)$ on the sonic line, i.e. for the velocity perturbation $\lambda' = \lambda u_\varphi$, as follows,

$$\begin{aligned} u_\varphi|_{AB} = {} & r_0 r p^{-r/3} (\tan\delta)^{1-r/3} (\tfrac{3}{4})^{2r/3} \frac{4r+1}{18\pi} \\ & \times \mathcal{B}(\tfrac{1}{6}, \tfrac{1}{6}) \mathcal{B}(\tfrac{1}{3}(2r+1), \tfrac{5}{6})(1-\psi)^{2r/3-1} + o((1-\psi)^{2r/3-1}), \end{aligned} \tag{3.61}$$

where

$$p = -(\gamma+1)(\tfrac{1}{2}(\gamma+1))^{2/(\gamma-1)} \lambda_\varphi|_A > 0,$$

and $\mathcal{B}(\alpha, \beta)$ is the classical beta-function.

A numerical simulation of the perturbation field in the decelerating flow (Section 4.1) reveals a perturbation wave, termed a δ-wave, which is incipient on the sonic line near the point A. This wave is similar to the compression wave emanating from the rear upper part of the sonic line in the flow with a local supersonic region (Figure 3.12). Formation of the δ-wave is accounted for by the perturbation which propagates in the supersonic flow downstream of the inlet, meets the sonic line, and partially reflects back into the supersonic region in the direction toward the lower wall. The expression (3.61) shows that the gradient of the velocity perturbation at the upstream endpoint A of the sonic line depends only on the angle δ made by the sonic line with the streamline and on the magnitude of the flow deceleration at that point.

If the shape of the sonic line changes in such a way that it becomes orthogonal to the streamline at A, $\delta \to \pi/2$, and $r < 3$, then $\tan\delta \to \infty$, and expression (3.61) tends to infinity. This corresponds to the numerical observation that the δ-wave approaches the characteristic line AA_1 and tends to a velocity discontinuity indicating formation of a shock wave in the linearized model (Figure 3.22b) similar to the bow discontinuity of the velocity field in the classical problem of supersonic flow past a thin airfoil. For $r \geqslant 3$, however, u_φ does not tend to infinity as $\delta \to \pi/2$, because the velocity perturbation (3.60) on the characteristic line AA_1 in the vicinity of point A is too small at large r.

3.6 FE and FD Treatment of Flow with Shock Waves

3.6.1 Finite-element approximations of transonic flow with shock waves

Suppose that a domain under consideration G consists of the subdomain G^+, in which flow is subsonic, and G^- where it is supersonic. The boundary between G^+ and G^- is formed by sonic lines and shock waves. The latter ones are characterized by jumps in the flow velocity and density. If the shock waves are weak, then the entropy changes across the shocks are small, and the flow may be described by the full-potential equation with respect to the velocity potential $\varphi(x, y)$.

Assume that the shock waves are piecewise-smooth curves and consider the jump conditions

$$\rho_- \frac{\partial \varphi}{\partial n}\bigg|_- = \rho_+ \frac{\partial \varphi}{\partial n}\bigg|_+, \tag{3.62}$$

$$\frac{\partial \varphi}{\partial s}\bigg|_- = \frac{\partial \varphi}{\partial s}\bigg|_+, \tag{3.63}$$

where '$-$' or '+' denotes the quantities in front of, or behind, the shock wave, respectively. By $\partial/\partial s$ and $\partial/\partial n$ we denote the derivatives with respect to the tangential and normal directions to the shock or to the boundary of the domain. Relations (3.62) and (3.63) must be satisfied by any weak solution of the full-potential equation in the form (2.38) or in the form of the system of continuity and the Bernoulli Equations (2.1), (2.2).

Consider, for example, a problem for flow in a channel G with impermeable walls and a local supersonic region over an airfoil placed on the lower wall. Then boundary conditions can be written in the form similar to (3.31) with $R \equiv 0$:

$$\left.\begin{array}{llll} \varphi = 0 & \text{at the inlet} & x = 0, & -1 < y < 1, \\ \rho\varphi_x = g_{\text{out}} & \text{at the outlet} & x = l, & -1 < y < 1, \\ \varphi_n = 0 & \text{on the walls}\,. & & \end{array}\right\} \tag{3.64}$$

Let V be the subspace of functions which belong to $W^{1,2}(G)$ and vanish at the inlet,

$$V = \{\varphi \in W^{1,2}(G);\ \varphi(0, y) = 0\}.$$

We multiply the continuity equation $(\rho\varphi_x)_x + (\rho\varphi_y)_y = 0$ by $\chi \in V$ and integrate over G. Then using Green's formula and boundary conditions (3.64), we obtain the weak formulation of this problem,

$$\int_G \rho\nabla\varphi \cdot \nabla\chi \,\mathrm{d}G = \int_{-1}^{1} g_{\text{out}}\chi\big|_{x=l}\,\mathrm{d}y, \qquad \forall\chi \in V. \tag{3.65}$$

Since the flow velocity cannot exceed

$$\lambda_{\max} = \left(\frac{\gamma + 1}{\gamma - 1}\right)^{1/2},$$

the solution $\varphi(x, y)$ is sought in the class of admissible functions

$$V^{\text{ad}} = \left\{\varphi \in V;\ \varphi_x^2 + \varphi_y^2 < \frac{\gamma + 1}{\gamma - 1}\right\}.$$

The physical model leading to the full-potential or the Euler equations for inviscid flow and to the weak formulation (3.65) does not include the important second fundamental principle of thermodynamics. Mathematically, these equations would allow solutions across which the entropy decreases and which would violate that principle. In order to get physical solutions, one has to enforce the increase in entropy across shocks as an additional constraint. That constraint, the entropy condition, implies that shocks must be compressive, i.e. they satisfy the inequality

$$|\nabla\varphi|\big|_{-} > |\nabla\varphi|\big|_{+}. \tag{3.66}$$

Instead of this inequality, Glowinski and Pironneau (1978) and Berger (1989) used a convenient selection principle, i.e. an inequality that is violated by expansion shocks. That inequality requires that the divergence of the flow field be bounded from above,

$$\Delta\varphi \leqslant m, \tag{3.67}$$

with $m \in \mathbb{R}$. For mathematical analysis of Problem (3.65), it is more suitable a weak version of (3.67). Multiplying (3.67) with a function $\eta \in C_0^{\infty}(G)$, $\eta \geqslant 0$, integrating over G and then using Green's formula, we obtain

$$-\int_G \nabla\varphi \cdot \nabla\eta \,\mathrm{d}G \leqslant m \int_G \eta \,\mathrm{d}G. \tag{3.68}$$

For piecewise-smooth solutions, inequality (3.68) is equivalent to (3.67).

Let us consider a system $\{G_h\}_{h\in(0,h_0)}$ of polygonal approximations of the domain G and decompositions $\mathcal{T}_h$ of $\overline{G}_h$ into closed triangular subsets, where h denotes the discretization parameter. We assume that the triangulations $\mathcal{T}_h$ are regular and quasi-uniform.

Formulation of a discretized problem corresponding to (3.65) is as follows. Find $\varphi_h \in V_h^{\text{ad}}$ such that

$$a(\varphi_h, \chi_h) = l(\chi_h), \quad \text{for all } \chi_h \in V_h, \tag{3.69}$$

where

$$a(\varphi_h, \chi_h) = \int_{G_h} \rho(\varphi_h) \cdot \nabla\varphi_h \nabla\chi_h \mathrm{d}G,$$

$$l(\chi_h) = \int_{-1}^{1} g_{\text{out}}\chi_h\Big|_{x=l} \mathrm{d}y,$$

and $V_h \subset V$, $V_h^{\mathrm{ad}} \subset V$ are the classes of continuous functions which are linear on every triangle $\tau_i \in \mathcal{T}_h$.

The discrete version of the selection condition (3.68) reads (Berger and Feistauer 1993):

$$-\int_{G_h} \nabla\varphi_h \nabla\eta_h \,\mathrm{d}G \leqslant m \int_G \eta \,\mathrm{d}G; \tag{3.70}$$

for all $\eta_h \in C^0(\overline{G})$ which are affine on each $\tau_i \in \mathcal{T}_h$, and $\eta_h \geqslant 0$ on G_h, $v_h = 0$ at $x = 0$, $x = l$.

Problem (3.69), (3.70) can be reformulated as the minimization problem (Glowinski and Pironneau 1978; Bristeau *et al.* 1979),

$$\mathcal{J}_h(\varphi_h) = \min_{\phi_h \in V_h^{\mathrm{ad}}} \mathcal{J}_h(\phi_h), \tag{3.71}$$

for the functional

$$\mathcal{J}_h(\phi_h) = \tfrac{1}{2}(\|\nabla\xi_h(\phi_h)\|^2_{L^2(G)} + \nu F_h(\phi_h)), \tag{3.72}$$

where the function $\xi_h(\phi_h) \in V_h$ is the solution to the discrete Poisson problem,

$$\int_G \nabla\xi_h \nabla\chi_h \,\mathrm{d}G = a(\phi_h, \chi_h) - l(\chi_h), \qquad \forall \chi_h \in V_h. \tag{3.73}$$

The penalty term F_h is defined as

$$F_h(\phi_h) = \frac{\nu}{2} \sum_{i=1}^{N} \frac{1}{h_i^{2\varepsilon}} \left[\left(-\int_{G_h} \nabla\phi_h \nabla\mu_i^h \,\mathrm{d}G - m \int_{G_h} \mu_i^h \,\mathrm{d}G \right)^+ \right]^2,$$

where $\nu > 0$, $m > 0$, $\varepsilon > 0$, $i = 1, \dots, N$, are arbitrary but fixed constants, μ_i^h are the hat functions for the given triangulation, and $(r)^+ = \max(r, 0)$ for $r \in \mathbb{R}$. Using the above representation for $F_h(\varphi_h)$ and the assumption that there exists the unique solution to the transonic flow problem in the weak form, which reduces to (3.65) in the case of flow in the channel, Berger (1989) proved the convergence of approximate finite-element solutions to the exact one. In order to establish the convergence, Berger generalized the entropy compactification results by Mandel and Nečas (1987) and Feistauer and Nečas (1985).

A general boundary-value problem for transonic flow in a bounded domain with mixed Dirichlet–Neumann conditions was studied by Berger and Feistauer (1993). The authors used a theory of variational crimes developed in the theory of elliptic equations in order to prove the convergence of finite-element approximations to the exact transonic solution of the full-potential equation.

For analysis of the exterior flow problem around an airfoil, Berger *et al.* (1997) introduced an artificial far-field boundary in the subsonic region. In the far field, the flow was approximated by the Prandtl–Glauert linearization. The latter leads via the Green representation theorem to an integral equation on the far-field boundary. This approach led to a coupled finite-element–boundary-element numerical method. The

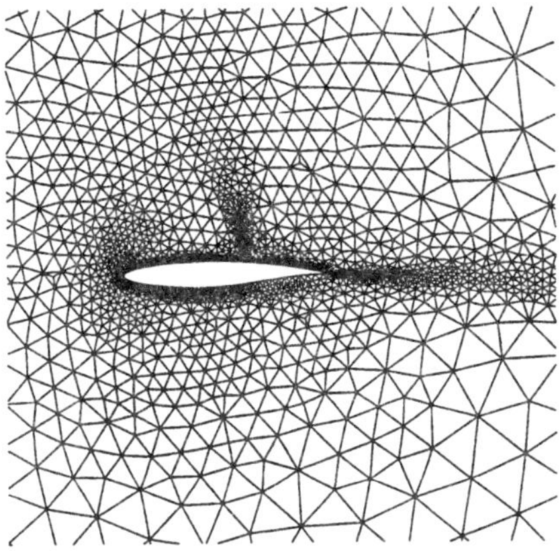

Figure 3.23 Details of an adaptive mesh for turbulent transonic flow. Courtesy of Ilinca *et al.* (1998). Reproduced by permission of the authors.

authors presented an error analysis and proved the convergence of FE approximations in the way outlined above under the assumption that a unique exact solution to the full-potential equation exists.

Gittel (1995) employed the variational approach in order to study boundary-value problems for transonic flow governed by a nonlinear system of first-order equations. Using compensated compactness arguments, he analysed the existence of solutions satisfying the entropy condition in the distributional sense and derived some properties of minimizers.

The stability of numerical solution of transonic flow with shock waves was analysed by Maršik and Daněk (1993), who derived a variational principle with dissipation, which was introduced through a physical viscosity. The authors presented a simplified relation describing the stabilizing influence of the dissipative process. The finite-element solutions of the potential equation were examined and compared with the formulation using the variational principle. The correspondence between the physical viscosity and artificial viscosity was shown both qualitatively and quantitatively. The necessary amount of artificial viscosity was investigated analytically and the results were compared with numerical experiments.

An important aspect of the practical use of a numerical method is the adaptation of an iterative numerical algorithm in which information from a previously calculated solution is used to make an improved calculation by changing the mesh or the trial functions. Generation of good-quality computational grids on complex geometries remains a difficult task for the routine application of CFD. In addition, generation of an efficient grid usually requires some prior knowledge of the flow behaviour in order to fit the grid resolution to the essential features of the flow field such as shock waves. In general, most adaptive methods fall into three broad categories: (1) mesh movement (r-refinement), (2) mesh enrichment (h-refinement), and (3) local solution enhancement (p-refinement). While the methods in the first two classes modify local mesh clustering in order to improve the solution accuracy (grid adaptation), those under the third category enhance the order of numerical approximation at locations where the solution undergoes abrupt variations (solution adaptation).

Göhner and Warnecke (1994) proposed a shock indicator based on the entropy inequality in the form of the selection condition (3.70). It was used to move nodes, in order to improve the alignment of the mesh with shock waves. In a subsequent paper, Göhner and Warnecke (1995) suggested an error indicator based on finite differences in the piecewise constant gradient field of a finite-element approximation. The indicator measures second-order distributional derivatives. The authors gave estimates for the approximate solution involving the error estimator, a residual and the error. This work was an extension of the approach worked out by Eriksson and Johnson (1988) and Verfürth (1989) for equations of elliptic type.

We would emphasize that computation of *viscous* transonic flow requires special effort and numerical algorithms to resolve delicate flow features such as boundary and shear layers. Figure 3.23 illustrates extreme clustering in the boundary layer and in the wake obtained with the algorithm by Ilinca *et al.* (1998). The authors incorporated turbulence via the k–ε model and employed the logarithms of k and ε as computational variables. The use of the logarithms guarantees positivity of turbulence variables throughout the domain. Because the logarithms vary more slowly than their arguments, the accuracy was improved near sharp fronts. Furthermore, accuracy in regions of low-turbulence level was also improved because divisions by small quantities were eliminated.

3.6.2 Convergence of finite-difference approximations

As pointed out in Section 2.4, for the time-dependent small disturbance equation

$$[K_{\text{sim}}\phi_x - \tfrac{1}{2}(\gamma+1)\phi_x^2]_x + \phi_{yy} - 2\phi_{tx} = 0,$$

Engquist and Osher split the initial–boundary-value problem into two differential problems and analysed them in the case when the flow is supersonic at the inlet and subsonic at the outlet of a channel. Following this approach, Fishelov (1988) treated a problem with subsonic inflow and outflow conditions under the Dirichlet condition at the inlet and the Neumann one at the outlet. He used a similar splitting of the problem into two ones. By omitting the factor 2, the first problem can be represented in the form

$$u_t = -[\tilde{K}(u)]_x, \tag{3.74}$$

$$u(1, y, t) = 0, \tag{3.75}$$

where $u = \phi_x$, $\tilde{K}(u) = K_{\text{sim}}u - (\gamma+1)u^2/2$. The second problem is

$$\phi_{tx} = \phi_{yy}, \tag{3.76}$$

$$\phi(-1, y, t) = 0, \tag{3.77}$$

$$\phi_y(x, \pm 1, t) = F_{\pm}(x), \tag{3.78}$$

where $F(x)$ describes the shape of the airfoil:

$$F_+(x) = 0, \qquad F_-(x) = \begin{cases} F'(x) & \text{if } |x| < x_0, \\ 0 & \text{if } |x| < x_0. \end{cases}$$

Both of the above problems are supplied with initial conditions.

Fishelov looked for a high-order scheme for the steady small disturbance equation. He focused attention on the x-direction, since changes spread much more slowly in the y-direction, and gave theoretical support to this approach using Chebyshev spectral methods. One may present both problems mentioned above in the form

$$u_t = G(u).$$

For the first problem,

$$G(u) = G_1(u) = \tilde{K}_x(u),$$

and for the second,

$$G(u) = G_2(u) = \int_{-1}^{x} u_{yy}\, \mathrm{d}x.$$

Step 1. *Discretization in time.* Fishelov discretized u_t in (3.74), (3.76) using finite differences and the modified Euler scheme,

$$u^{n+1/2} = u^n + \tfrac{1}{2}\tau G(u^n), \qquad u^{n+1} = u^n + \tau G(u^{n+1/2}).$$

Let $L(\tau)$ be the operator which acts on u^n to yield u^{n+1}:

$$L(\tau)u^n = u^{n+1} = u^n + \tau G(u^n + \tfrac{1}{2}\tau G(u^n)).$$

$L_1(\tau)$, $L_2(\tau)$ are defined as $L(\tau)$ with G_1, G_2 replacing G. According to Gottlieb (1972), the discretization in time

$$u^{n+1} = L_1(\tfrac{1}{2}\tau)L_2(\tfrac{1}{2}\tau)L_2(\tfrac{1}{2}\tau)L_1(\tfrac{1}{2}\tau)u^n \tag{3.79}$$

is second-order accurate, even in the nonlinear case. Therefore, one can use it to approximate the time-dependent solution as well, and if higher-order accuracy in time is desired, one may use higher-order Runge–Kutta schemes.

Step 2. *Discretization in the space variable x.* It is sufficient to illustrate the discretization of $\partial u/\partial x$ and $\int_{-1}^{x} u(\xi)\, \mathrm{d}\xi$.

Let $P_N u$ be the Chebyshev pseudospectral projection of u on the subspace of polynomials of degree N or less, i.e.

$$u_N(x, y) = P_N u(x, y) = \sum_{n=0}^{N} a_n(y) T_n(x),$$

where

$$u_N(x_i, y) = u(x_i, y), \qquad x_i = \cos\frac{\pi i}{N}, \quad 0 \leqslant i \leqslant N.$$

Fishelov discretized $\partial/\partial x$ by differentiating $P_N u$, and then denoted the resulting operator by L_N. Hence

$$L_N u = P_N \frac{\partial}{\partial x} P_N u = \sum_{n=0}^{N} a_n(y) T_n'(x) = \sum_{n=0}^{N} b_n(y) T_n(x), \tag{3.80}$$

where

$$b_N(y) = 0, \qquad b_{N-1}(y) = 2N a_N(y),$$

and

$$\begin{aligned} \bar{c}_k b_k(y) &= b_{k+2}(y) + 2(k+1) a_{k+1}(y), \qquad 0 \leqslant k \leqslant N-2, \\ \bar{c}_0 &= \bar{c}_N = 2, \\ \bar{c}_j &= 1, \qquad 1 \leqslant j \leqslant N-1. \end{aligned}$$

The operator $L_N u$ is applied for $y = y_j = \cos(\pi j/M), 0 \leqslant j \leqslant M$.

Next, integration is done in a similar way,

$$I_n = P_N \int_{-1}^{x} P_N u \, \mathrm{d}\xi = P_N \sum_{n=0}^{N} a_n(y) \int_{-1}^{x} T_N(\xi) \, \mathrm{d}\xi = P_N \sum_{n=0}^{N+1} d_n(y) T_n(x).$$

By integrating the recurrence formula

$$2T_n(x) = \frac{T_{n+1}'(x)}{n+1} + \frac{T_{n-1}'(x)}{n-1},$$

one finds

$$\begin{aligned} d_{N+1} &= \frac{a_N}{2(N+2)}, \qquad d_N = \frac{a_{N-1}}{2(N+1)}, \\ d_n &= \frac{1}{2} \frac{a_{n-1} - a_{n+1}}{n}, \qquad 3 \leqslant n \leqslant N-1, \\ d_2 &= \tfrac{1}{4} a_1 - \tfrac{1}{4} a_3, \qquad d_1 = a_0 - \tfrac{1}{2} a_2, \end{aligned}$$

and d_0 is chosen such that

$$\sum_{n=0}^{N+1} d_n(y) T_n(-1) = 0.$$

Step 3. *Discretization in the space variable y.* For the discretization of $\partial^2/\partial y^2$, Fishelov used two types of schemes.

Spectral approximation in y (Scheme A). As described above, one can discretize $\partial^2/\partial y^2$ by a Chebyshev pseudospectral method. For a fixed x, define $P_M u(x, y)$ to be the Chebyshev pseudospectral projection of $u(x, y)$ onto the subspace of polynomials of degree less than or equal to M, i.e.

$$P_M u = \sum_{m=0}^{M} e_m(x) T_m(y).$$

Define $D_M^{(1)}(y)$ as follows,

$$D_M^{(1)}(y) = P_M \frac{\partial^2}{\partial y^2} P_M u(x, y)$$
$$= \sum_{m=0}^{M} e_m(x) T_m''(y) = -\sum_{m=0}^{M} \tilde{K}_m(x) T_m(y), \tag{3.81}$$

where (see Gottlieb and Orszag 1977, p. 16)

$$\tilde{K}_M = \tilde{K}_{M-1} = 0,$$
$$c_m \tilde{K}_m = -\sum_{\substack{p=m+2,\\ p+m \text{ even}}}^{M} p(p^2 - m^2) a_p, \qquad 0 \leqslant m \leqslant M-2,$$

and

$$c_0 = 2, \qquad c_m = 1, \quad 1 \leqslant m \leqslant M-2.$$

Now the semidiscrete spectral scheme for (3.74)–(3.78) can be derived as follows. Denote by U the approximation to u and by Φ the approximation to ϕ, where

$$\Phi = I_N U.$$

The semidiscrete approximation to (3.74), (3.75) is

$$\frac{\partial U}{\partial t} = -L_N(x) f(U), \tag{3.82}$$
$$U(-1, y, t) = 0, \tag{3.83}$$

and for (3.76)–(3.78) we obtain the following scheme:

$$\frac{\partial \Phi}{\partial t} = I_N D_M^{(1)}(y) \Phi,$$
$$L_M(y) \Phi(x, \pm 1, t) = P_N F_{\pm}(x).$$

Finite-difference approximation in the space variable y (Scheme B). Using finite differences for approximating $\partial^2/\partial y^2$, we have

$$u_{yy} \approx D_M^{(2)}(y) u(x, y) = \frac{u(x, y_{j+1}) - 2u(x, y_j) + u(x, y_{j-1})}{h_2^2},$$

where $h_2 = 2/M$, $y_j = 1 - h_3 j$, $1 \leqslant j \leqslant M-1$. The operator $D_M^{(2)}(y)$ is applied for $x = x_i = \cos(\pi i/N)$, $0 \leqslant i \leqslant N-1$.

For (3.74), (3.75) we use the same discretization as in (3.82), (3.83) and for (3.76)–(3.78) we obtain the following semidiscrete scheme (scheme B),

$$\Phi_t = I_N(x) D_M^{(2)}(y)\Phi,$$
$$\frac{\Phi(x_i, 1) - \Phi(x_i, 1 - h_2)}{h_2} = F_+(x_i), \qquad 0 \leqslant i \leqslant N - 1,$$
$$\frac{\Phi(x_i, -1 + h_2) - \Phi(x_i, -1)}{h_2} = F_-(x_i), \qquad 0 \leqslant i \leqslant N - 1,$$

where $\Phi = I_N U$. This scheme has spectral accuracy in x and is of second order in y.

The reason Fishelov looked for other schemes rather than the spectral one for approximating $\partial^2/\partial y^2$ is the asymptotic behaviour of the eigenvalues of $D_M^{(1)}(y)$, where

$$D_M^{(1)}(y) = P_M(P_M)_{yy}.$$

The largest eigenvalue of $D_M^{(1)}(y)$ grows like $O(M^4)$ when M grows to infinity (see Gottlieb and Orszag 1977, p. 115). This implies a restriction on τ of the form

$$\tau = O(1/M^4).$$

For M of the order 10, one may still use spectral differentiation with respect to y and get spectral accuracy. But if we wish to increase M, it is preferable to use a finite-difference approximation, in which

$$\tau = O(1/M^2).$$

An alternative is to use an implicit scheme for the time discretization. In this case the scheme is unconditionally stable but one should solve a full system of linear equations at every time step.

Fishelov showed that the spatial operators of the schemes A and B have eigenvalues with nonpositive real parts. Moreover, stability was proved for the second operator.

Mostrel (1989) worked out new second-order accurate finite-difference approximations for a class of nonlinear differential equations of mixed type, which includes the low-frequency transonic small disturbance Equation (2.104) and the two-dimensional (2D) full-potential equation. In distinction to the above schemes, which are type-dependent, Mostrel proposed a special kind of upwinding uniformly in all regions. The resulting stencil, the same in all regions, was of seven-point bandwidth, with four points upwind and two points downwind. For the small disturbance Equation (2.102), the scheme is implemented via a time-stepping algorithm; the inclusion of flux limiters keeps the total variation nonincreasing and eliminates spurious oscillations near shocks. Global linear stability, total variation diminishing and entropy stability aspects were examined under assumption of convergence of the finite-difference solutions.

For the steady full-potential Equation (2.38), Morice and Viviand (1980) and Osher *et al.* (1985) studied a new discrete entropy inequality which rules out expansion shocks. This inequality is enforced across a shock wave if and only if the usual criterion (3.66) of the velocity magnitude decreasing across the shock is valid. Osher *et al.* (1985) and Osher (1986) described a class of conservative difference approximations for the full-potential equation which satisfy that discrete entropy inequality and have sharp and steady shocks. The presented examples involved the Engquist–Osher scheme, Godunov's scheme, and general fixes of Murman's scheme. The concept of flux biasing was discussed and shown to be more convenient than the density biasing, as well as having the same truncation error. Numerical experiments by Hafez *et al.* (1984) confirmed the worth of the approach.

In the 1980s, strategies for improving the accuracy of finite-difference stencils were based on the approximation of smooth solutions outside shock regions and the suppression of potentially catastrophic oscillations near them. Total variation diminishing schemes reduce to first-order accuracy at shocks to suppress the oscillations, but excessively damp small-scale flow features. Essentially nonoscillatory (ENO) schemes employ an adaptive stencil to avoid interpolation across discontinuities. The original ENO schemes were proposed in a cell-averaged framework (Harten *et al.* 1987). Soon afterward they were extended to finite-difference formulations, which are significantly faster in multiple space dimensions (Shu and Osher 1988, 1989). At each grid point the ENO algorithm selects, from several candidates, the particular stencil that has the smoothest local data, and interpolation across discontinuities is thereby avoided. These first-generation ENO schemes are inefficient on vector architectures because of the logical operations in the stencil selection algorithm. Furthermore the stencil choice can be sensitive to slight perturbations in the data, leading to a lack of robustness, and in some situations, loss of accuracy. While some of these problems are alleviated by slight modifications (Shu 1990), the integer shifting between stencils remains a source of difficulties for accuracy, steady-state convergence, and vectorization of the original ENO schemes.

Second-generation ENO schemes use a weighted combination of all the candidate fluxes (Liu *et al.* 1994; Jiang and Shu 1996). These weighted ENO schemes do not suffer from the integer shifting. If the local solution is smooth each candidate is assigned an optimal weight; then the weighted combination of candidate stencils corresponds to the optimal stencil. But if a candidate contains a discontinuity it is assigned a very small weight, and does not contribute to the numerical flux. In this way interpolation across shocks is avoided.

For more details on FD, FE and FV methods for solving the Euler and Navier–Stokes equations, we refer the reader to the books by Habashi and Hafez (1995), Ferziger and Peric (1995), Warsi (1999) and Wesseling (2001). A brief review of CFD techniques was given by Jameson (1995). Advances in the finite-volume high-resolution schemes for solving the Euler equations were recently presented by Sonar and Süli (1998).

We remark that for shock waves which propagate with variable velocity, the majority of high-resolution schemes degrade to the first-order accuracy in the vicinity of shocks. That problem was thoroughly analysed by Carpenter and Casper (1997), Engquist and Sjogreen (1998) and Ostapenko (2000).

3.7 Viscous Transonic Flow

3.7.1 Simplified models studied with methods of functional analysis

In the nearsonic approximation, steady flow of viscous gas is governed by the third-order equation

$$\varepsilon(\varphi_{xx} + \varphi_{yy})_x + (\gamma + 1)(1 - \varphi_x)\varphi_{xx} + \varphi_{yy} = 0, \tag{3.84}$$

if the streamline stresses are predominant as compared with transverse ones. This equation is obtained from the Navier–Stokes equations by detaining terms with the streamline stresses and the convective derivative. Equation (3.84) can serve, for example, for description of the structure of shock waves or asymptotic far-field pattern of transonic flow past an airfoil (Niewland and Spee 1973; Ryzhov 1978).

The term $\varepsilon(\varphi_{xx} + \varphi_{yy})_x$ is similar to the singular perturbation used in Equation (1.11) for the proof of solvability of the basic problem. Moreover, it coincides with the singular perturbation used by Vragov (1977) and his group for the study of boundary-value problems for linear multidimensional equations of mixed type. When treating the direct problems for transonic flow in the previous sections, we actually used such a viscous perturbation for the analysis of solvability of the problems. However, we considered shock-free flows, therefore, the employed method was a mathematical tool rather than consideration of the actual viscosity, because in this case the transverse viscous stresses in the boundary layer are of primary importance. For Equation (3.84), Sevost'yanov (1973) and Glazatov (1997) studied a few boundary-value problems in bounded domains or with periodic boundary conditions, however, their practical applications remained unclear.

Gamba and Morawetz (1996) tackled steady two-dimensional inviscid irrotational compressible flow using the artificial viscosity approach as follows. They introduced the artificial viscosity term into the Bernoulli law and analysed the system of equations

$$\operatorname{div}(\rho \cdot \nabla\varphi) = 0, \tag{3.85}$$

$$|\nabla\varphi|^2 + \frac{\gamma+1}{\gamma-1}\left(\frac{\rho}{\rho_0}\right)^{\gamma-1} = \frac{\gamma+1}{\gamma-1} + \varepsilon\Delta\varphi f(|\nabla\varphi|), \tag{3.86}$$

with respect to the velocity potential $\varphi(x, y)$ and the density $\rho(x, y)$ in the physical plane. The domain under consideration D was bounded by four smooth curves Γ_i, $i = 1, 2, 3, 4$, of the class $C^{2,\delta}$, and at corner points of D the inner angles were equal to $\pi/2$. Such a domain has the property that there is a conformal mapping of D onto a rectangle of auxiliary plane. Boundary conditions for the viscous system (3.85), (3.86) were imposed as follows:

(1) $\varphi = 0$, $\nabla\varphi \cdot \boldsymbol{n} < 0$, $\rho = h(x, y)$, at the inlet Γ_1,

(2) $|\nabla\varphi| = g(x, y)$ at the outlet Γ_3,

(3) $\varphi_n = 0$ on the walls $\Gamma_2 \cup \Gamma_4$, where n is the outer normal to the boundary.

Gamba and Morawetz established solvability of the problem for the viscous model and proved the following conjecture. If

$$g \in C^{1,\delta},\ 0 < g^2 < \frac{\gamma+1}{\gamma-1}, \quad h \in C^{1,\delta},\ 0 < h < \rho_0,$$

then there exists a solution $(\rho^\varepsilon, \varphi^\varepsilon) \in C^{1,\delta}(\overline{D}) \cap C^{2,\delta}(\overline{D})$ for sufficiently small $\varepsilon \leqslant \varepsilon_0$, which is subject to the bounds

$$0 < k_\varepsilon \leqslant |\nabla\varphi^\varepsilon| \leqslant K^*, \qquad 0 \leqslant \varphi^\varepsilon \leqslant K^*, \quad 0 < l_\varepsilon \leqslant \rho^\varepsilon \leqslant L^*,$$

with constants K^* and L^* independent of ε. The first step of the proof were iterations on the continuity equation and the viscous Bernoulli one, and obtaining *a priori* estimates in appropriate functional spaces. The second step was construction of a nonlinear compact mapping from a Banach space into itself which admits a fixed point that represents the desired solution.

Because of the validity of the uniform upper bounds, there exists a convergent subsequence $(\varphi^\varepsilon, \rho^\varepsilon)$ with a limit (φ^0, ρ^0). However, it has not been proved that this limit is a weak solution to a boundary-value problem for the inviscid system. In order to obtain a limit which is a solution to the inviscid system, Morawetz (1995) proposed to adapt a method of compensated compactness, which has been mainly used in studies of hyperbolic problems. However, in order to apply this method, one needs the inequality

$$0 < \delta_1 \leqslant |\nabla\varphi^\varepsilon|^2 \leqslant \frac{\gamma+1}{\gamma-1} - \delta_1,$$

which means that the velocity $|\nabla\varphi^\varepsilon|$ must be uniformly bounded below from zero and above from the cavitation speed. Also, the flow deflection angle must be bounded: $|\theta^\varepsilon| \leqslant M$. These bounds have not been proved.

Similar difficulties arose in the existence theorems formulated by other authors. Feistauer and Nečas (1988) used a vanishing viscosity approach based on the Navier–Stokes equations and formulated an existence theorem for the inviscid model. But they needed to assume that divergence of the viscous velocity field satisfies uniform bounds in ε. Klouček and Nečas (1990) and Klouček (1994) used the method of stabilization with an artificial time variable to obtain the inviscid solution. They also needed to assume uniform bounds with respect to the velocity and the entropy in order to pass to the limit.

The complete system of equations describing 2D unsteady viscous compressible flow consists of the continuity equation, Navier–Stokes equations and energy equation:

$$\left.\begin{aligned}
\frac{\partial \rho}{\partial t}+\frac{\partial \rho u}{\partial x}+\frac{\partial \rho v}{\partial y}&=0,\\
\frac{\partial \rho u}{\partial t}+\frac{\partial \rho u^2}{\partial x}+\frac{\partial \rho u v}{\partial y}&=-\frac{\partial p}{\partial x}+\frac{\partial \tau_{11}}{\partial x}+\frac{\partial \tau_{21}}{\partial y},\\
\frac{\partial \rho v}{\partial t}+\frac{\partial \rho u v}{\partial x}+\frac{\partial \rho v^2}{\partial y}&=-\frac{\partial p}{\partial y}+\frac{\partial \tau_{12}}{\partial x}+\frac{\partial \tau_{22}}{\partial y},\\
\frac{\partial e}{\partial t}+\frac{\partial e u}{\partial x}+\frac{\partial e v}{\partial y}&=\operatorname{div}(-p\boldsymbol{v}+k\nabla T)\\
&\quad+\frac{\partial(\tau_{11}u+\tau_{12}v)}{\partial x}+\frac{\partial(\tau_{21}u+\tau_{22}v)}{\partial y}.
\end{aligned}\right\} \tag{3.87}$$

Here

$$\tau_{11}=-\tfrac{2}{3}\mu\operatorname{div}\boldsymbol{v}+2\mu\frac{\partial u}{\partial x},\qquad \tau_{22}=-\tfrac{2}{3}\mu\operatorname{div}\boldsymbol{v}+2\mu\frac{\partial v}{\partial y},$$

$$\tau_{12}=\tau_{21}=\mu\left(\frac{\partial u}{\partial y}+\frac{\partial v}{\partial x}\right).$$

From thermodynamics we have

$$e=\rho(c_v T+\tfrac{1}{2}(u^2+v^2)),\qquad p=(\gamma-1)c_v\rho T. \tag{3.88}$$

In (3.87), (3.88) the notation used is as follows: t, time; x and y, Cartesian coordinates in $\mathbb{R}^2$; ρ, density; u and v, components of the flow velocity in the directions x and y, respectively; p, pressure; T, absolute temperature; e, total energy; τ_{ij}, components of the viscous part of the stress tensor; $\gamma > 1$, Poisson adiabatic constant; c_v, specific heat at constant volume; k, heat conductivity; μ, viscosity coefficient. The quantities c_v, k, μ are positive constants. The outer volume forces and heat sources are neglected. System (3.87), (3.88) is to be equipped with the initial conditions and boundary conditions which characterize the behaviour of the flow field on the boundary.

Lions (1998) considered the system of compressible Navier–Stokes equations together with the isentropic relation $p=\vartheta\rho^\gamma$. He looked for global solutions $(\rho, \boldsymbol{v})$ of the following system,

$$\rho_t+\operatorname{div}(\rho\boldsymbol{v})=0, \tag{3.89}$$

$$(\rho v_i)_t+\operatorname{div}(\rho\boldsymbol{v}v_i)-\mu\Delta v_i-\xi(\operatorname{div}\boldsymbol{v})_{x_i}+(\vartheta\rho^\gamma)_{x_i}=\rho f_i, \tag{3.90}$$

$1 \leqslant i \leqslant N$, where ρ is a nonnegative density, $\boldsymbol{v}$ is a vector-valued in $\mathbb{R}^N$ function that corresponds to the flow velocity, $\mu > 0$, $\mu+\xi > 0$, $\vartheta > 0$ and $\gamma \in (1, \infty)$. Lions mainly considered three situations.

1. The case when Equations (3.89), (3.90) are set in the whole space.
2. The case of the Dirichlet boundary condition when (3.89), (3.90) hold in a domain $\Omega \times (0, T)$ for some fixed T and $\boldsymbol{v}=\boldsymbol{0}$ on $\partial\Omega \times (0, T)$.

3. The periodic case when (3.89), (3.90) hold in $\mathbb{R}^N \times (0, T)$, and it is required that all the data and unknowns be periodic in each x_i $(1 \leqslant i \leqslant N)$ with period $T_i \in (0, \infty)$.

The initial conditions read

$$\rho|_{t=0} = \rho_0, \qquad \rho\boldsymbol{v}|_{t=0} = \boldsymbol{m}_0 \quad \text{in } \Omega,$$

where ρ_0, $\boldsymbol{m}_0$ are given functions on Ω.

The general strategy for construction of the global weak solutions was as follows: by convenient approximations, one builds sequences of approximate solutions or, more precisely, sequences of solutions to approximated problems. The key issue was the passage to the limit which, at least formally, should lead to solutions of the original problem. This passage to the limit always requires some form of compactness in order to be able to handle the nonlinearities. This is why considerable effort was made to obtain compactness results for the compressible isentropic Navier–Stokes equations.

Lions (1998) also obtained some results on the regularity of the global weak solutions. On the other hand, it is worth mentioning the recent examples by Vaigant (1994), who demonstrated the formation of singularities in finite time for 'smooth' solutions $(\rho, \boldsymbol{v})$ of (3.89), (3.90). Specifically, if $\gamma < 1 + 1/N$ (and thus if $1 \leqslant \gamma < 2$ when $N = 2$), one can find $f \in L^q(\Omega \times (0, T))$ for some $q > N$ such that the local in time unique solution $(\rho, \boldsymbol{v})$ of (3.89), (3.90) with the Dirichlet boundary condition blows up in finite time.

Novotny (1998) proved a general compactness of sequences of weak solutions to Equations (3.89), (3.90). He studied the summability, the regularity and the compactness of different quantities in the flow field. He thus got results similar to those of P.-L. Lions by a different method. The main tool in the proofs was the 'supersonic' method of decomposition, which is a global version of the similar method used earlier for studying compressible flows near the equilibrium states.

The Euler equations can be obtained from (3.89), (3.90) by omitting the viscous terms and f_i:

$$\left.\begin{aligned} \rho_t + \operatorname{div}(\rho\boldsymbol{v}) &= 0, \\ (\rho v_i)_t + \operatorname{div}(\rho\boldsymbol{v}v_i) + (\vartheta\rho^\gamma)_{x_i} &= 0, \end{aligned}\right\} \tag{3.91}$$

in $\mathbb{R}^N \times (0, \infty)$, where $N \geqslant 1$, $\vartheta > 0$, $\gamma > 1$. This system is complemented by the following initial conditions:

$$\rho|_{t=0} = \rho_0 \geqslant 0, \qquad \rho\boldsymbol{v}|_{t=0} = \boldsymbol{m}_0 \quad \text{in } \mathbb{R}^N.$$

The system (3.91) of partial differential equations has a history going back more than two centuries. A few classical facts on the above system are as follows.

(i) If ρ_0 and $\boldsymbol{m}_0$ are smooth enough, then there exists a maximum time interval $[0, T_c)$ on which there exists a unique smooth solution $(\rho, \boldsymbol{v})$ of (3.91) for $0 \leqslant t \leqslant T_c$. In addition, if $T_c < \infty$ (and this is the case in general), $(\rho, \boldsymbol{v})$ becomes discontinuous as t goes to T_c.

(ii) If we allow for discontinuous solutions, for instance, solutions $(\rho, \boldsymbol{v}) \in L^\infty$ satisfying (3.91) in the sense of distributions, then the solutions are neither unique nor stable. More precisely, one can extract sequences of such solutions which converge weakly to functions which do not satisfy (3.91).

(iii) In order to restore the stability of solutions and (possibly) the uniqueness, one should impose further restrictions, namely an entropy inequality, on bounded solutions of (3.91).

The reader is referred to Lax (1957, 1973) for more details regarding these results.

Up to now, difficulties with providing the entropy inequality are the primary reason for a very slow progress in rigorous studies of the discontinuous solutions to the compressible Euler Equations (3.91), as well as to the full-potential equation, in space dimensions greater than one. In the preface to his book, Lions (1998) wrote, '. . . we briefly discuss the compressible Euler equations in section 8.7, making some rather wild speculations more or less motivated by our analysis of the compressible Navier–Stokes equations'.

Kweon and Kellogg (1997) studied a boundary-value problem for the compressible Navier–Stokes equations in a bounded two-dimensional domain assuming that the flow is barotropic, i.e. the density is a function of the pressure only: $\rho = \rho(p)$. Nonzero velocities were prescribed on the boundary, and the density was prescribed on the inflow part of the boundary. Particular attention was paid to the boundary points where the velocity vector was tangent to the boundary since these gave rise to singularities in the solution. However, the existence of the unique solution to the problem was proved under the assumption of a sufficiently large viscosity coefficient μ which is not relevant for transonic applications.

Liu (2000) analysed finite-element solutions to the compressible Navier–Stokes equations and performed an error analysis in the case of barotropic two- and three-dimensional unsteady flow. For the continuous problem it was assumed that the exact solution exists, the density is strictly positive, and both the velocity and density, together with certain first-order derivatives, are uniformly bounded. The implicit Euler scheme was used to discretize the problem in time. A standard conforming Galerkin finite-element scheme in space was stabilized by applying the streamline diffusion technique to the continuity equation. The main result was concerned with the convergence of the fully discretized solution to the solution of the continuous problem. Moreover, an *a priori* error estimate of optimal order and uniform in time was obtained.

Kellogg and Liu (2000) studied a similar problem by rewriting the continuity equation in the form in which a dependent variable is the logarithm of the density. The advantage of this new formulation is that the duality between the gradient of the pressure in the momentum equation and the divergence of the velocity in the continuity equation is preserved.

The main unclear aspect of these results, as well as of the results by Lions obtained for the compressible Navier–Stokes, is their practical relevance, since the isentropic or

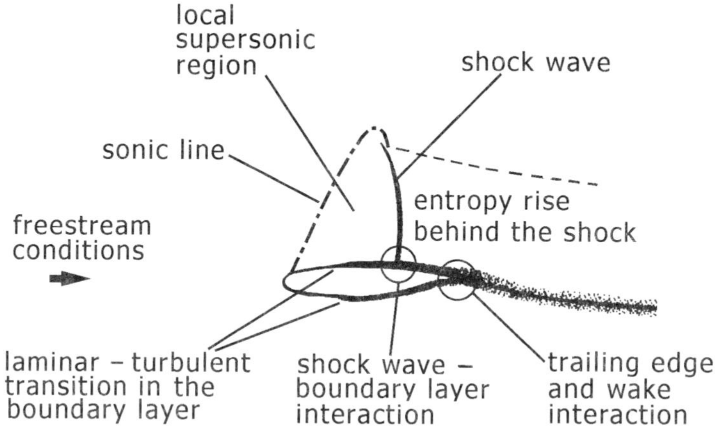

Figure 3.24 A scheme of transonic flow at a high Reynolds number in the case of the attached boundary layer.

barotropic assumption appears to be incompatible with the real viscous flow because of the dissipation of the kinetic energy.

3.7.2 Asymptotic theory of viscous–inviscid interaction

The complexity of the physical phenomena that occur in viscous compressible flow is revealed by both experimental observations and the asymptotic analysis of the governing equations. At high Reynolds numbers, viscous effects are confined to boundary layers, shock waves and wakes. If the boundary layer remains attached to the airfoil, the inviscid flow is then the correct first approximation to the flow field (see Figure 3.24). To take into account the influence of the boundary layer, one can use a direct iteration procedure, wherein high-Reynolds-number flow is assumed, allowing the flow field to be divided into two regions: an outer region governed by the full-potential or the Euler equations which permit strong shocks and rotational flow, and an inner or viscous-layer region, which is governed by Prandtl's equations. The Prandtl equations for steady compressible flow in the laminar boundary layer are obtained from (3.87) under the assumption $\partial p/\partial y \equiv 0$, where y is aligned with the normal to the airfoil, and minor simplifications of the viscous terms:

$$\left.\begin{aligned} \frac{\partial \rho u}{\partial x} + \frac{\partial \rho v}{\partial y} &= 0, \\ \frac{\partial \rho u^2}{\partial x} + \frac{\partial \rho u v}{\partial y} &= -\frac{\mathrm{d}p}{\mathrm{d}x} + \frac{\partial}{\partial y}\left(\mu \frac{\partial u}{\partial y}\right), \\ \frac{\partial e u}{\partial x} + \frac{\partial e v}{\partial y} &= -\frac{\partial p u}{\partial x} + \frac{\partial}{\partial y}\left(k \frac{\partial T}{\partial y}\right) + \mu \left(\frac{\partial u}{\partial y}\right)^2, \end{aligned}\right\} \tag{3.92}$$

where $p(x)$ is determined by the external inviscid flow, and

$$e = \frac{p}{\gamma - 1} + \rho \frac{u^2 + v^2}{2}, \qquad T = \frac{p}{\rho R}.$$

The system (3.92) is endowed with the boundary conditions:

$$u = 0, \quad v = 0, \quad \partial T/\partial y = T_{\mathrm{w}}(x) \quad \text{on the airfoil ;} \tag{3.93}$$

$$u = u_{\mathrm{e}}(x), \quad T = T_{\mathrm{e}}(x) \quad \text{at the edge of the boundary layer } y = \delta(x). \tag{3.94}$$

In the 'direct' coupling method, the inviscid flow solution provides the pressure $p(x)$ at the edge of the boundary layer. If a solution of the system (3.92)–(3.94) is found, e.g. with an integral boundary-layer solver, then the displacement thickness δ^* can be calculated via the formula (Fletcher 1991, Section 16)

$$\delta^*(x) = \int_0^{\delta} \left(1 - \frac{\rho u}{\rho_{\mathrm{e}} u_{\mathrm{e}}}\right) \mathrm{d}y,$$

and it can be added to the airfoil geometry to provide the boundary condition for the next iteration in the outer region.

With increasing freestream Mach number M_∞, viscous effects become more important. The increase of M_∞ from low subsonic to nearsonic values often leads to flow separation. Typically, the boundary layer tends to separate from the surface either as a result of an incident shock wave (at transonic speed) or due to a high angle of attack and an adverse overall pressure gradient (at subsonic speed). Numerical results illustrating transonic flow separation behind the shock wave impinging on the boundary layer at the end of the local supersonic region are shown in Plate 1. Boundary-layer separation leads to a decrease in the aerodynamic performance and an increase in drag and therefore fuel consumption. In case of turbomachines, separation can occur on the suction side of blades of compressors and turbines. The flow unsteadiness associated with separation (buffet) leads to buffeting and fatigue damage which limits the durability of aircraft structures and turbomachines.

In the presence of a large adverse pressure gradient, the steady two-dimensional boundary-layer Equations (3.92) develop a singularity at the streamwise location at which the skin-friction vanishes (Goldstein 1948). This result is interpreted usually as an indication that the physics involved in the flow development can no longer be described simply by the concept of a noninteractive boundary-layer driven by an external flow. The study of Stewartson (1970a) indicated that this singularity is not removable in the sense that it cannot be smoothed out on a shorter length-scale around the point of separation.

When separation takes place, or the state of the boundary layer is pre-separated, the above-mentioned way for an iterative search of the solution fails, and the viscous flow near the airfoil and the external inviscid flow become interplaying. Thus a new approach is needed for predicting strong inviscid/viscid interaction phenomena such as viscous-layer separation, shock/boundary-layer interaction and trailing-edge/near-wake interaction over an airfoil/wing and in blade passages, particularly at off-design operating conditions.

A significant advance towards the development of a local-interaction theory for laminar flows was made by Lighthill (1953), who suggested dividing the boundary layer into two regions, namely an outer inviscid but rotational layer in which the flow is only slightly disturbed from the incoming boundary-layer profile, and a thin viscous sublayer in which the velocity is reduced to zero on the wall to satisfy the no-slip condition.

The problem of high-Reynolds-number laminar separation has undergone a systematic analysis through the asymptotic triple-deck theory. It was formulated simultaneously by Neiland (1969) and Stewartson and Williams (1969) for the self-induced separation in supersonic flow, and by Messiter (1970) for incompressible fluid flow near a trailing edge of a flat plate. It was shown that in many cases of practical interest, for sufficiently small separation-causing disturbances, the structure of the separation region can be represented analytically by a three-layer model which was named the 'triple-deck'. Three regions were identified – the inner, middle and outer decks – all having characteristic longitudinal length of order $\varepsilon^3 L$ and normal height of the order $\varepsilon^5 L$, $\varepsilon^4 L$, and $\varepsilon^4 L$, respectively. L is the characteristic length of the geometry and ε is the small parameter in the perturbation expansion related to the large characteristic Reynolds number by the relation $\varepsilon^8 = 1/Re$, $Re = v_\infty L/\nu_\infty$, v_∞ is the freestream velocity, ν is the kinematic viscosity. These three regions or decks are characterized by viscous, inviscid–rotational, and potential flow conditions, respectively, and their structure and properties are described to first order in the perturbation parameter ε by the fundamental triple-deck problem. This consists of the bottom deck, boundary-layer-like equations coupled to a pressure interaction law relating the viscous displacement thickness growth and the inviscid outer deck pressure gradient, according to linear airfoil theory.

It turned out that the triple-deck interaction region, while being small, plays a key role in many fluid dynamics phenomena.

Solutions to the triple-deck equations are important for two reasons: first, they provide rational, complete solutions for high Reynolds-number flows where the classical boundary-layer approach fails (e.g. in regions where separation and reattachment occur); second, they offer a reliable test base for evaluation of more comprehensive viscous flow models (e.g. numerical algorithms for the Navier–Stokes or interacting boundary-layer equations).

Both separation and stability exhibit some peculiarities when the external velocity is close to the speed of sound.

Flow over a bump

A fundamental problem is the flow over a flat plate which features a small bump situated downstream of the leading edge (Stewartson 1970b, 1971; Messiter 1970). The description of the boundary-layer flow, based upon the triple-deck theory, showed how the presence of the bump generated an interaction between the inviscid region just outside the layer and the viscous region near the bump. The problem reduces to that in the lower deck, where the motion is essentially viscous with an induced pressure gradient.

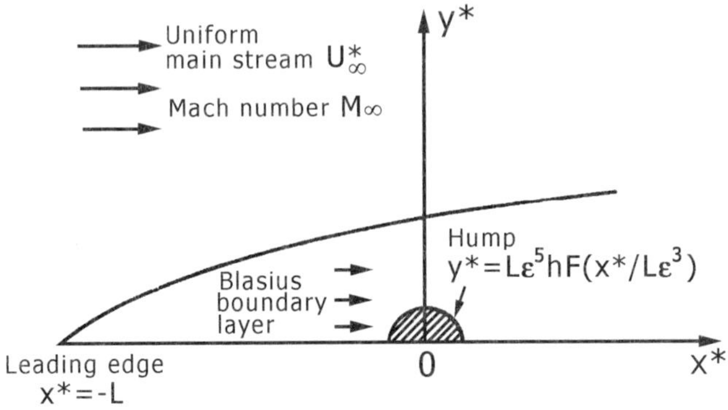

Figure 3.25 Laminar boundary layer over a small hump on a flat plate. Reproduced from Smith (1973) by permission of Cambridge University Press.

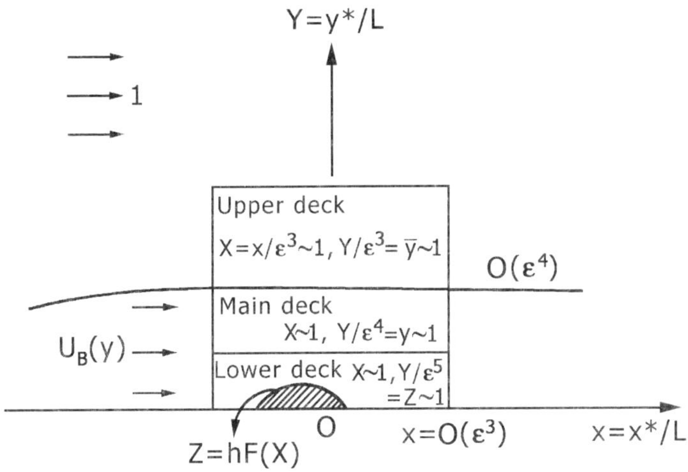

Figure 3.26 The triple-deck structure of laminar flow over a hump. Reproduced from Smith (1973) by permission of Cambridge University Press.

Smith (1973) considered the bump profiles

$$y^*/L = \varepsilon^5 hF(x^*/L\varepsilon^3), \tag{3.95}$$

where $\varepsilon = Re^{-1/8}$, the parameter h was initially of order one and the function F was such that $hF(X)$ was of order one or less for all $X = x^*/L\varepsilon^3$ (see Figures 3.25 and 3.26). 'The expansion and subsequent contraction of the boundary-layer thickness caused by the presence of the hump (3.95) produce first-order perturbations in the upper deck just outside the boundary layer that interact with and serve to maintain the lower-deck flow. The triple-deck structure, based on the matching and overall consistency of asymptotic expansions in three regions, the upper, main and lower decks, incorporates this feedback effects and also we are able to satisfy the boundary conditions on the hump and those of matching upstream and outside the boundary layer' (Smith 1973). Asymptotic expansions yield the system governing the flow in the lower deck

$$\left.\begin{aligned} UU_X + VU_Y &= -P_X + U_{YY}, \\ U_X + V_Y = 0, \qquad & P_Y = 0, \end{aligned}\right\} \tag{3.96}$$

with the boundary conditions

$U = V = 0$ on $Y = hF(X)$ (no-slip on the bump),

$U_Y \to 1$ as $X \to -\infty$ (matching with the Blasius solution upstream),

$U \sim Y + A(X)$ as $Y \to \infty$ (matching with the main deck),

where, for subsonic main flow,

$$A_{XX} = -\frac{1}{\pi}\int_{-\infty}^{\infty}\frac{P_t \, dt}{X - t},$$

or, for supersonic flow,

$$A(X) = -\int_{\infty}^{X} P(t)\, dt.$$

Linearized solutions for both supersonic and subsonic main streams demonstrated the consistency of the model by the match achieved between these solutions and the external inviscid flow. In the special case of very small bump, the solution was obtained by considering the effects of a suitably defined point disturbance within the lower deck. Similarity solutions for the decay of the velocity perturbations in the far-wake, $x^*/L\varepsilon^3 \gg 1$, were found to be in agreement with the full linearized solutions, while an analysis of the near-wake of the point disturbance, $x^*/L\varepsilon^3 \ll 1$, showed that the effect of the main stream remains a primary influence in the local flow. This influence is independent of the subsonic or supersonic nature of the main stream. Also, Smith calculated the first-order forces and couples on the general $L\varepsilon^3$ by $L\varepsilon^5$ bump and demonstrated that the couples of forces due to streamwise and normal stresses, respectively, contribute equally to the total couple on the bump. On the other hand, the force normal to the main-stream flow is zero, while the drag is given to first order by the undisturbed boundary-layer contribution.

The description was extended to laminar flow over small bodies by considering bumps of length of order unity. Perturbation solutions to the boundary-layer equations were obtained. This made it possible to examine the implied efflux from the layer and to show that pressure feedback effects cannot be neglected as one approaches the region $x^*/L \sim \varepsilon^3$. It is the triple-deck theory that resolves this difficulty.

A similar problem for laminar flow over a bump on a flat plate was studied by Korolev (1983) and Ryzhov (1989), who concentrated on computation of the separated zones behind and in front of the bump (Figure 3.27). The solution to the equations of

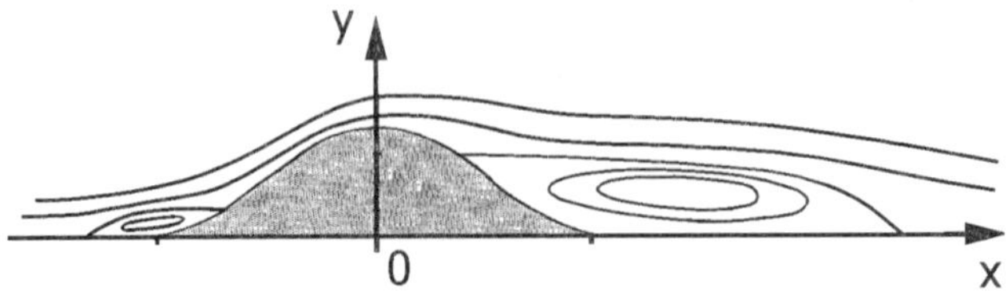

Figure 3.27 A scheme of laminar flow separation behind and in front of a smooth bump.

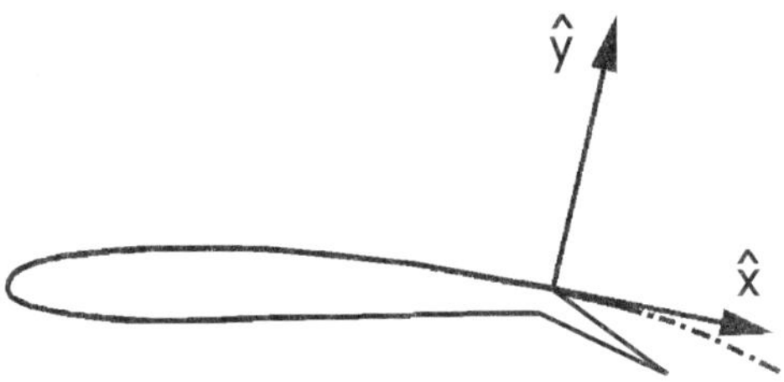

Figure 3.28 Flow separation induced by a deflection flap. Reprinted from Ruban and Turkyilmaz (2000) by permission of Cambridge University Press.

the triple-deck theory was shown to depend on the similarity parameter

$$K_\infty = \frac{M_\infty^2 - 1}{\varepsilon^2 2^{4/5} C^{1/5} \lambda_s^{2/5} (T_w^*/T_\infty^*)^{4/5}},$$

where T_w^* is the plate temperature, T_∞^* is the temperature of the free stream, λ_s is the coefficient of skin friction, and C is the Chapman constant occurring in the linear viscosity law. Korolev and Ryzhov obtained numerical solution for $K_\infty = \pm 3$ and the bump shape $y'_w(x) = h_0(1 + \cos x)$, $-\pi \leqslant x \leqslant \pi$, where y' is a nondimensional coordinate normal to the plate. In the case of the supersonic freestream flow ($K_\infty = +3$), a principal separation occurred in front of the bow of the bump, whereas in the case of the subsonic main stream ($K_\infty = -3$) a salient feature of the flow was development of a large recirculation bubble on its back side. Thus, the size of the separation zone was much greater in the second case than in the first one.

The theoretical knowledge gained from the asymptotic approaches was used in recent developments of numerical interacting boundary-layer schemes. Bartels and Rothmayer (1994), for example, extended a quasi-simultaneous interacting boundary-layer method with alternating direction explicit acceleration to include the nonlinear weak-shock branch of a nonlinear transonic viscous–inviscid interaction. They studied the case of supersonic main flow over an expansion ramp and a parabolic bump and mentioned the development of a growing separation region behind the bump as $M_\infty \to 1$.

Trailing edge flow

Asymptotic analysis of transonic flow near the trailing edge of a flat plate was performed by Bodonyi and Kluwick (1982, 1998). In this problem the inviscid part of the flow remains uniform everywhere except in a small vicinity of the trailing edge, and the boundary layer on the plate surface is described by the compressible version of the Blasius solution. The interaction region, which forms near the trailing edge, has conventional triple-deck structure. The interaction may be described as follows. In the viscous sublayer the flow velocity is relatively small and for this reason the lower deck exhibits very high sensitivity to pressure variations. Even a small pressure rise along the wall may cause significant deceleration of the flow. That is why the streamlines change their shape being displaced from the wall. The slope of the streamlines is then transmitted through the main part of the boundary layer to the potential flow in the upper deck. The potential flow plays the role of converting the perturbations of the slope into the perturbations of pressure. The latter, being transmitted back to the lower deck, cause thickening of the viscous sublayer, and the process repeats. The middle deck plays a passive role in the interaction process. It does not contribute to the displacement effect of the boundary layer, and it does not change the pressure perturbations while transmitting them from the upper deck to the viscous sublayer.

Ruban and Turkyilmaz (2000) focused on the separation of the laminar boundary layer from a convex corner made by an airfoil and a flap in transonic flow (Figure 3.28). They demonstrated that the viscous–inviscid interaction changes dramatically when instead of the uniform flow outside the interaction region a real transonic flow separating from a rigid surface is considered. It appear that the flow separation is accompanied by a very strong pressure gradient acting upon the boundary layer upstream of the separation. This pressure gradient causes the velocity profile in the boundary layer to take a form which is significantly different from that observed in the subsonic or supersonic flows.

> It is shown that the flow in a small vicinity of the separation point is governed, as usual, by strong interaction between the boundary layer and inviscid part of the flow. Outside the interaction region the Kármán–Guderley equation describing transonic inviscid flow admits a self-similar solution with the pressure on the body surface being proportional to the cubic root of the distance $-x$ from the separation point, and the pressure gradient $\mathrm{d}p/\mathrm{d}x \sim (-x)^{-2/3}$. Analysis of the boundary layer driven by this pressure shows that as the interaction region is approached the boundary layer splits into two parts, the near-wall viscous sublayer whose thickness may be estimated as $y \sim Re^{-1/2}(-x)^{5/12}$, and the main body of the boundary layer where the flow is locally inviscid. It is remarkable that contrary to what happens in subsonic and supersonic flows, the displacement effect of the boundary layer is primarily due to the inviscid part. The contribution of the viscous sublayer proves to be negligible to the leading order. Consequently, the flow in the interaction region is governed by the inviscid–inviscid interaction. The longitudinal

extent of this region is estimated as $\Delta x = O(Re^{-3/7})$. To describe this flow one needs to solve the Kármán–Guderley equation for the potential flow region outside the boundary layer; the solution in the main part of the boundary layer was found in an analytical form, thanks to which the interaction between the boundary layer and external flow might be expressed via the corresponding boundary condition for the Kármán–Guderley equation. Formulation of the interaction problem involves one similarity parameter which in essence is the Kármán–Guderley parameter suitably modified for the flow at hand. The solution of the interaction problem has been constructed numerically.

(Ruban and Turkyilmaz 2000)

3.7.3 Shock-wave–boundary-layer interaction

When a shock wave impinges on the boundary layer, in the area of shock-wave–boundary-layer interaction the flow parameters change abruptly in both the streamwise and normal directions. Therefore, basic assumptions of the classical theory of the boundary layer fail to be true. Available experimental data make it possible to describe the interaction pattern as follows.

The existence of a shock wave results in a strong pressure rise along the wall. The pressure rise acts upstream through the subsonic part of the boundary layer and invokes steep thickening or even separation of the boundary layer. This in its turn influences the shape and strength of the shock. The pattern of interaction depends significantly on the flow regime in the boundary layer.

For a laminar boundary layer, if the Mach number in the supersonic zone in front of the shock is close to one, then the pressure jump across the shock is small and then the shock wave is not of sufficient strength to cause separation. The interaction is weak, and the sonic line is placed not too deep within the boundary layer. The thickening of the boundary layer extends both upstream and downstream of the shock impingement point thereby causing local changes in the external flow due to the outward displacement of the streamlines. The boundary layer either remains attached or separates over a short distance which terminates by reattachment terminating a short separation bubble. After the reattachment, flow may remain laminar or change for turbulent regime. Behind the shock there is an expansion fan similar to that in the problem of flow over a convex wall. At the reattachment point the incident flow hits the wall at a nonzero angle with respect to the tangent direction, hence, a new shock appears which may cause further boundary-layer separations.

With increasing Mach number in front of the shock wave, the strength of the shock increases. This causes a considerable thickening of the boundary layer, which leads to a long-scale separation (Plate 1). It was found that under laminar flow conditions the separation point moves forward along the body rather rapidly as the strength of the shock wave is increased. Compression waves arise at the point of abrupt thickening of the boundary layer in the same way as in supersonic flow over a concave wall. At

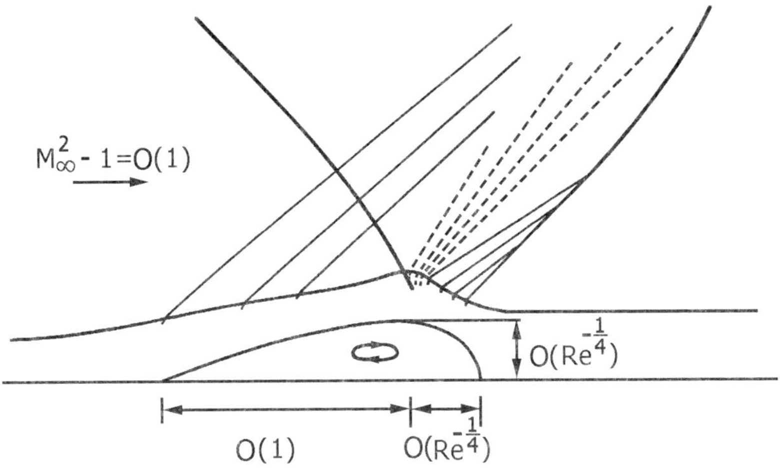

Figure 3.29 A scheme of shock-wave–boundary-layer interaction. After Bodonyi and Smith (1986). Copyright © 1986 Elsevier Science. Reproduced with permission.

a distance from the wall, the compression waves coalesce into one or several oblique shocks.

In a phenomenological sense, it is evident that the upstream pressure rise must balance the rate of growth of the boundary layer ahead of the shock wave. The appropriate scaling laws for transonic interactions were studied by Brilliant and Adamson (1973), who considered the shock-wave–boundary-layer interaction for sufficiently weak oblique shock waves such that the boundary layer remains unseparated throughout the interaction region. A perturbation analysis of Adamson and Feo (1975) and Melnik and Grossman (1976) showed that the boundary-layer structure is multilayered in the vicinity of the interaction.

As discussed in a review paper by Adamson and Messiter (1980), increases in the shock wave strength result in the sonic line descent to the lower part of the boundary layer, and the interaction becomes strong with flow separation occurring. A broad turbulent wake then results along with a substantial increase in the airfoil drag and loss of lift. The wave drag engendered by a weak shock wave is small but develops rapidly with increasing freestream Mach number. Usually, the total drag coefficient increases, the lift coefficient decreases and the centre of pressure moves aft as the shock wave moves rearward across the airfoil upper surface. These trends need not be monotonic with increasing freestream Mach number, particularly when lower surface shock waves form on a lifting airfoil.

Bodonyi and Smith (1986) examined the response of the laminar boundary layer beneath a supercritical transonic flow to an incident oblique shock wave as the shock strength increases from that of a weak incident shock involving a localized separated flow region towards a shock with strength of order one and a rather large-scale separated-flow region (Figure 3.29).

> It is found that for relatively mild shocks both the separation and reattachment processes take place within the triple-deck structure, which has a longitudinal length-scale of order $Re^{-3/10}$ in transonic flow. As the shock strength increases, however, the separation and reattachment processes tend to become dissociated. The separation process remains within the triple-deck structure and is pushed further and further upstream of the shock-impingement point in the limit of an order one shock strength. Also, a rather long constant pressure plateau region develops and finally reattachment occurs within a relatively short region. For further details concerning this large-scale separation problem the interested reader is referred to Smith (1985). Scaling laws have been proposed for this reattachment region. However, it must be stressed that they are only tentative as the effects of the backward jet in the vicinity of the reattachment point are not fully understood as yet. Nevertheless, the solution found to date are of interest in our attempt to understand the complicated flow processes which occur during a shock-wave boundary-layer interaction.
>
> (Bodonyi and Smith 1986)

The problem is more complicated in case of the turbulent boundary layer. The location of the sonic line within the boundary layer is of importance in dictating the type of interaction. The sonic line location is a function of the freestream Mach number and the boundary layer velocity profile. The velocity profile is, in turn, a function of the pressure gradient under which the boundary layer had developed before encountering the interaction. For a weak interaction, the sonic line resides in the large-eddy region of the boundary layer, and the shock wave has little influence on the turbulent energy production in the wall layers. This situation arises when $|M_\infty^2 - 1| = O(\varepsilon)$ and the flow is properly transonic.

Agrawal and Messiter (1984) studied formation of an oblique shock wave within an unseparated turbulent boundary layer caused by a shallow concave (compression) corner. Analytical solutions for the local mean-velocity and pressure distributions were found for the transonic range which agree favourably with available experimental data.

Fundamental and well-documented experiments by Délery (1989) were aimed at the physical understanding of transonic shock-wave–boundary-layer interaction and comparison of various turbulence models. Figure 3.30 shows a tracing of Mach isolines in the two-dimensional flow over a bump in a channel. The figure confirms the existence of a strong interaction leading to the formation of a large separated region with the accompanying lambda shock system. The turbulence models incorporating a history effect were shown to provide a good agreement with the experiment. Also, Délery established that the structure of the flow resulting from shock-induced interactions in a 3D channel is far more complicated than that of a nominally 2D interaction. Examination of skin-friction line patterns revealed a complex topology containing a large number of singular points and separation lines.

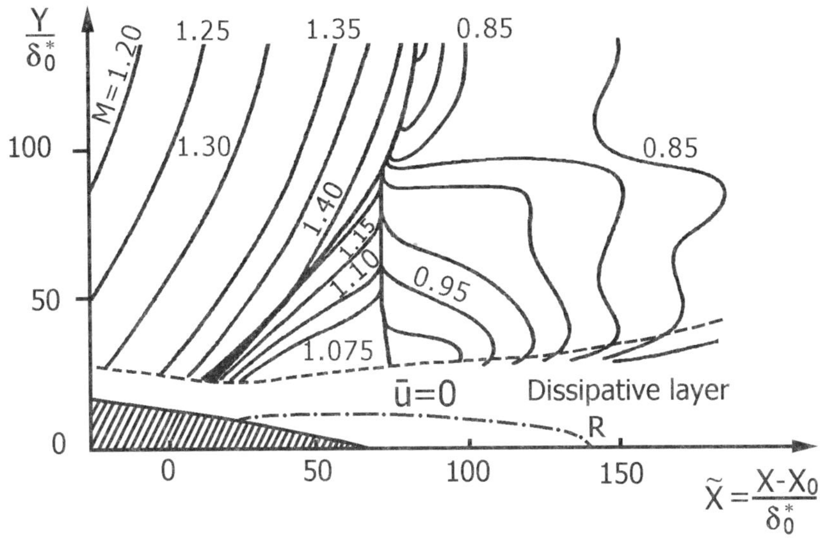

Figure 3.30 Transonic shock wave–turbulent boundary layer interaction. Mach number isolines obtained in experiments of Délery (1989, Fig. 2b). Reproduced by permission of Springer and the author.

3.7.4 Unsteady viscous flows

Self-excited oscillations

Separation of the boundary layer can lead to the development of self-excited shock oscillations or shock buffet for perfectly rigid wings/airfoils (McDevitt *et al.* 1976). This is a periodic shock motion involving alternating separation and reattachment of the boundary layer. The unsteady airloads generated are quite large, they occur for narrow ranges of the freestream Mach number and angle of attack, and have characteristic frequencies which can be near those involved in flutter.

As experiments demonstrate (see Mabey (1989) and references therein), for a thick biconvex airfoil at zero angle of attack and Mach numbers $0.74 < M < 0.82$, a shock wave forms downstream of the maximum thickness. The flow is almost steady, with a small, symmetric separation downstream of the shock. For Mach numbers from $0.82 < M < 0.86$ the flow is periodic, with shocks moving upstream and downstream on alternative surfaces of the airfoil. The chordwise pressures are correlated strongly in both space and time. For Mach numbers $M > 0.86$ the flow is again steady, with a separation extending from the shock to the trailing edge. The upper boundary of the self-excited oscillations actually corresponds to the flow conditions when the shock is placed near the airfoil trailing edge.

Even an attached turbulent boundary layer produces local pressure fluctuations on the airfoil. This excitation is confined to high frequencies and has small length-

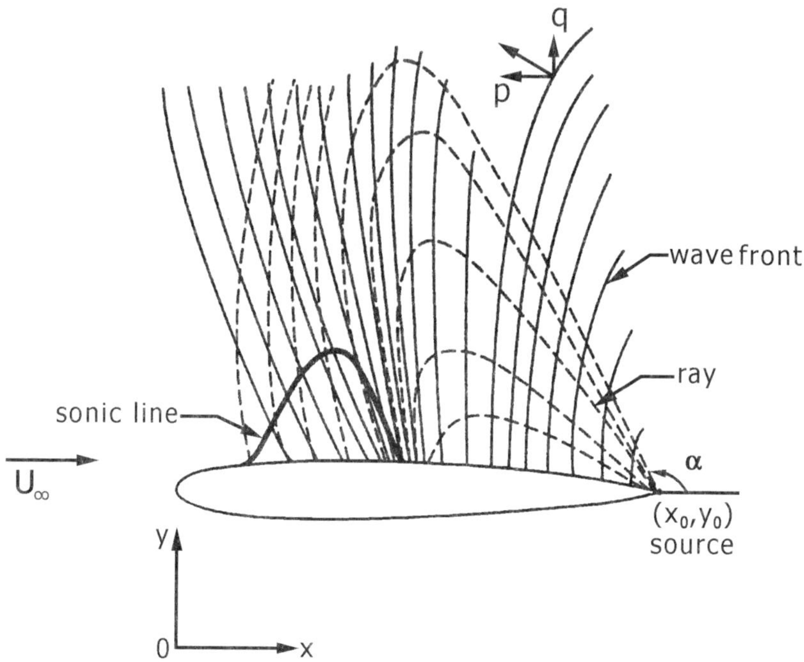

Figure 3.31 A scheme of wavefronts and rays propagating in transonic flow from a source at the trailing edge of an airfoil. Courtesy of Lee *et al*. (1994). Reproduced by permission of the authors.

scales relative to the airfoil thickness. If the excitation is integrated in space and time across the airfoil, there is no net unsteady force. For higher speeds the coupling between the steady and time-dependent flows modifies the surface pressure fluctuations in a complicated manner. Initially, the thickening of the attached boundary layer and its intermittency causes a small shock oscillation, which in turn modifies the boundary-layer thickness and further increases the local pressure fluctuations (Mabey 1989). The pressure at the trailing edge provides a link between the top and bottom surfaces of the airfoil. The requirement of constant mean pressure at the trailing edge and the wake development impose the condition that these oscillations be in antiphase on opposite surfaces, as illustrated by measurements of McDevitt and Okuno (1985).

The onset of the large-amplitude oscillatory shock motion in the case of a perfectly rigid airfoil is associated with the boundary-layer separation. Separated flow is strongly unstable and can act as a source of acoustic perturbations in the surrounding inviscid field. Thus, the shock oscillations comprise the boundary-layer excitation of disturbances and the feedback through the inviscid region maintaining the oscillations. Current understanding of the basic mechanisms leading to the flow instability is incomplete. However, spark schlieren photographs of flow fields indicate clearly the presence of upstream-moving waves originating at the trailing edge and near-wake region. They are associated with wake fluctuations due to unsteady shock motions and termed 'Kutta waves' by Tijdeman (1977). A scheme of wavefronts and rays propagating in transonic flow from a source at the trailing edge is shown in Figure 3.31. The feedback through the subsonic region is of paramount importance for accounting for the self-sustained oscillations.

Several explanations of the mechanisms of the shock wave oscillations have been proposed by a few authors in their interpretation of experimental and computational results obtained for different flow conditions. However, there is no clear understanding of why the oscillations exist in the specific ranges of freestream parameters. Moreover, full details of the role of viscosity are still not clear. An attempt to approach the problem of self-excited oscillations theoretically using the asymptotic theory of self-similar solutions in the transonic range has been recently made by Lifshitz (2000). We shall continue the discussion of shock buffet in Section 4.4.

Laminar–turbulent transition

Bowles and Smith (1989, 1993) analysed the development of instabilities in the boundary layer with the triple-deck theory and identified the mechanisms of the laminar–turbulent transition in the transonic range behind three major processes: (a) substantial external-flow deceleration, (b) rapid boundary-layer thickening, (c) three-dimensional nonlinear interactions.

In two-dimensional flow, a transonic range was found to possess essentially the same linear and nonlinear properties as those which prevail in the pure subsonic or supersonic regimes, with some notable exceptions, e.g. at higher amplitudes. An alternative range of local Mach numbers closer to unity, however, exposes two substantially different aspects, namely, external shock effects and internal separation effects, corresponding to (a), (b) above.

> The nonlinear linkage between (a), (b) provides a by-pass mechanism (as in buffeting or shock flutter) which is potentially powerful locally and could readily induce transition. Likewise, the 3D influence... corresponding to (c) above turns out to be very important, in particular yielding substantial differences between linear and nonlinear behaviour and between 2D- and 3D-based predictions, in the transonic range. Thus new forms of three–wave resonance, of strong 3D nonlinear growth with focusing, and of vortex–wave interactions, are found throughout the transonic regime, along with a distinct effect produced by slight 3D warping of an otherwise 2D input disturbance upstream. This last is perhaps the finding of most interest and may be of significance for the nacelle configuration outlined at the start of this section as well as in other transonic-flow applications. Both slight warping, i.e. the low-spanwise-wavenumber content of the input, and enhanced 3D input, at much higher spanwise wavenumbers, provoke spatial focusing and amplification fast

nonlinearity, in contrast with the spanwise wavenumbers in-between and with 2D theory.

(Bowles and Smith 1993)

Ryzhov (1989, 1993) studied the issues of stability and laminar–turbulent transition with a similar technique and distinguished between two transonic ranges: (i) $M_\infty^2 - 1 = O(\varepsilon^{-8/5})$, and (ii) $M_\infty^2 - 1 = O(\varepsilon^{-8/9})$, where as usual $\varepsilon = Re^{-1/9}$, and Re is the Reynolds number based on the freestream velocity, kinematic viscosity, and a length based on the global scale. In range (i), the perturbation potential governing the upper deck can be reduced to a nonlinear, unsteady, second-order partial differential equation, the same as the equation used in unsteady transonic small disturbance calculations, whereas the canonical form of the lower-deck equation (Prandtl's viscous wall sublayer equation) is identical to that of the standard triple-deck theory. Ryzhov showed that certain transformations of the time and longitudinal coordinates, and substitution of asymptotic expansions into the Navier–Stokes equations reduce them to the standard Equations (3.96) for an incompressible laminar boundary layer. The absence of time implies that the viscous flow in the wall region is capable of adjusting instantaneously to low-frequency perturbations propagating through the external potential region. To elucidate how this adjustment occurs, one needs to study processes in the greatest part of the boundary layer. Thus the lower deck in range (i) is quasi-steady, responding passively to the upstream propagating, nonlinear 'receding' wave system; whether this wave system may receive amplification through the interaction is unclear. Detailed study for range (i) involved only examples of two low-lying bumps in the steady case at $M_\infty < 1$.

In range (ii), which is not so close to $M_\infty = 1$, the equation governing the unsteady flow in the upper deck can be linearized, and the lower-deck equation system must now include an unsteady term. This formulation makes use of time-dependent oscillator eigenmodes for instability analysis of an insulated boundary layer in transonic flow over a flat plate. The analysis indicated a temporal/absolute instability in the transonic range, which becomes progressively worse as the wavenumber k of the eigenmode becomes unbounded. Ryzhov (1993) regarded the breakdown as a consequence of the problem being ill-posed. His analysis indicated strongly that the contribution of the normal pressure gradient in the main deck, omitted as higher-order effect from most triple-deck theories, can provide a remedy and exerts the necessary stabilizing influence. For the neutral stability, Ryzhov found that inclusion of the term $\partial p/\partial y$ in the formulation would cause a second pair of critical k and frequency to emerge, at least for an $M_\infty < 1$ in range (ii). Whereas the scaling analysis and conclusion in this aspect of the extension remain to be more firmly established, the importance of $\partial p/\partial y$ ought not to be denied, since it is an important ingredient of the Orr–Sommerfeld equations. However, the study of compressible boundary layer instability cannot be complete without considering the 'inflectional instability' modes as well as the additional spanwise wave component (Smith 1982).

Recent developments of the topics outlined were made by Smith *et al.* (1998a,b). Analysis of 2D and 3D perturbations in axisymmetric boundary layers at transonic speeds of the main stream was performed by Savenkov (1999).

Further studies using the asymptotic technique are to clarify these issues by developing the appropriate high Reynolds number models within the framework of viscous–inviscid interaction theory and validating these models against finite Reynolds number interactive flow and full Navier–Stokes simulation. Also, theoretical models for the acoustic noise generation as a result of flow instability interacting with spatial inhomogeneity in the boundary layer are still awaiting their development.

4

Numerical Simulation of Transonic Flow Patterns

In this chapter we discuss mainly transonic flow with a local supersonic region and focus on the analysis of formation and propagation of shock waves. The knowledge of the structure of the flow field may contribute to further rigorous mathematical studies of transonic flow problems. Also, the understanding of the flow structure is of practical importance as it may facilitate the development of advanced concepts for the airfoil/wing design and flow control aimed at higher aerodynamic performance and stability of flight.

4.1 Formation of Shock Waves in the Local Supersonic Region

4.1.1 A shock wave terminating the supersonic region

The first attempt of a numerical study of the shock wave which terminates the local supersonic region was presumably made by Nocilla (1957/58a,b, 1986), who used the hodograph technique and approximate solutions of Tomotika and Tamada. However, Nocilla assumed the existence of an expansion characteristic, which met the origin of the shock wave inside the supersonic region, in order to match the solutions obtained in two subdomains in the vicinity of the origin.

Murman (1974) used early versions of conservative and nonconservative finite-difference schemes in order to solve the von Kármán equation and simulate transonic flow over the airfoil constituted by two parabolic arcs of thickness/length ratio 0.06 at the freestream Mach number $M = 0.851$. He considered symmetric flow and employed a fine mesh to resolve the flow details in the end of the supersonic region. The computational domain was a rectangle $-0.2875 < x < 1.1125$, $0 < y < 3.0$, and the fine mesh cell sizes were $\Delta x = 0.001$, $\Delta y = 0.0025$. Far-field asymptotic boundary conditions were prescribed on the left, right and upper sides of the domain.

The flow field obtained by Murman is shown in Figure 4.1a, and the shock region is enlarged in Figure 4.1b. Above the shock origin P with the coordinates $x_P \approx 0.72$, $y_P \approx 0.14$, the deceleration is continuous. At point P, the compression waves

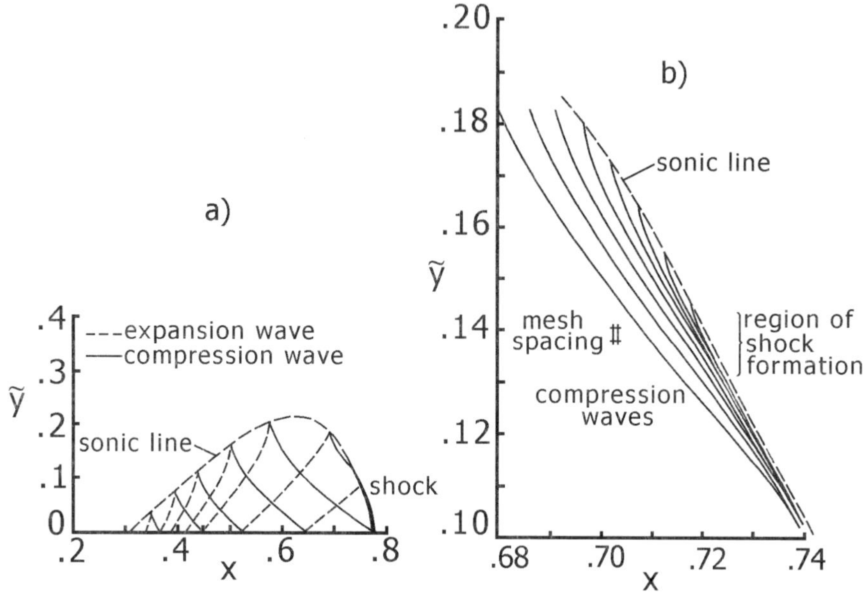

Figure 4.1 Formation of a shock wave in the end of the local supersonic region. Calculations of Murman (1974).

emanating from the sonic line coalesce to form a shock wave. The shock is enhanced as it approaches the airfoil.

The pattern shown in Figure 4.1b illustrates the second concept of shock wave formation pointed out in Figure 3.12. However, the location of the downstream endpoint B of the sonic line, which is expected to lie on the shock below the origin P, remained unclear because of the insufficient resolution of the employed mesh.

Numerical confirmations of this pattern with a shock wave terminating the local supersonic region were then presented in many papers. For example, Figure 4.2 demonstrates the results of numerical simulation of a transonic flow over the NACA-0012 airfoil at the freestream Mach number $M_\infty = 0.8$ and the angle of attack $\alpha = 0.125°$ obtained by Delanaye *et al.* (1997) with a finite-volume scheme using a quadratic reconstruction on adaptive unstructured meshes. The endpoint of the sonic line appears to lie on the shock wave slightly below its origin embedded in the supersonic region.

A well-known example of flow with a local supersonic region was introduced by Ni (1982). He considered two-dimensional, inviscid, steady flow through a channel $\{(x, y) : 0 < x < 3,\ 0 < y < 1\}$ with parallel walls and a circular-arc airfoil of height 0.1 placed on the lower wall:

$$y_1(x) = -1.2 + [1.69 - (1.5 - x)^2]^{1/2} \quad \text{at } 1.0 < x < 2.0. \tag{4.1}$$

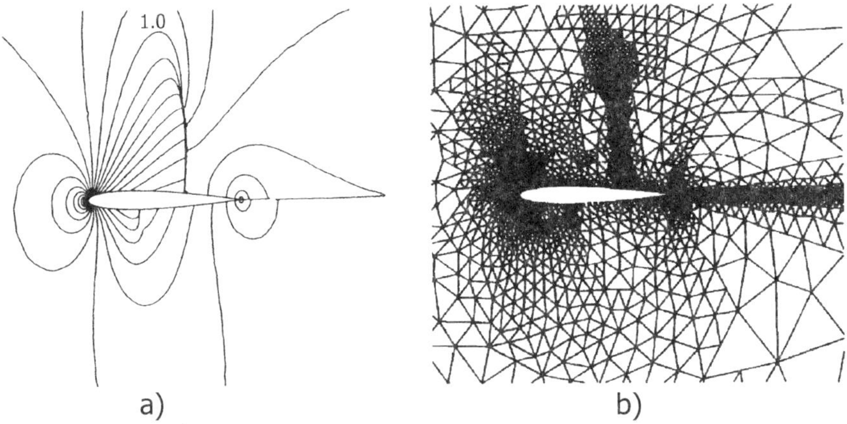

Figure 4.2 Flow over the NACA 0012 airfoil at $M_\infty = 0.8$ and the angle of attack $\alpha = 0.125°$. (a) Mach number isolines (0–1.4, $\Delta M = 0.05$), (b) the last adapted mesh. Computations of Delanaye *et al.* (1997). Copyright © 1997 AIAA. Reproduced with permission.

The boundary conditions for the system of continuity, Euler and energy equations are the entropy, enthalpy, the deflection angle $\theta = 0$ prescribed at the inlet of the channel, and the static pressure p_{out} fixed at the outlet. Specifically, the value of p_{out} was chosen according to the isentropic relation $p_0/p = [1 + (\gamma - 1)M^2/2]^{\gamma/(\gamma-1)}$ with $M'_{out} = 0.675$. The slip condition was utilized on the solid walls.

Apart from Ni (1982), the above problem was solved numerically by many authors (see, for example, Eidelman *et al.* 1983; Hirsh 1990, p. 395). Figure 4.3b demonstrates the Mach number isolines in transonic flow over the airfoil (4.1) in the channel obtained with an ENO2 scheme on a nonuniform mesh of 401×171 grid points and the cell sizes $\Delta x = \Delta y = 0.004$ in the supersonic region. Details of the method are described below in Section 4.3. The shock region is enlarged in Figure 4.3a. Again, one can observe the merging of compression waves near the rear part of the sonic line and the formation of a shock wave with the origin at $x \approx 1.70$, $y \approx 0.41$.

4.1.2 Shock wave due to the small local curvature of an airfoil

A number of numerical studies of transonic flow over different airfoils revealed peculiar wavy configurations of the boundary of the supersonic region at freestream velocities smaller than the on-design value. However, the nature of such configurations was not clearly accounted for.

Pfenninger *et al.* (1986) studied the issue of the shape of the airfoil which maximizes M_{design} and delays the laminar–turbulent transition. The authors suggested to avoid excessively high supersonic zones with unsatisfactory off-design characteristics by allowing only moderate Mach numbers (not greater than 1.10) in the central part of the supersonic region. Following this idea and using the Navier–Stokes equations,

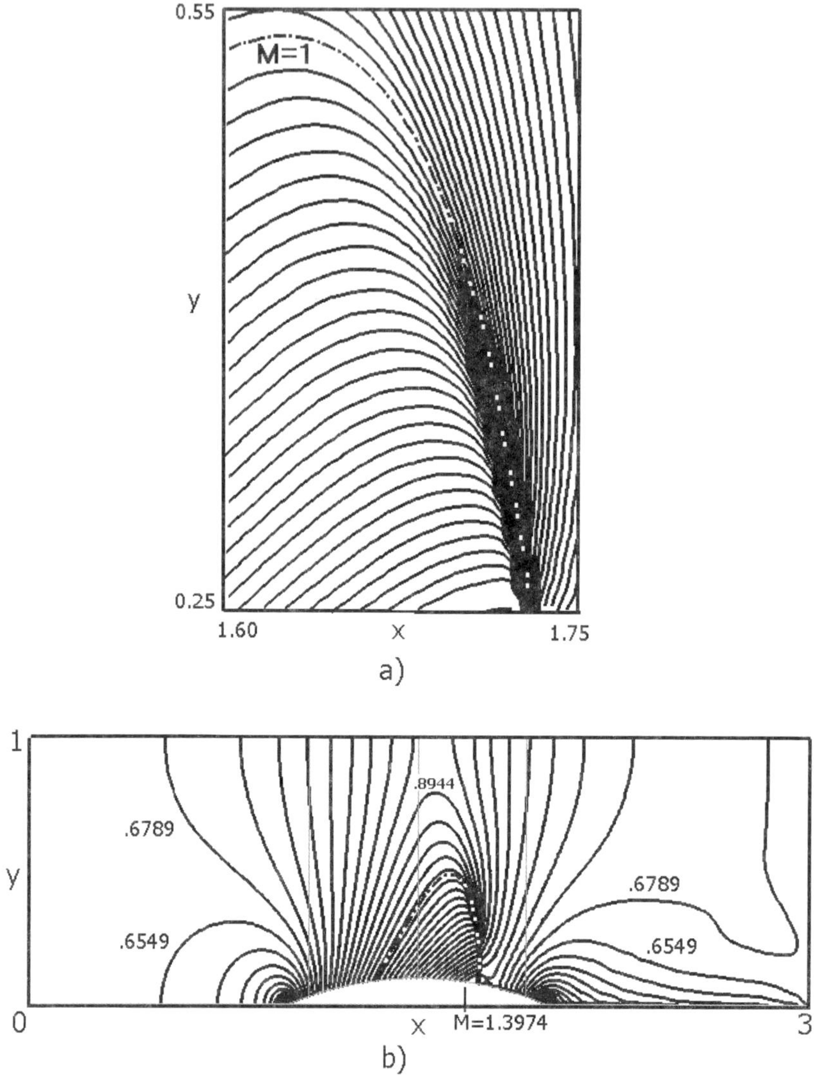

Figure 4.3 Mach number contours over the 10% circular-arc airfoil (4.1) in a channel for $M'_{\text{out}} = 0.675$.

Pfenninger *et al.* designed the shock-free airfoil X63T18S for $M_{\text{design}} = 0.7815$ and the lift coefficient $C_L = 0.524$. At on-design conditions, the flow over the airfoil was smooth and the supersonic region was convex, as shown in Figure 4.4a.

This airfoil revealed, however, a strong sensitivity of the flow pattern to small variations in the freestream Mach number, the angle of attack, and the shape of the

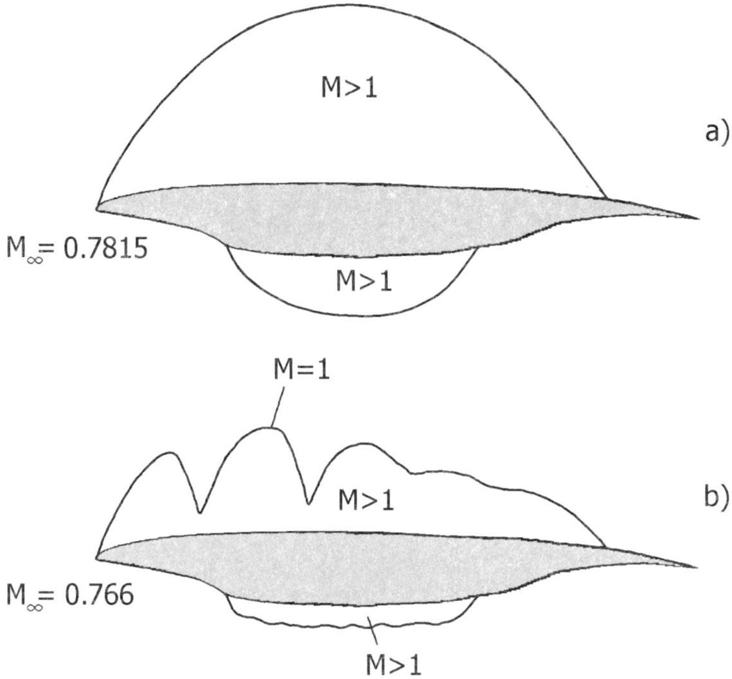

Figure 4.4 The shape of the supersonic region over the X63T18S airfoil: (a) $M_{\text{design}} = 0.7815$, (b) $M_\infty = 0.7660$. Computations of Pfenninger *et al.* (1986). Copyright © 1986 AIAA. Reproduced with permission.

airfoil. In particular, computations demonstrated wavy configurations of the supersonic region at the off-design freestream Mach number $M_\infty = 0.7660 < M_{\text{design}}$ and the same lift coefficient C_L (Figure 4.4b). By comparing Figure 4.4a and Figure 4.4b, one can notice that the small decrease, about 1.5%, of the freestream Mach number (that corresponds to a decrease less than 5 m s^{-1} in the freestream velocity) and the accompanying small decrease of the angle of attack to support C_L = const. yield a crucial change of the flow pattern. In fact, the computations demonstrated that a weak gust may influence the flow pattern drastically. Pfenninger *et al.* (1986) used the term 'double shock formation' to describe the obtained flow field, however, details of the flow field were not presented, and the locations of shock waves in the physical plane were not pointed out.

In order to clarify the flow structure, we recomputed a front part of the inviscid supersonic region shown in Figure 4.4b using the same location of the sonic line and the airfoil shape as in the paper of Pfenninger *et al.* (1986). A shock-fitting method of characteristics and a mesh of 220 × 100 was employed for the fragment of the supersonic region depicted in Plate 2. Across the shock waves we used the standard

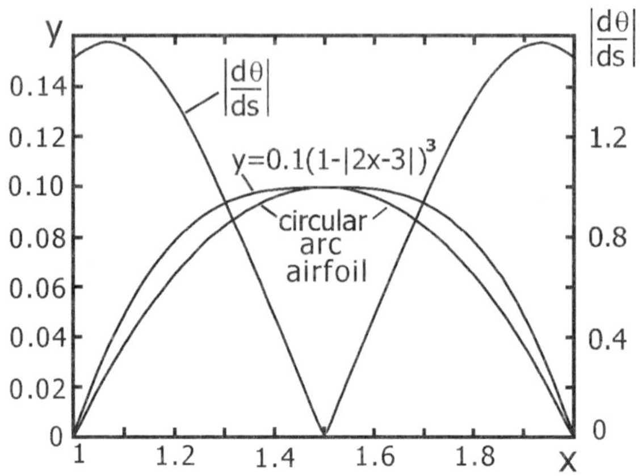

Figure 4.5 The shape of the airfoil (4.4) and its curvature at $n = 3$, $y_{\max} = 0.1$. The circular-arc configuration is also represented for comparison.

oblique shock relations

$$\tan \beta_2 = \tan \beta_1 \frac{1 + \frac{1}{2}(\gamma - 1)M_1^2 \sin^2 \beta_1}{\frac{1}{2}(\gamma + 1)M_1^2 \sin^2 \beta_1}, \tag{4.2}$$

$$\lambda_2^2 = \lambda_1^2 \cos^2 \beta_1 + \frac{(1 - ((\gamma - 1)/(\gamma + 1))\lambda_1^2 \cos^2 \beta_1)^2}{(1 - \cos^2 \beta_1)\lambda_1^2}, \tag{4.3}$$

which follow from the Rankine–Hugoniot conditions. In these relations, λ_1, β_1 and λ_2, β_2 are the velocity magnitude and the angle made by the velocity with the shock wave in front of the shock and behind it, respectively. The computations revealed a strong compression wave which appears near the airfoil arc $0.090 < x < 0.108$ (due to its small curvature) and forms a shock wave in the supersonic region. The shock wave propagates towards the sonic line, and then reflects back to the airfoil. Therefore, the descending portions of the boundary of the supersonic region in Figure 4.4b are to be interpreted as shock waves. In the regions of the second and subsequent reflections from the sonic line, the shocks are considerably smeared by the physical and numerical viscosity.

At the same time, at slightly higher freestream Mach numbers $0.770 < M_\infty < 0.7815 = M_{\text{design}}$, the descending portions of the supersonic region in the flow patterns obtained by Pfenninger *et al.* (1986) (e.g. Fig. 8(c) in their paper) are the sonic lines whose indentations are caused by the compression wave originated near the airfoil arc $0.090 < x < 0.108$ and multiply reflected from the sonic line and the airfoil.

As a simple example illustrating the direct formation of a shock wave due to the small local curvature of an airfoil, we considered transonic flow in the channel $\{(x, y) : 0 < x < 3,\ 0 < y < 1\}$ with the airfoil

$$y(x) = y_{\max} \cdot (1 - |2x - 3|^n) \quad \text{at } 1.0 < x < 2.0, \tag{4.4}$$

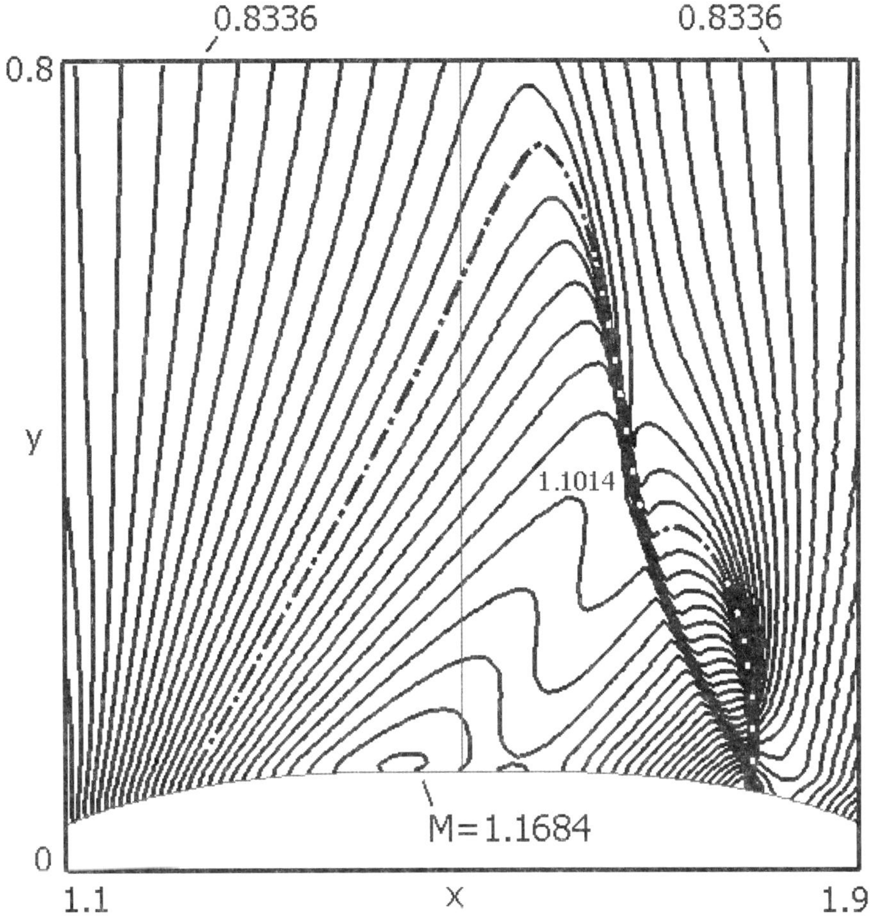

Figure 4.6 Mach number isolines over the airfoil (4.4) ($y_{max} = 0.1$, $n = 3$, $M'_{out} = 0.675$).

placed on the lower wall. The airfoil configuration for $n = 3$ and $y_{max} = 0.1$ is plotted in Figure 4.5. Obviously, its curvature vanishes at $x = 1.5$. In order to simulate the flow over the profile (4.4) we employed the same code based on the ENO2 scheme for solving the Euler equations as used above for computation of the flow over the profile (4.1). Plate 3 and Figure 4.6 show the steady-state solution obtained for $M'_{out} = 0.675$ on the mesh with cell sizes $\Delta x = \Delta y = 0.004$ in the supersonic region. It can be seen that there is a strong compression wave, which originates at the centre of the airfoil and yields a shock wave in agreement with Nikolsky–Taganov's theory discussed in Section 3.3. The shock propagates away from the airfoil and meets another shock formed near the top of the supersonic region due to the strong deceleration there.

Another simple example of shock formation due to the small local curvature of the airfoil is shown in Plate 7. In this case we introduced the straight segment into

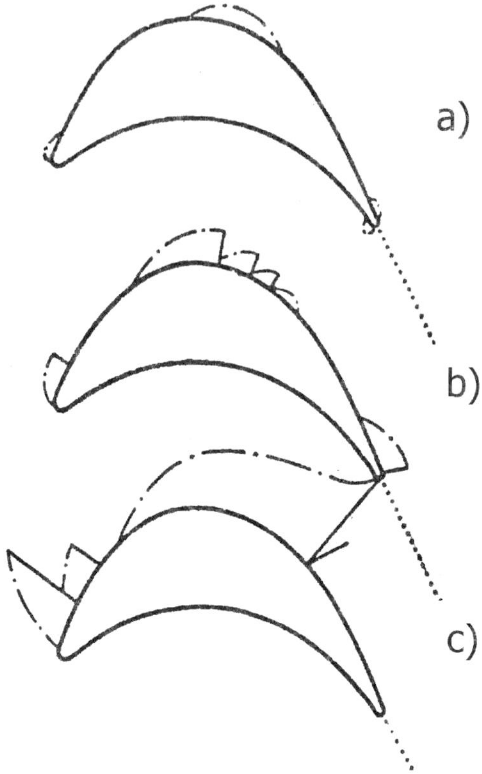

Figure 4.7 A scheme of flow past turbomachinery blades suggested by Stepanov (1962) for interpretation of his experimental study.

the circular-arc configuration (4.1) at $1.45 < x < 1.50$, while the other boundary conditions were the same as those for Ni's flow (Figure 4.3).

We notice that a flow pattern with a concave portion of the boundary of the local supersonic region developing at a Mach number smaller than the on-design one was revealed in the well-known paper by Murman and Cole (1971), which was not only pioneering in numerical methods for computation of transonic flow, but also gave a hint at the manner of the arising of strong compression waves and the structure of the local supersonic region at freestream Mach numbers smaller than the on-design value.

Sawtooth-like configurations of the supersonic region were observed in the 1950s in an experimental study of transonic flow between turbomachinery blades by Stepanov (1962, p. 222), who suggested a scheme of location of shock waves and made an attempt to analyse the flow pattern with increasing subsonic Mach number at the inlet (Figure 4.7). Though the obtained experimental patterns were not interpreted adequately, the work of Stepanov (1962) was probably the first one in which complex

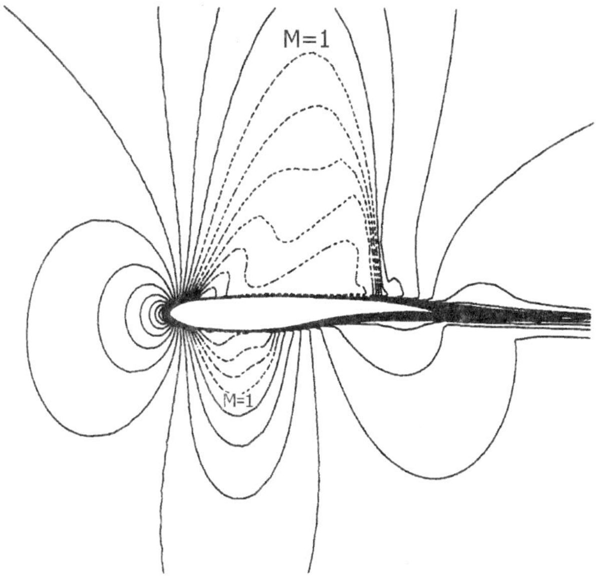

Figure 4.8 Mach number contours obtained in computations of Yamamoto and Inoue (1995) when solving a design problem. Reproduced by permission of the authors.

sawtooth configurations of the supersonic region at off-design flow conditions were observed.

Among recent papers, we mention the one by Yamamoto and Inoue (1995), who performed computations of the flow over airfoils with a genetic algorithm for aerodynamic shape optimization based on the Navier–Stokes equations. An example of solving a C_L/C_D maximization problem for transonic flow over the airfoil at $M_\infty = 0.8$ demonstrated a compression wave which was generated near the nose of the airfoil at $x \approx 0.17$ (Figure 4.8). However, the compression wave was not strong enough to distort the shape of the sonic line.

4.1.3 Shock waves due to a short perturbation of an airfoil shape

Johnson and Bachalo (1980) performed an experimental study of transonic flow past a symmetric NACA 64A010 airfoil for three angles of attack with the freestream Mach number fixed at 0.8.

> The cases studied included a weak-shock-wave–boundary-layer interaction, an interaction of medium strength with mild separation, and an interaction of sufficient strength to produce a shock-induced stall situation. Two nonintrusive optical techniques, laser velocimetry and holographic interferometry, were used to characterize the flows. The results include Mach number contours and flow angle distributions in the inviscid flow

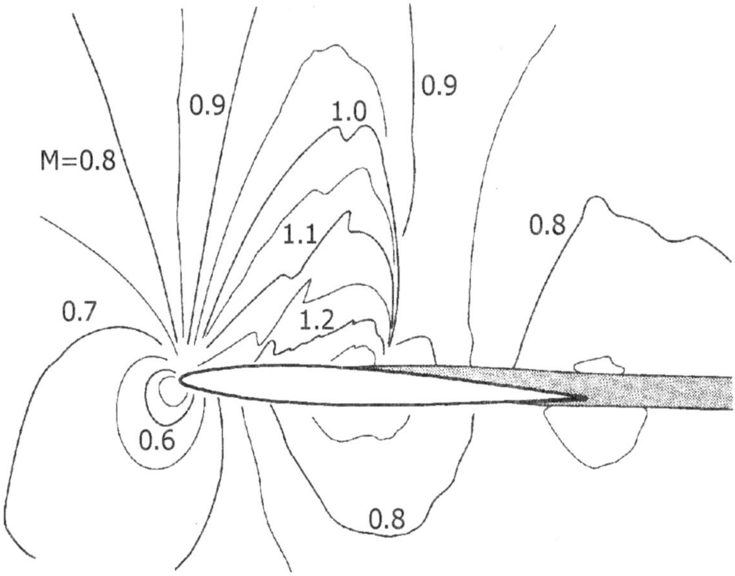

Figure 4.9 Mach number contours over the NACA 64A010 airfoil obtained from interferograms of Johnson and Bachalo (1980).

> regions, and turbulent flow properties, including the turbulent Reynolds stresses, of the upper surface viscous layers, and of the near-wake.
>
> (Johnson and Bachalo 1980)

Among the three angles of attack investigated, $\alpha = 0$, 3.5 and 6.2°, most pronounced compression waves arising near the airfoil were observed at $\alpha = 3.5$ (Figure 4.9). It can be seen that a compression wave generated near the airfoil at $x \approx 0.17$ is followed by an expansion wave, which in its turn is followed by another compression wave arisen at $x \approx 0.22$. Such a pattern obviously stems from a short deformation, a small bump, occurred on the airfoil specimen, whose concave parts generate two compression waves while the central convex part yields an expansion fan in agreement with our discussion of Section 3.3.

Below we consider the channel $\{(x, y) : 0 < x < 3,\ 0 < y < 1\}$, the same as in Figure 4.3, whereas the shape of the airfoil placed on the lower wall is different from (4.1). The baseline airfoil was designed to be shockless at the on-design Mach number $M_{\text{out}} = M_{\text{design}} = 0.675$ and the vanishing deflection angle θ at the inlet. The fictitious gas method mentioned in Section 3.1 was used to calculate the shockless configuration $y_0(x)$.

The shockless airfoil $y_0(x)$ is defined by arrays of about 300 Cartesian coordinates at $1.27 \leqslant x \leqslant 1.7415$, while outside this segment the shockless shape coincides with the circular arc (4.1). The maximum thickness of the shockless airfoil is $y_{0,\max} =$

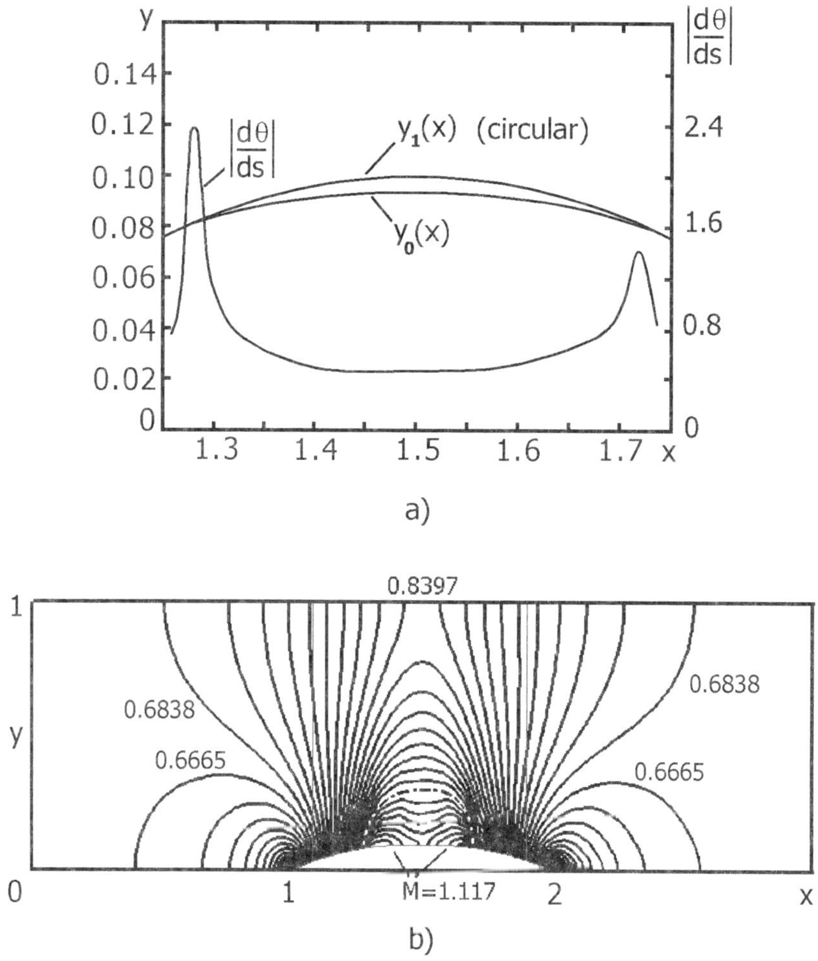

Figure 4.10 (a) The shapes of the shockless $y_0(x)$ and circular-arc $y_1(x)$ airfoils, and the curvature $|\mathrm{d}\theta/\mathrm{d}s|$ of $y_0(x)$. (b) Mach number isolines over the shockless airfoil.

0.0938 at $x = 1.4992$ (Figure 4.10a). The obtained shock-free flow field is shown in Figure 4.10b.

Kuz'min (1994a) explored the smooth short perturbation

$$\Delta y = f(x) = h \cdot \sin^4\left(\pi \frac{x - x_1}{x_2 - x_1}\right) \quad \text{at } x_1 \leqslant x \leqslant x_2, \tag{4.5}$$

superimposed on the shockless airfoil $y_0(x)$. For computation of the flow field, the domain under consideration was decomposed into subsonic and supersonic regions. In the subsonic region, the full-potential Equation (2.39) was written in nonuniform,

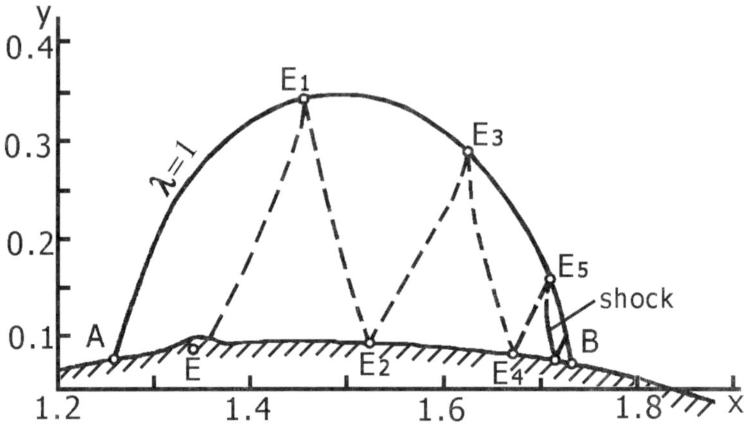

Figure 4.11 Formation of a microshock in the end of the supersonic region due to the perturbation (4.5) of height $h = 0.3 \times 10^{-5}$ imposed on the shockless airfoil at $1.30 < x < 1.38$. The height of the perturbation is exaggerated.

curvilinear, orthogonal coordinates (ξ, η) such that $\eta = 1$ coincided with the upper wall of the channel, while $\eta = 0$ was aligned with the lower wall and the shockless airfoil. Central differences were employed in the subsonic region, and the system of algebraic equations was solved with an implicit approximate-factorization scheme. In the supersonic region, the solution was obtained by a second-order shock-fitting method using a marching scheme for computation of the solution along left-going characteristics from the airfoil to the sonic line/shock step by step in the downstream direction. The solutions obtained in the subsonic and supersonic regions were smoothly and iteratively coupled on the sonic line, while the shock relations (4.2), (4.3) were employed across the shock waves. A characteristic mesh of about 300×80 was used in the computations.

For perturbation (4.5) with $x_1 = 1.30$, $x_2 = 1.38$, and $h = 0.3 \times 10^{-5}$, the obtained flow is shown in Figure 4.11, where E indicates the midpoint of the perturbation. There is a very weak shock (microshock) formed by the coalescence of right-running characteristics downstream of the third reflection of the perturbation from the sonic line (point E_5).

If the amplitude h of the perturbation is increased from $h = 0.3 \times 10^{-5}$ to 0.15×10^{-4}, then the shock origin jumps from E_5 to the vicinity of the first reflection of the flow perturbation from the sonic line. In this case the shock is formed by the coalescence of the right-running characteristics below the point E_1. In fact, the shock is engendered by the compression wave emanated from the tail of the perturbation (4.5) and reflected from the sonic line (Kuz'min 1994a). The shock incipient near E_1 propagates towards the airfoil and is reflected from it as another shock.

A further increase in the amplitude h entails a shift of the shock origin to the point E, an increase in the shock strength, and the appearance of a second shock formed by the compression wave emanated from the bow of perturbation (4.5). Since the amplitudes h under consideration are very small, the perturbation (4.5) resides actually in the boundary layer, where the inviscid equations are not valid. However, estimates based on the triple-deck theory have shown that the inclusion of viscous effects does not influence significantly the pattern of shock formation. This can be explained by the fact that the resulting streamline perturbation at the edge of the boundary layer over the short bump (4.5) is similar in shape and size to the bump itself.

For computation of flow over bumps (4.5) with greater h, we employed the above-mentioned shock-capturing ENO2 code, as the shock-fitting technique becomes very complex in case of multiple shocks arising in the flow.

Plate 8 shows the calculated flow pattern over the airfoil $y_0(x)$ with the superimposed perturbation (4.5) for $M'_\infty = 0.675$ and $h = 2 \times 10^{-4}$, $x_1 = 1.37$, $x_2 = 1.40$. Formation of two shocks due to the bow and the tail of the perturbation (4.5) agrees with the scheme proposed in Figure 3.15. Both shocks reach the sonic line, reflect back, and propagate towards the airfoil. However, the first shock becomes considerably weakened after meeting the expansion fan emanating from the top of the perturbation (4.5).

A similar pattern was obtained for $h = 3 \times 10^{-4}$, $x_1 = 1.35$, $x_2 = 1.40$. The longer bump obviously needs a larger height h to conserve the curvature and to provide the same order of indentations in the shape of the supersonic region.

Plate 9 demonstrates the influence of the smaller height $h = 1.5 \times 10^{-4}$ of the bump (4.5) under the fixed parameters $x_1 = 1.37$, $x_2 = 1.40$, $M'_{\text{out}} = 0.675$. One can estimate the amplitudes h which lead to the direct arising of shock waves over the perturbation (4.5) according to the asymptotic formula (3.24). From the profile $y_0(x)$ (Figure 4.10a), it can be inferred that $|d\theta/dx| \approx 0.48$ for $x_1 = 1.37 < x < x_2 = 1.40$. Since $x_2 - x_1 = 0.03$, the estimate (3.24) yields $h > 0.19 \times 10^{-4}$.

We notice that in the examples shown in Plates 8 and 9, the oblique shocks propagating in the supersonic region are smeared out after a few reflections from the sonic line and the airfoil because of the effect of numerical viscosity. Another feasible cause of the vanishing of the oblique shocks is their weakening by the expansion waves emanating from the convex airfoil (Figure 3.11b). In the end of the supersonic region, a terminating shock forms according to the scheme discussed in Section 4.1. In order to explore the influence of the mesh size and numerical viscosity, we recomputed the flow represented in Plate 9 on a mesh of 801×341. The obtained results are shown in Plate 4. The fine mesh proved to give a better resolution of details of the shock waves reflection from the sonic line, though there is some evidence that the solution is on the verge of instability.

The shockless flow depicted in Figure 4.10b turned out to be more sensitive to the perturbation (4.5) or to the modification of the airfoil $y_0(x)$ by a straight segment as compared with Ni's circular-arc airfoil. Introduction of a straight segment of the same length as in Plate 7 into the airfoil $y_0(x)$ resulted in two strong expansion fans

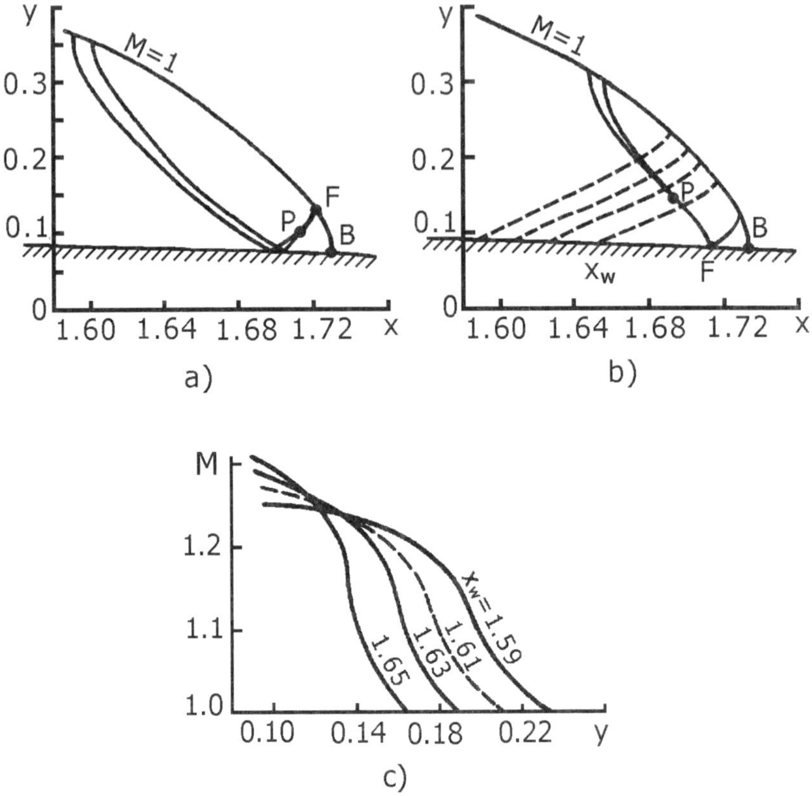

Figure 4.12 Formation of a shock wave over the profile (4.6) at $M'_{\text{out}} = 0.675$: (a) $p = 0.2$, (b) $p = 0.5$, (c) Mach number distributions along the characteristic lines shown in (b) by the dashed curves.

emanating from the beginning and end of the straight segment, and a shock wave originating by the central part of the segment. The shock wave propagated to the sonic line and, having reached it, produced a crucial change of the flow pattern.

Complex structures of the local supersonic region over special wavy airfoils with a few concave arcs have been revealed recently by Hafez and Guo (1999a,b). The structures agree well with the concepts presented above. We shall discuss the results of Hafez and Guo in Section 4.6 in connection with nonunique numerical solutions.

4.1.4 Accumulative effect

The accumulative phenomenon in the end of the local supersonic region can be observed, e.g. in numerical simulation of transonic flow over the airfoil

$$y(x) = p \cdot y_1(x) + (1 - p) \cdot y_0(x), \tag{4.6}$$

where the functions $y_1(x)$ and $y_0(x)$ correspond to the circular and shockless shapes given by (4.1) and pointed out in Figure 4.10a, respectively. At small values of the parameter p, the airfoil (4.6) is close to the shockless one, while at $p = 1$ it coincides with the circular arc. The right-hand side of (4.6) can be represented in the form

$$y(x) = y_0(x) + p \cdot [y_1(x) - y_0(x)],$$

hence, the profile (4.6) can be considered as a result of the superposition of the extended perturbation $p \cdot [y_1(x) - y_0(x)]$ on the shockless airfoil $y_0(x)$.

Using a shock-fitting algorithm, Kuz'min (1994b) calculated the flow over the airfoil (4.6) at small p and verified the importance of the accumulative effect at small deviations of the airfoil shape from the shockless configuration. Figure 4.12 demonstrates the coalescence of compression waves into a shock wave in the end of the supersonic region. When $p = 0.5$ (Figure 4.12b) the compression waves emanating from the sonic line are not as strong as in the case of the circular-arc airfoil (Figure 4.3), and the shock origin is shifted from the sonic line to the airfoil. Figure 4.12c shows Mach number distributions along left-going characteristic lines. It can be seen that the steepening flow gradients in the narrowing part of the supersonic region eventually lead to the formation of a shock wave. The originating shock is reflected from the airfoil as another shock which propagates towards the sonic line.

When $p = 0.2$ (Figure 4.12a) the compression waves emanating from the sonic line are too weak for direct shock formation before meeting the airfoil surface. However, having been reflected from the surface, the compression waves become stronger due to the accumulative effect, and this finally results in the formation of a weak shock wave.

A refined numerical analysis of the accumulative effect in transonic decelerating flow can be carried out on the basis of the linearized full-potential Equation (2.56) written in the streamline coordinates,

$$\left[\frac{(\rho\lambda)_\lambda}{\rho} u_\varphi\right]_\varphi + (\rho^2 u_\psi)_\psi = 0, \tag{4.7}$$

where $u(\varphi, \psi)$ is the perturbation of the given potential φ and ψ is the stream function.

An example of smooth decelerating transonic flow is given by the solution of the von Kármán equation

$$\lambda(\varphi, \psi) = 1 - 0.12[\varphi - 0.06(\gamma + 1)\psi^2].$$

By shifting the origin to the point $\varphi = 1.054$, $\psi = -4.0$ and retaining the notation for the velocity potential and the stream function, we obtain the velocity field

$$\lambda(\varphi, \psi) = 1 - 0.12(\varphi - 0.144\psi^2 + 1.152\psi - 1.250). \tag{4.8}$$

Let us choose the streamlines $\psi = 0$ and $\psi = \psi_2 = 1.294\,553$ to coincide with the lower and upper walls of the channel

$$G = \{(\varphi, \psi) : 0 < \varphi < \varphi_2,\ 0 < \psi < \psi_2\},$$

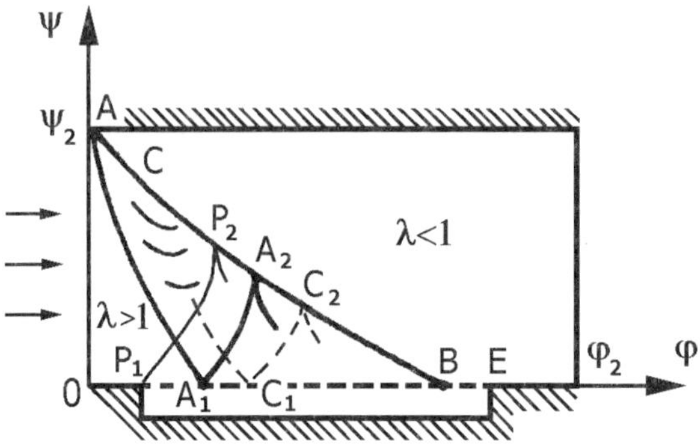

Figure 4.13 Characteristic points and δ-wave in the decelerating transonic flow.

where $\varphi_2 = 2.5$, and λ varies from 1.15 at the origin to 0.7 at the upper right corner. Both u and u_φ are to be prescribed at the inlet $\varphi = 0$, where the flow velocity is supersonic. Kuz'min (1992a,b) treated two versions of the inlet conditions:

$$u = 0, \quad u_\varphi = 0.1 \qquad \text{at } \varphi = 0, \tag{4.9}$$

or

$$u = 0, \quad u_\varphi = 0.1 \sin^2(\tfrac{1}{2}\pi(1 - \psi/\psi_2)) \quad \text{at } \varphi = 0, \tag{4.10}$$

where the derivative u_φ is linked with the velocity perturbation λ' by the relation (2.61): $\lambda' = \lambda u_\varphi$. The condition $u_\psi = 0$ at the outlet, where the flow is subsonic, implies that the angle of the velocity vector inclination to the φ-axis vanishes.

On the lower wall, we prescribe the oblique derivative condition

$$u_\psi - R(\varphi)u_\varphi = 0 \quad \text{at } \psi = 0, \tag{4.11}$$

while the condition $u_\psi = 0$ on the upper wall is the standard slip condition for inviscid flow. Relation (4.11) implies an active control governed by Darcy's law for the transpiration velocity across the perforated surface as discussed in Section 3.4. For the coefficient R we choose

$$R = 0.2 \quad \text{for } 0.11 \leqslant \bar{\varphi} \leqslant 0.53, \tag{4.12}$$

or

$$R = 1.0 \quad \text{for } 0.11 \leqslant \bar{\varphi} \leqslant 0.53, \tag{4.13}$$

where $\bar{\varphi} = \varphi/\varphi_2$ is the longitudinal coordinate measured in terms of the channel length. The value $R = 1.0$ in (4.13) corresponds to the 13.6% perforation of the wall and to the porosity coefficient $\sigma \approx 0.63$ in the Darcy condition written in the standard form (3.29). Outside the perforated portion of the lower wall $\psi = 0$ we have $R(\varphi) \equiv 0$.

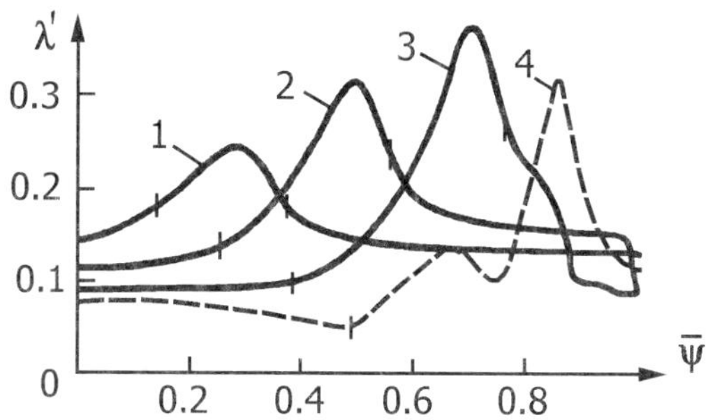

Figure 4.14 Velocity perturbation at cross-sections φ = const.: graphs 1 for $\bar{\varphi} = 0.06$, 2 for $\bar{\varphi} = 0.11$, 3 for $\bar{\varphi} = 0.17$, 4 for $\bar{\varphi} = 0.22$.

The above problem was solved numerically with a method of decomposition of the computational domain into subsonic and supersonic subdomains. In the former one, the solution was obtained on a uniform grid of 101×81 with the approximate factorization method, while in the supersonic subdomain the solution was found with a finite-difference marching scheme from the inlet in the downstream direction up to the point B. The step size in the φ-direction was $\Delta\bar{\varphi} = 5 \times 10^{-4}$, while that in the ψ-direction was $\Delta\psi = \Delta(\psi/\psi_2) = 1/80$. Both solutions were iteratively coupled on the sonic line. Verifications included computations on refined meshes with the cell size two and four times smaller than that of the basic mesh. Figure 4.13 shows points A_1 and A_2, which are determined by the characteristic lines emanated from point A of the sonic line, and the point P_1 which is the beginning of the perforated portion of the wall; $\bar{\varphi}_{A_1} = 0.256$.

Figure 4.14 demonstrates the obtained velocity perturbation λ' at cross-sections φ = const. in the case of boundary conditions (4.9), (4.13). The vertical dashes on the plots indicate the points of intersection with the sonic line and the bow characteristic AA_1. It can be seen that a perturbation wave emanates from the sonic line in the vicinity of its upstream endpoint and propagates in the downstream-downward direction.

Figure 4.15 shows the calculated velocity perturbation λ' on the sonic line. Curve 1 corresponds to the case (4.10), (4.12) and reveals a maximum C, indicating the wave which originates near point A of the sonic line. This is followed by a minimum P_2, arising due to the discontinuity of the velocity perturbation at the beginning of the perforation P_1 (this discontinuity can be removed by using a smooth coefficient $R(\varphi)$, which provides a continuous raise of the perforation.) There is then a maximum at A_2, which separates the influence of different physical phenomena along the arcs P_2A_2 and A_2C_2. The minimum C_2 shows the δ-wave after it has been reflected from the wall. The maximum K is caused by the accumulative effect.

Curve 2 in Figure 4.15 illustrates the solution in case of the boundary conditions (4.10), (4.13). Due to the increase in R from 0.2 to 1.0, the minimum at P_2 deepens, while that at A_2 disappears. The accumulative effect also disappears (being damped by the perforation). The decaying oscillations are due to the reflection of the strong

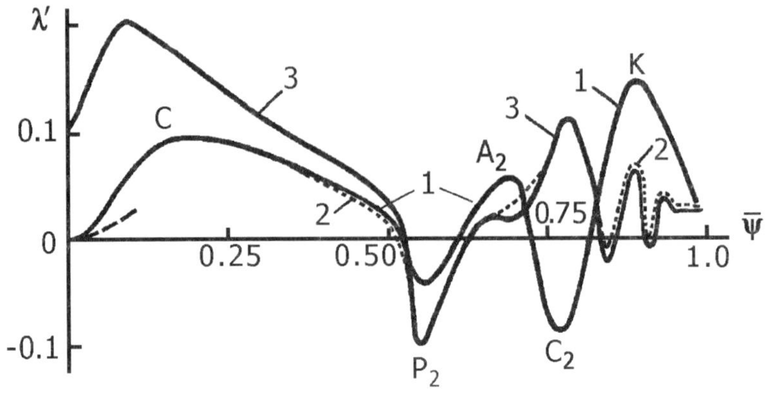

Figure 4.15 Velocity perturbation on the sonic line: graphs 1 for the conditions (4.10), (4.12), 2 for (4.10), (4.13), 3 for (4.9), (4.13).

discontinuity that arose at the point P_1, where the perforation starts, from the wall and the sonic line.

Curve 3 (the case (4.9), (4.13)) is qualitatively the same as 2. This is explained by the fact that, though a different condition at the inlet yields the solution with the discontinuous second-order derivatives at the point A and a different amplitude of the initial wave, the perturbation propagating in the supersonic region is damped by the perforated wall and so has hardly any effect on the subsequent behaviour of the solution.

Calculations were also performed for a wall with a short perforated segment: $R = 1.0$ for $0.28 < \bar{\varphi} < 0.53$. The computed values of λ' on the sonic line for this case have a strong minimum at the point C_2, due to the fact that the point C_1 on the lower wall lies upstream of the perforated zone, and so the δ-wave that arose on the sonic line is not absorbed in the vicinity of the point C_1 but, on the contrary, after being reflected from the solid wall at point C_1, is increased in accordance with the classical theory. The reflected perturbation is augmented by a velocity discontinuity that arose at the upstream edge of the perforation. However, subsequent oscillations ultimately decay owing to the damping effect of the perforation.

For all boundary conditions considered, the velocity perturbation λ' is continuous on the leading characteristic AA_1, and the behaviour of λ' near the beginning A of the sonic line is independent of any change in the conditions at the channel outlet. This is accounted for by the fact that the conditions in the outlet section can affect the values of the solution near the point A only by means of perturbations which propagate upstream in the subsonic region; yet the cross-sectional area of this subsonic region and the velocity of the upstream propagation of perturbations tend to zero as the section approaches the point A. For the same reason, the behaviour of the curves near the point A is in fact independent of the kind of perforation or the length of the perforated part of the wall.

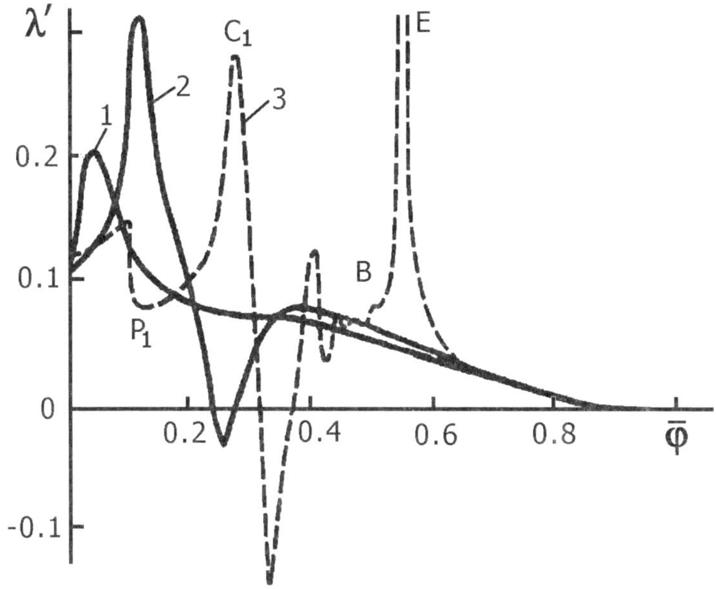

Figure 4.16 Velocity perturbation at cross-sections ψ = const.: graphs 1 for $\bar{\psi} = 0.125$, 2 for $\bar{\psi} = 0.5$, 3 for $\bar{\psi} = 1.0$.

Velocity perturbations at longitudinal sections ψ = const. of the channel for the boundary conditions (4.9), (4.13) are shown in Figure 4.16, where $\bar{\psi} = \psi/\psi_2$. The graphs confirm that λ' is continuous on the leading characteristic AA_1.

Similar results are obtained for the vanishing perturbation of the velocity magnitude at the inlet section ($u_\varphi \equiv 0$) and nonzero perturbation of the velocity inclination angle. The δ-wave here turns out to have a different shape, whereas the perforation prevents the accumulative effect and ensures a stable flow as before. Thus, transonic decelerating flow (4.8) is linearly stable with respect to the types of perturbation considered if the perforation condition (4.11) with $R = 1$ holds on a sufficiently long part of the wall containing the point B. On the other hand, if the perforated segment is not sufficiently long or the coefficient R is small, then the solution develops a strong singularity at point B in accordance with the Busemann–Guderley concept discussed in Section 3.3.

4.2 Interaction of a Shock Wave with the Sonic Line

4.2.1 Self-similar solutions describing patterns of interaction

A technique of self-similar solutions was developed by Frankl (1945, 1955), Guderley (1957) and Landau and Lifshits (1987) for the analysis of a number of transonic flow

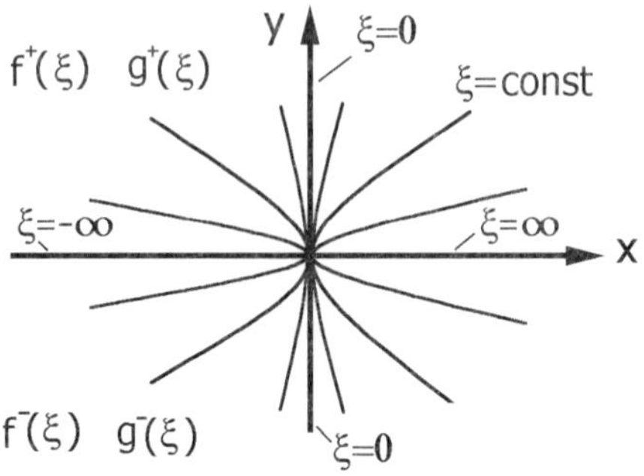

Figure 4.17 Curves ξ = const. in the (x, y)-plane.

problems, including the interaction of a weak singularity with the sonic line. Kuz'min (1996) employed this technique for the study of the patterns of shock-wave–sonic-line interaction and predicted patterns involving either a cusp or inflection point of the sonic line. Subsequent numerical studies have confirmed the occurrence of the patterns.

Suppose that the flow velocity λ and the deflection angle θ depend on the coordinates (x, y) as follows,

$$\lambda(x, y) = 1 + \frac{f^{\pm}(\xi) \cdot |y|^{2n-2}}{\gamma + 1}, \qquad \theta(x, y) = \frac{g^{\pm}(\xi) \cdot |y|^{3n-3}}{\gamma + 1}, \tag{4.14}$$

where $n > 1$ is a parameter, $\xi = x/|y|^n$ is the self-similar variable, the upper (lower) superscript corresponds to the half-plane $y > 0$ $(y < 0)$ of the physical plane (x, y), respectively (Figure 4.17). By substituting expressions (4.14) into the system (2.9) equivalent to the von Kármán equation, one finds that the function $f^{\pm}(\xi)$ satisfies the ordinary differential equation

$$(f - n^2\xi^2)\frac{\mathrm{d}^2 f}{\mathrm{d}\xi^2} + \left(\frac{\mathrm{d}f}{\mathrm{d}\xi}\right)^2 + n(3n-5)\xi\frac{\mathrm{d}f}{\mathrm{d}\xi} - 2(2n-3)(n-1)f = 0, \tag{4.15}$$

while the function $g^{\pm}(\xi)$, which governs the deflection angle θ, is expressed in terms of $f^{\pm}(\xi)$ through the relation

$$g^{\pm}(\xi) = \frac{\pm 1}{3(n-1)}\left[2n(n-1)\xi f^{\pm} + (f^{\pm} - n^2\xi^2)\frac{\mathrm{d}f^{\pm}}{\mathrm{d}\xi}\right]. \tag{4.16}$$

It can be seen from (4.14) that positive values of $f^{\pm}$ correspond to the supersonic velocities, while negative $f^{\pm}$ refer to the subsonic ones. Therefore, in the case of flow of the form (4.14), the location of the sonic line is determined by the ξ-coordinate defined as a root of the equation $f^{\pm}(\xi) = 0$.

The vicinity of the x-axis, that is a narrow infinite strip enclosing the x-axis in the plane (x, y), corresponds to a vicinity of the infinitely remote point $\xi = \pm\infty$

(Figure 4.17). The behaviour of the functions $f^{\pm}(\xi)$ at $\xi \to \pm\infty$ is described by the asymptotic expansion

$$f^{\pm}(\xi) = C_1|\xi|^{2-2/n} \pm C_2|\xi|^{2-3/n} + o(|\xi|^{2-3/n}), \qquad (4.17)$$

which follows from the condition of the continuity of λ and θ on the x-axis and the assumed dependence of $\lambda(x, 0)$ and $\theta(x, 0)$ on the powers of x.

If the configuration of a shock wave located in the lower half-plane $y < 0$ is described by the equation $\xi = \xi_s$, that is $x = \xi_s(-y)^n$, then the limits f_1^- and f_2^- of the function f^- as $\xi \to \xi_s \pm 0$ satisfy the relations

$$f_1^- + f_2^- = 2n^2\xi_s^2, \qquad (4.18)$$

$$\left(\frac{\mathrm{d}f^-}{\mathrm{d}\xi}\right)_1 + \left(\frac{\mathrm{d}f^-}{\mathrm{d}\xi}\right)_2 = -10n(n-1)\xi_s, \qquad (4.19)$$

which are a consequence of the Rankine–Hugoniot jump conditions. To analyse the behaviour of the integral curves of Equation (4.15) near the parabola $f = n^2\xi^2$, we notice that this equation can be reduced to a first-order differential equation in a (F, Ψ)-plane by the change of variables

$$f(\xi) = \xi^2 F(\eta), \qquad \frac{\mathrm{d}F}{\mathrm{d}\eta} = \Psi(\eta), \qquad \eta = \log|\xi|$$

(see Lifshitz 1965). On the line $F = n^2$, this first-order equation has two singular points: $F = n^2$, $\Psi = -n(n+1)$ and $F = n^2$, $\Psi = -6n(n-1)$. Hence, due to the relation $\Psi \equiv \mathrm{d}F/\mathrm{d}\eta = \xi^{-1}\,\mathrm{d}f/\mathrm{d}\xi - 2\xi^{-2}f$, there are two directions (apart from the vertical direction $\xi = \text{const.}$), along which the integral curves of Equation (4.15) can approach the singular parabola $f = n^2\xi^2$:

$$\frac{\mathrm{d}f}{\mathrm{d}\xi} = n(n-1)\xi, \qquad (4.20)$$

$$\frac{\mathrm{d}f}{\mathrm{d}\xi} = -2n(2n-3)\xi. \qquad (4.21)$$

For constructing the first example of transonic flow, we choose $n = 1.7$ and the initial conditions

$$f^+ = -0.5, \qquad \frac{\mathrm{d}f^+}{\mathrm{d}\xi} = -2n(2n-3)\xi \quad \text{at } \xi = 0.5, \qquad (4.22)$$

which determine the integral curve of Equation (4.15) emanating from the point $\xi = 0.5$ of the singular parabola $f = n^2\xi^2$ as the separatrix of the saddle point (4.21). A numerical solution of Problem (4.15), (4.17)–(4.19), (4.22) was found as follows. Equation (4.15) with respect to $f^-(\xi)$ was integrated numerically under initial data (4.22) from $\xi = 0.5$ to a sufficiently large $\xi_\infty > 0$. Then the asymptotic expansion (4.17) provided the initial data for $f^+(\xi)$ at $\xi = \xi_\infty$. After that we integrated Equation (4.15) in the negative direction of the ξ-axis from ξ_∞ to $-\xi_\infty$ and, thus, found $f^+(\xi)$

on the segment $[-\xi_\infty, \xi_\infty]$. Then, using again expansion (4.17), we obtained the initial data for $f^-(\xi)$ at $\xi = -\xi_\infty$, and integrated Equation (4.15) from $-\xi_\infty$ in the positive direction of the ξ-axis up to the intersection of the integral curve in the plane (f, ξ) with the parabola $f = n^2\xi^2$ at a point E. The computations demonstrated that the integral curve approaches this point in the direction of the node (4.20).

After that, we chose a coordinate ξ_r of the reflected shock from the interval $0.5 < \xi_r < 0.574$ and used the limit values of f^- and $df^-/d\xi$ at $\xi = \xi_r + 0$ in order to find f^- and $df^-/d\xi$ at $\xi = \xi_r - 0$ with relations (4.18), (4.19). Integration was then performed in the negative direction of the ξ-axis up to the intersection with the parabola $f = n^2\xi^2$ at a point F. Finally, the coordinate of the incident shock wave ξ_i was chosen in such a way that, in the integration from point E in the positive direction of the ξ-axis, the solution enters the same point F. To ensure the uniqueness of the solution, we imposed the condition of infinite differentiability of the integral curve $f^-(\xi)$ emanating from the node E in the direction (4.20) at $\xi \geqslant \xi_E$.

Figure 4.18a depicts the solutions of Problem (4.15), (4.17)–(4.19), (4.22) for different values of the coordinates of the incident and reflected shock waves. For instance, at $\xi_r = 0.55$ (curve 2) the coordinate of the incident shock wave is equal to $\xi_i = -1.0725$. The corresponding function $g^\pm(\xi)$ obtained with (4.16) is represented in Figure 4.18b.

Figure 4.18c shows the flow pattern in the physical plane. It can be seen that point O, at which the shock is reflected from the sonic line, is an inflection point of the sonic line composed by the convex arc $x = -1.24y^{1.55}$, $y > 0$ and the concave one $x = 5.94|y|^{1.55}$, $y < 0$. On the characteristic line E_*O, which is located upstream of the incident shock wave, the first-order derivatives of the flow velocity are continuous by virtue of the employed continuation through the point E. On the characteristic line F_*O located downstream of the incident shock wave, the first-order derivatives of the velocity components become infinite, although the components u and v themselves are continuous.

As computations demonstrate, when the coordinate ξ_r of the reflected shock wave approaches 0.574, the corresponding coordinate ξ_i of the incident shock wave approaches -1.157, the strength of the incident shock tending to zero. At $\xi_r = 0.574$ we obtain a Gor'kov–Pitaevskii solution with a weak singularity reflected from the sonic line in the form of a shock wave (see Landau and Lifshits 1987).

On the other hand, at $\xi_r \to 0.5$ the strength of the reflected shock wave tends to zero. Moreover, the integral curve f^- emanated from the point G in the direction of the separatrix (4.21) enters this point from the left-hand side again in the direction of the separatrix. In the limit case $\xi = 0.5$ we obtain a solution with a shock wave which ends on the sonic line without formation of a reflected singularity. However, the flow velocity behind the incident shock is then subsonic, so that a second subsonic zone appears, and the original flow scheme is violated. The appearance of the second subsonic zone can be prevented by specifying the self-similarity parameter value $n = 1.52$ under the same initial conditions (4.22).

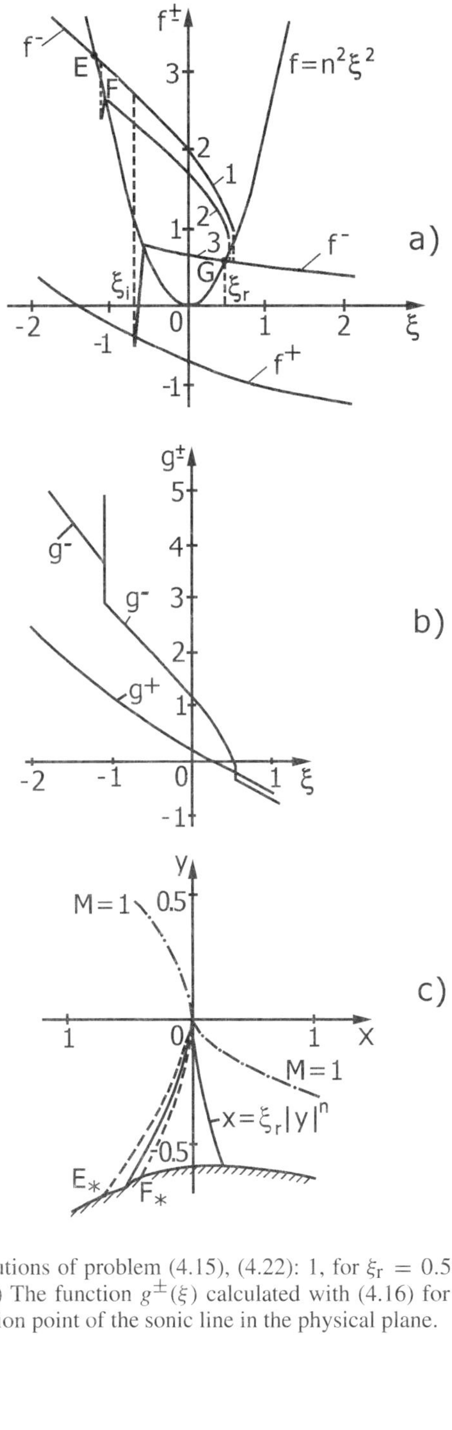

Figure 4.18 (a) Solutions of problem (4.15), (4.22): 1, for $\xi_r = 0.574$; 2, for $\xi_r = 0.550$; 3, for $\xi_r = 0.500$. (b) The function $g^{\pm}(\xi)$ calculated with (4.16) for $\xi_r = 0.550$. (c) Flow pattern with an inflection point of the sonic line in the physical plane.

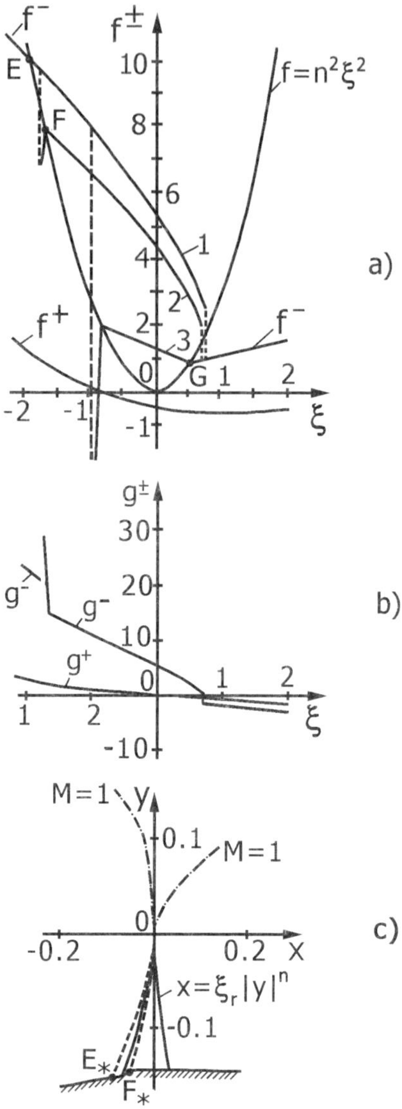

Figure 4.19 (a) Solutions of problem (4.15), (4.23): 1, for $\xi_r = 0.76$; 2, for $\xi_r = 0.70$; 3, for $\xi_r = 0.55$. (b) The function $g^{\pm}(\xi)$ calculated with (4.16) for $\xi_r = 0.70$. (c) Flow pattern with a cusp of the sonic line in the physical plane.

For constructing a second example of transonic flow, we considered $n = 1.7$ and the initial conditions

$$f^{+} = -0.5, \qquad \frac{\mathrm{d}f^{+}}{\mathrm{d}\xi} = -0.3 \quad \text{at } \xi = 0. \tag{4.23}$$

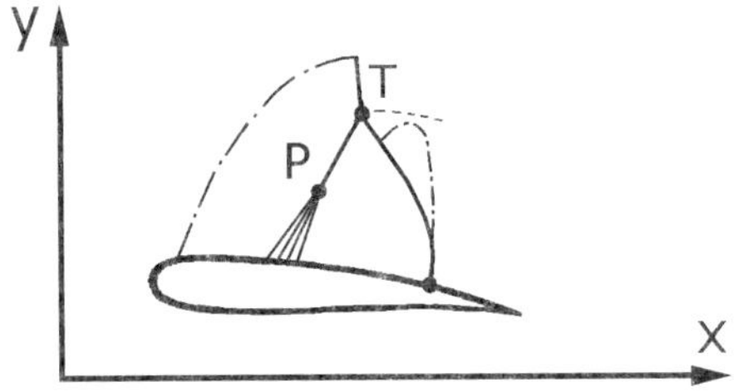

Figure 4.20 The qualitative pattern of the Mach reflection of a shock wave from the sonic line.

Problem (4.15), (4.17)–(4.19), (4.23) was solved as follows. With the initial data (4.23), Equation (4.15) was numerically integrated from 0 to a sufficiently large $\xi_\infty > 0$, and also from 0 to $-\xi_\infty$. Thus the solution f^+ was obtained on the segment $-\xi_\infty \leqslant \xi \leqslant \xi_\infty$. Then the initial data for the function f^- at $\xi = \pm\xi_\infty$ were determined using expansions (4.17), and f^- was calculated by solving Equation (4.15) from $\xi = -\xi_\infty$ in the positive direction of the ξ-axis up to the intersection with the parabola $f = n^2\xi^2$ at a point E, and also from $\xi = \xi_\infty$ in the negative direction up to the intersection with the parabola at a point G.

The solutions of Problem (4.15), (4.17)–(4.19), (4.23) at different values of the coordinates of the reflected shock wave ξ_r are shown in Figure 4.19a. The corresponding function $g^\pm(\xi)$ obtained with (4.16) for $\xi_r = 0.70$ is demonstrated in Figure 4.19b.

The flow pattern in the physical plane is represented in Figure 4.19c. The sonic line is described by the formulae $x = -0.80y^{1.7}$, $y > 0$ and $x = 8.03y^{1.7}$, $y > 0$. As $\xi_r \to 0.76$ the strength of the incident shock wave tends to zero, so that in the limit a Gor'kov–Pitaevskii-type solution is recovered, although unlike Figure 4.18 the sonic line has a cusp instead of the inflection point.

Depending on initial conditions, the strength of the shock reflected from the sonic line may be either greater or smaller than that of the incident shock. The above self similar solutions can be interpreted as the ones describing transonic flows over a special family of wall shapes with a concave portion of the wall contour, which leads to formation of the incident shock wave. Moreover, there must exist a point F_* of the contour, at which the tangent to the contour has a jump.

4.2.2 The Mach reflection of a shock wave from the sonic line

We present now a construction of a nonself-similar solution which describes shock–sonic-line interaction involving the Mach configuration of shock waves with a triple point (Kuz'min 1996). The scheme of interaction is shown in Figure 4.20. It is similar to a Mach reflection of a shock wave from the solid wall. We shall validate the scheme by solving numerically a special problem for the full-potential equation in the subsonic region and then a problem in the supersonic region. The approach is

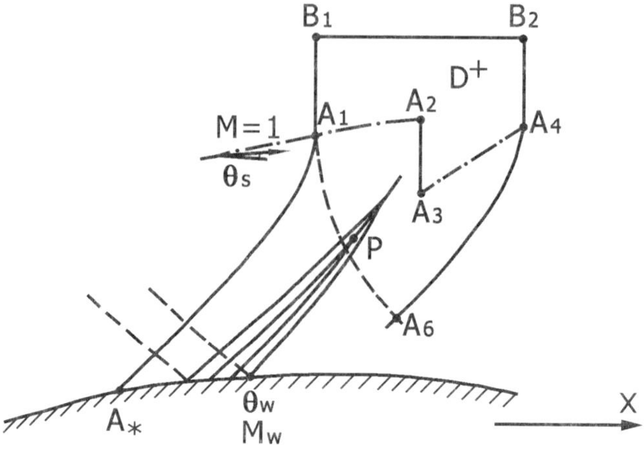

Figure 4.21 Domain D^+ in which the boundary-value problem (4.24), (4.25) for subsonic flow is formulated.

similar to the construction of particular solutions for the shock-free airfoil design discussed in Section 3.1.

We start by formulating a boundary-value problem in the subsonic region. Once it has been solved, it is easy to construct the flow in the supersonic region using the obtained deflection angle θ on the sonic lines, and also the angle θ and the Mach number M behind the shock waves.

In the (x, y)-plane we fix the points

$$A_1(0.40, 0.35), \quad A_2(0.42, 0.36), \quad A_3(0.42, 0.21),$$
$$A_4(0.44, 0.34), \quad B_1(0.40, 0.45), \quad B_2(0.44, 0.45)$$

(see Figure 4.21). Let D^+ be the hexagonal domain bounded by the segments A_1B_1, B_1B_2, B_2A_4, A_3A_4, A_2A_3 and the arc A_1A_2 of the parabola $y = 0.36 - 25(x - 0.42)^2$. We set the boundary conditions:

$$\left.\begin{array}{ll} \lambda = 1 & \text{on arc } A_1A_2 \text{ and segment } A_3A_4, \\ \lambda = 24.5(0.44 - x) + 24.75(x - 0.40) & \text{on segment } B_1B_2, \\ \lambda = (100/11)(0.42 - y) + 9(y - 0.34) & \text{on segment } B_2A_4, \\ \lambda = 1 - (0.36 - y)(y - 0.21)/1.625 & \text{on segment } A_2A_3, \\ \theta = 0.001\,76 + 0.0343y - 0.049y^2 & \text{on segment } A_1B_1, \end{array}\right\} \tag{4.24}$$

where λ is as usual the ratio of the flow velocity to the critical speed of sound.

Due to the prescribed $\lambda = 1$ on the segment A_3A_4 and the arc A_1A_2, these are arcs of the sonic line, while the flow is subsonic on the portions B_1B_2, B_2A_4, and A_2A_3 of the boundary.

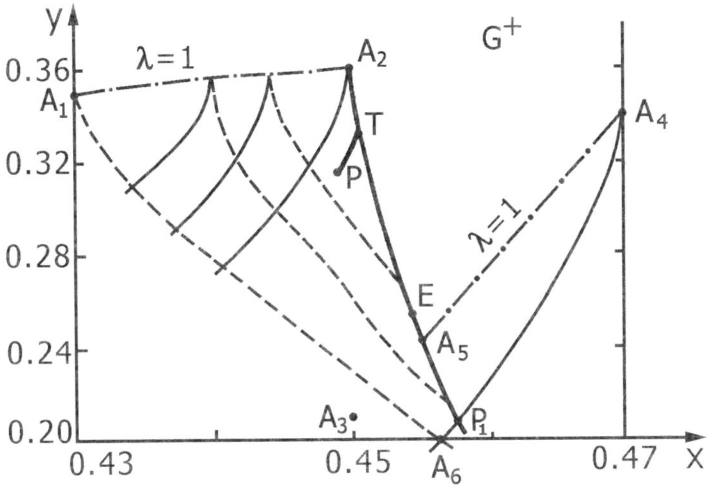

Figure 4.22 Details of the construction of supersonic flow upstream of the shock waves A_2T and TA_5.

The flow in D^+ is governed by the full-potential Equation (2.38), which can be represented in the form

$$[(\gamma+1)(\varphi_x^2-1)+(\gamma-1)\varphi_y^2]\varphi_{xx}+4\varphi_x\varphi_y\varphi_{xy} \\ +[(\gamma+1)(\varphi_y^2-1)+(\gamma-1)\varphi_x^2]\varphi_{yy}=0, \qquad (4.25)$$

with $\varphi_x = \lambda\cos\theta$, $\varphi_y = \lambda\sin\theta$. This equation was approximated using central differences for the derivatives of φ, and the solution of Problem (4.24), (4.25) was found with an approximate factorization scheme on an 81×51 Cartesian grid. The obtained solution $\varphi(x, y)$ yielded the velocity λ and the angle θ in the domain D^+ and over its boundary.

We now construct two curves TA_2 and TA_5 in D^+ near segment A_2A_3 and ignore the computed subsonic flow in the region between the curves and A_2A_3. The first curve is prescribed by the equation $x = 1.42 + 0.08(0.36 - y)^{1.5}$ for $0.334 \leqslant y \leqslant 0.36$, and the second one by a third-order polynomial which passes through the points T(0.420 33, 0.334), A_5(0.4252, 0.2438) and points (0.4209, 0.316), (0.4225, 0.280). We assign the curves TA_2 and TA_5 to shock waves separating the downstream subsonic flow from the upstream supersonic one (see Figure 4.22). The angle between the tangents to these shock waves at the point T is 9.9×10^{-3}.

The flow parameters downstream of the shock waves are defined by the solution of Problem (4.24), (4.25); these are shown by the dashed lines in Figure 4.23. Using the oblique shock relations (4.2), (4.3), one can then determine the parameters immediately ahead of the shock waves; these are represented in Figure 4.23 by the solid lines.

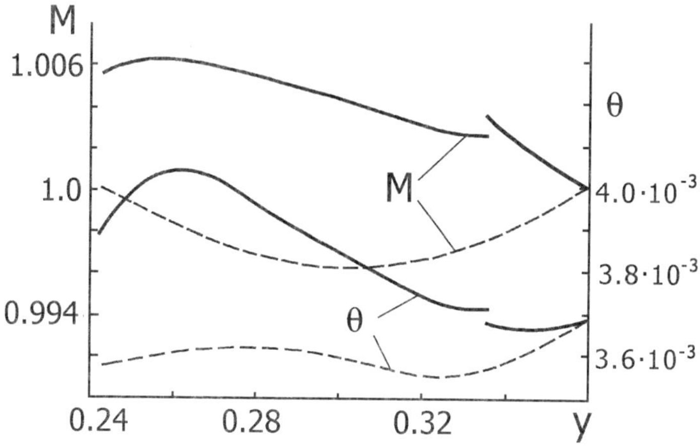

Figure 4.23 The Mach number and the angle θ behind (dashed lines) and ahead (solid lines) of the shock waves TA_2 and TA_5.

The supersonic flow in the region upstream of shock waves TA_2 and TA_5 and in the region underneath the sonic line A_4A_5 is calculated by the method of characteristics. In Figure 4.22 the fine lines show characteristics of the first family, the dashed lines those of the second one; the bold lines show the shock waves, and A_5P_1 is the continuation of the shock wave TA_5. The incident shock PT is constructed via several characteristic mesh steps from point T to the point P using the oblique shock relations. Also, by solving the Cauchy problem with the initial data on the line A_1A_2, one finds supersonic flow in a triangular domain bounded by A_1A_2 and the characteristics of the first and second family emanating from points A_2 and A_1, respectively. In this way supersonic flow is constructed in the entire domain above the characteristics A_1A_6 and A_4A_6 (see Figure 4.22).

A wall shape that corresponds to the obtained flow field can be designed, e.g. by assigning the following distribution of the Mach number along the characteristic line of the first family A_1A_* emanated from point A_1:

$$M(y) = 0.48(0.35 - y) - 6.656 + 7.656e^{(0.35-y)^4}, \qquad y \leqslant 0.35.$$

This makes it possible to construct supersonic flow in the domain below the characteristics A_1A_* and A_1A_6 by solving Goursat's problem with initial data on those characteristics. The streamline passing through the point (0.345,0.09) can be taken for the wall contour. The graph of the function $y = y_w(x)$ which describes this contour is shown in Figure 4.24, which also displays the coordinates x_*, x_2, x_5 and x_6 of points of the wall at which characteristics of the first family emanated from points A_1, A_2, A_5, A_6, respectively, end. In addition, the distributions of the Mach number M_w and the difference $\theta_w - \zeta(M_w)$ along the wall are plotted. It can be seen that this difference

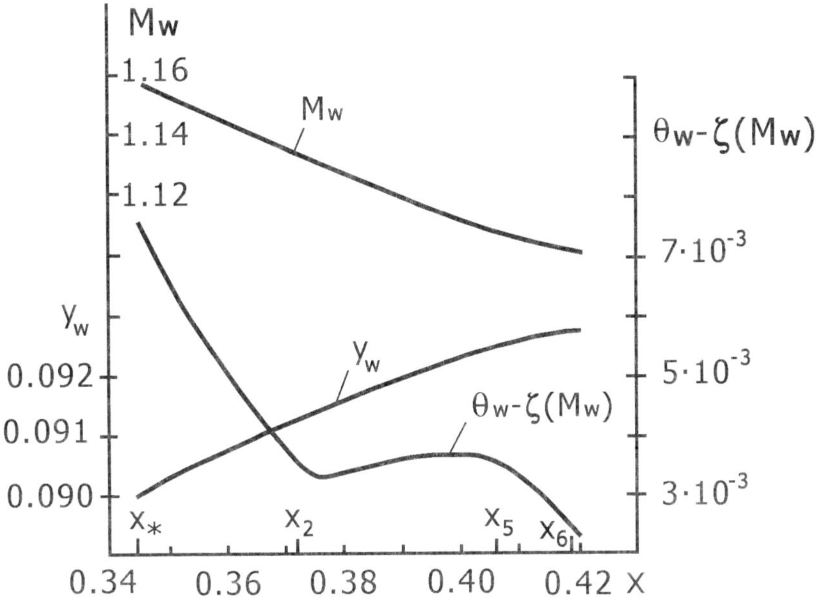

Figure 4.24 The contour $y_W(x)$ of the designed wall and distributions of the Mach number and the difference $\theta - \zeta(M)$ along it.

increases slightly in the interval $0.375 < x < 0.400$ in accordance with what was said in Section 3.3 concerning the conditions under which shock waves arise when the local curvature of the airfoil is small.

The weakening of the shock wave TEP_1 along the portion EP_1 and the acceleration of flow across the sonic line A_5A_4 are caused by the expansion waves emanating from the wall.

4.2.3 Discussion

Numerical simulation of transonic flow over a circular-arc airfoil with a straight segment introduced at $1.45 < x < 1.50$ demonstrated a pattern (Plate 7) which looks like the pattern of the shock-wave–sonic-line interaction with a cusp of the sonic line. The latter was obtained in the example of the self-similar solution depicted in Figure 4.19.

Computational results for the flow over a short bump superimposed on the shockless airfoil shown in Plates 4, 8 and 9 reinforce the example of the nonself-similar solution which describes shock–sonic-line interaction involving the Mach configuration of shock waves with a triple point. Also, flow over the polynomial airfoil (4.4) with $n = 3$ and vanishing curvature at the centre (Plate 3) demonstrates a pattern close to the Mach reflection of a shock wave from the sonic line. The specific shape of the sonic line is explained by the fact that the velocity losses across two oblique shocks below the triple point are smaller than those across the Mach stem above it.

For obtaining a numerical example of shock-wave–sonic-line interaction with the inflection point of the sonic line, we employed the profile (4.4) with $n = 2.5$ which has a greater curvature in the vicinity of the midpoint as compared with that for $n = 3$. Hence, compression waves emanating from the centre of the profile are not so strong. Still the curvature vanishes at the centre, consequently, a weak shock wave arises and interacts with the sonic line. The computed flow field is represented in Plate 10. It corresponds to the example of self-similar solution depicted in Figure 4.18.

The nonself-similar solution shown in Figure 4.22 demonstrates that transonic flow with a local supersonic region over a smooth boundary may contain a contact discontinuity (emanating from the triple point T, see Figure 4.20). This yields a conclusion which should be taken into consideration in rigorous mathematical studies of the solvability of transonic flow problems. The conclusion is that the choice of the Sobolev space $W^{1,2}(G)$ for the sought-after solution $\varphi(x, y)$ of the full-potential equation would be incompatible with the oblique shock relations (4.2), (4.3) if the triple point of shock waves occurred in the flow. Indeed, the oblique shock relations necessarily lead to a contact discontinuity of the flow velocity and, hence, discontinuity of the potential $\varphi(x, y)$ along the streamline emanating from the point T. Consequently, the potential cannot have generalized derivatives (in distinction to the flow with shock waves but without contact discontinuities). An analysis of the solvability of boundary-value problems in this case is to be done in a wider functional class.

In the isentropic flow, the difference $\theta - \zeta(M)$ is constant along characteristics of the first family as was discussed in Section 3.1. Let us show that, in the nearsonic approximation, this remains true if the characteristics intersect a weak shock behind which the velocity is supersonic, and the angle θ decreases. We estimate the variation of the quantity $\theta - \zeta(M)$ across a shock wave such as A_5P_1 in Figure 4.22. One can easily show that, in the nearsonic approximation, $\theta-\zeta(M) \approx \theta-\frac{2}{3}(\gamma+1)^{1/2}(\lambda-1)^{3/2}$ and therefore the variation of this quantity is equal to

$$\delta[\theta - \zeta(M)] \approx \delta\theta - (\gamma + 1)^{1/2}(\lambda - 1)^{1/2}\delta\lambda.$$

The last expression vanishes by virtue of $\delta\theta < 0$, $\delta\lambda < 0$ and the known equation of a shock polar (see, for example, Landau and Lifshits 1987). These considerations yield a qualitative conclusion regarding the location of the arc A_4A_5 of the second sonic line behind the triple point T: it must lie downstream of the point E which indicates a local maximum of $\theta - \zeta(M)$ on the shock wave A_2P_1. To prove this, we assume that the opposite situation takes place, i.e. that E lies downstream of A_5. Then the angle θ_+ would increase along some part of A_4A_5, since values of $\theta - \zeta(M)$ are constant along characteristics of the first family. That increase would conflict with the theorem about the decreasing angle θ_+ along the sonic line (see (3.17)). The numerical solution presented in Figure 4.24 confirms that conclusion perfectly.

We notice that in the vicinity of the meeting point A_2 of the shock wave and the sonic line, the nonself-similar numerical solution depicted in Figure 4.22 is close to

Frankl's example of a shock wave orthogonal to the sonic line at their meeting point (Frankl 1961). At the same time, we admit the possibility of a pattern in which the shock wave does not meet the sonic line in the direction orthogonal to the sonic line, but is formed by the coalescence of compression waves emanated from the sonic line near its top as sketched in Figure 3.13. This scheme is possibly in better agreement with the flow field shown in Plate 3, where the continuously increasing strength of compression waves preceding the compression wave emanated from the centre of the profile yields the strong deceleration of the flow near the top of the sonic line.

4.3 Off-design Conditions and Flow Control

4.3.1 Dependence of flow pattern on the Mach number

In this section, we continue studying inviscid transonic flow in the channel $\{(x, y) : 0 < x < 3,\ 0 < y < 1\}$ with an airfoil placed on the lower wall. We use the relaxation approach in order to obtain a steady-state solution. The governing system consists of the continuity, Euler and energy equations which follow from the system (3.87) if the viscosity and heat-conductivity terms are neglected:

$$\left.\begin{aligned} \frac{\partial \rho}{\partial t} + \frac{\partial \rho u}{\partial x} + \frac{\partial \rho v}{\partial y} &= 0,\\ \frac{\partial \rho u}{\partial t} + \frac{\partial (\rho u^2 + p)}{\partial x} + \frac{\partial \rho u v}{\partial y} &= 0,\\ \frac{\partial \rho v}{\partial t} + \frac{\partial \rho u v}{\partial x} + \frac{\partial (\rho v^2 + p)}{\partial y} &= 0,\\ \frac{\partial e}{\partial t} + \frac{\partial (e + p) u}{\partial x} + \frac{\partial (e + p) v}{\partial y} &= 0,\\ e = \rho(c_v T + \tfrac{1}{2}(u^2 + v^2)), \qquad c_v T &= \frac{p}{\rho(\gamma - 1)}, \end{aligned}\right\} \tag{4.26}$$

where p, ρ, u, v are the static pressure, density, and components of the velocity in x- and y-directions, respectively; e is the total energy per unit volume, T is the absolute temperature.

For subsonic velocities at the inlet and outlet of the channel, the necessity of prescribing three quantities at the inlet and one quantity at the outlet as boundary conditions for system (4.26) was justified with general reasons in a number of papers (see Fletcher 1991; Eidelman *et al.* 1983). In case of irrotational isentropic flow, system (4.26) of four differential equations reduces to the second-order full-potential equation. Consequently, instead of the three conditions, only one condition is required at the inlet. This perfectly agrees with the rigorous considerations of Section 3.4 of this book.

For the system (4.26), one typically specifies the entropy, enthalpy and $v = 0$ at the inlet. These quantities determine the stagnation pressure p_0. In order to provide the on-design Mach number at the outlet for the airfoil $y_0(x)$ pointed out in Section 4.1, one must ensure that $p_0/p_{\text{out}} = [1 + (\gamma - 1)M_{\text{out}}^2/2]^{\gamma/(\gamma-1)} = 1.356\,940\,4$, which corresponds to $M_{\text{out}} = 0.675$.

That is why we specify the *on-design* boundary conditions as follows:

$$\left.\begin{array}{ll} \text{at the outlet } x = 3: & \text{the pressure } p_{\text{out}} = 72\,218.0 \text{ N m}^{-2}, \\ \text{at the inlet } x = 0: & \text{the velocity component } v = 0,\ p/\rho^{\gamma} = 73\,340.042, \\ & \text{the total enthalpy} \\ \multicolumn{2}{c}{\dfrac{u^2+v^2}{2} + \dfrac{\gamma}{\gamma-1}\dfrac{p}{\rho} = 278\,850.0 \text{ m}^2\text{ s}^{-2}, \qquad \gamma = 1.4,} \end{array}\right\} \tag{4.27}$$

where the density ρ is measured in kg m^{-3}. The accepted values of the total enthalpy and the ratio p/ρ^{γ}, which determines the entropy, yield the stagnation pressure $p_0 = 97\,995.522$ N m^{-2}. Then using p_0 and p_{out}, one can verify with the relation $p_0/p_{\text{out}} = [1 + (\gamma - 1)M_{\text{out}}^2/2]^{\gamma/(\gamma-1)}$ that $M_{\text{out}} = 0.675$ at the outlet.

Isentropic relations also make it possible to find $\rho_{\text{out}} = 0.989\,047\,98$ kg m^{-3} and $(u_{\text{out}}^2 + v_{\text{out}}^2)^{1/2} = 215.815\,11$ m s^{-1}, the stagnation density $\rho_0 = 1.229\,995\,8$ kg m^{-3}, and the sound speed $a_{\text{out}} = 319.726\,08$ m s^{-1}.

The system (4.26) is to be endowed with initial conditions for all dependent variables and the usual slip condition on the walls of the channel.

The solution of the obtained initial–boundary-value problem was computed with a finite difference essentially nonoscillatory scheme ENO2 based on the characteristic flux difference splitting method by Yang (1990). The scheme maintains the second-order accuracy throughout even in regions of flow nonsmoothness and at local extrema. We used a nonuniform Cartesian mesh with constant cell sizes in the central region of the channel, where flow is supersonic, and increasing cell sizes in the subsonic region near the inlet, outlet, and the upper wall (Kotov and Kuz'min 2000). A typical mesh was 401×171 (with $\Delta x = \Delta y = 0.004$ in the central region), though some computations were performed on a mesh of 801×341 (with $\Delta x = \Delta y = 0.002$). The code achieved a steady-state solution in 5×10^3 to 12×10^4 time steps depending on the mesh and initial data used.

The accuracy of the method was verified by the direct recomputation of the shock-free flow obtained earlier with the fictitious gas method (Figure 4.10b) and also by computation of transonic flow with a shock wave over the circular 10% arc airfoil (4.1) and comparison with results obtained in other works (Ni 1982; Eidelman *et al.* 1983; Hirsh 1990, p. 395). The verification has shown the robustness of the developed code and its capability to resolve weak shock waves of strength less than 1%.

At *off-design* conditions, we prescribed p_{out} different from 72 218.0, while the conditions at the inlet were kept the same. The imposed p_{out} can be related to a quantity M'_{out} via the isentropic relation. However, at off-design conditions M'_{out} is different from the real Mach number at the outlet near the lower wall because of the entropy increase across the shock waves in the supersonic region and, consequently, the violation of the isentropic relations in the region downstream of the shock waves.

The numerical simulation demonstrated that shock waves appear in different parts of the supersonic region depending on the off-design Mach number at the outlet of the channel. At $M'_{\text{out}} = 0.673 < M_{\text{design}} = 0.675$, computations on the fine mesh of 801×341 revealed a pattern with $\mathcal{M}$-shaped sonic line (Plate 5). The small curvature of the shockless airfoil at $1.37 \leqslant x \leqslant 1.40$ yields a strong compression wave which propagates toward the sonic line and causes the first indentation in the shape of the sonic line. Then the compression wave, having been reflected from the sonic line, forms a shock wave with the origin at $x \approx 1.43$, $y \approx 0.19$. The shock propagates to the airfoil and is reflected from it as another shock, which reaches the sonic line and yields the second indentation.

Plate 6 shows a flow pattern obtained at a smaller Mach number $M'_{\text{out}} = 0.671$. The shock wave arisen near the sonic line at $x \approx 1.41$ is now strong enough to split the supersonic region into two parts. By comparing Plates 5 and 6, one can notice that the crucial change of the flow pattern occurs under a very small variation, only 0.3%, of the prescribed M'_{out}, which corresponds to the variation of less than 1 m s^{-1} in the flow velocity at the outlet. Both patterns persisted regardless of the changes in the initial conditions and employment of different meshes.

At the Mach number $M_{\text{out}} = 0.680 > M_{\text{design}} = 0.675$, computations showed an increase in the height of the supersonic region and formation of a shock by the coalescence of compression waves emanated from the rear upper part of the sonic line in agreement with the scheme discussed in Section 3.3.

Another airfoil we treated was the one obtained by a modification of the K-2 Grumman airfoil in the nose area. The modification was introduced to simplify the mesh-generation process. Plate 13 demonstrates the results of transonic flow computation over this airfoil placed on the lower wall of the same channel. The mesh of 401×171 with $\Delta x = \Delta y = 0.004$ in the supersonic region was used. Again we see that at Mach number $M'_{\text{out}} = 0.772$ smaller that the on-design value, a compression wave arises near the nose of the airfoil, and at $M'_{\text{out}} = 0.770$ it becomes strong enough to split the supersonic region into two separate bubbles.

Cosentino (1983) discussed an example of a low-speed high-lift general aviation airfoil redesigned to be shock-free at transonic conditions. The design procedure coupled an interactive boundary-layer calculation with the fictitious gas method for finding a solution of the full-potential equation. The on-design Mach number was $M_{\text{design}} = 0.7725$, and the lift coefficient $C_{\text{L}} = 0.50$. At $M_\infty = 0.76$ the computations revealed the splitting of the supersonic region into two bubbles, each of them being terminated with a weak shock wave.

A wavy configuration of the boundary of the supersonic region similar to those shown in Figure 4.4b was obtained by Kraiko and P'yankov (2000). They applied the fictitious gas method for the design of engine nacelles and obtained a shockless profile for $M_{\text{design}} = 0.64$ and the angle of attack $\alpha = 2°$ (Figure 4.25a). The decrease of the Mach number from 0.64 to 0.62 led to the appearance of indentations in the shape of the supersonic region and to the flow pattern represented in Figure 4.25b.

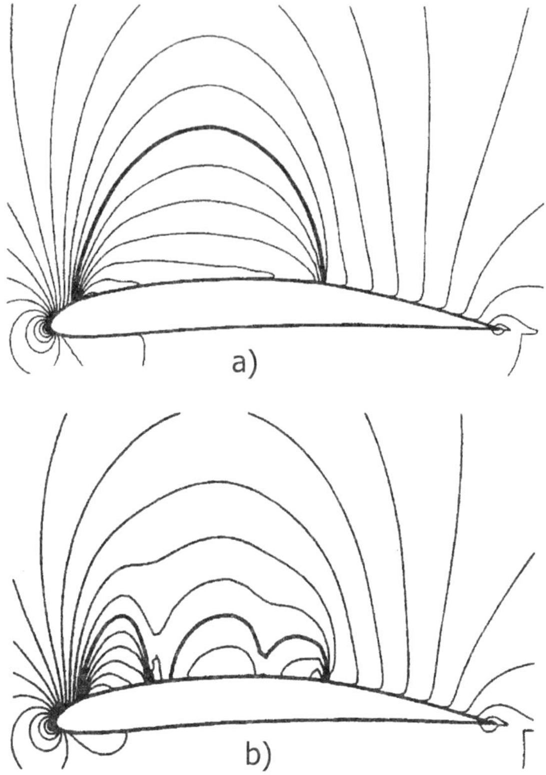

Figure 4.25 Flow over a nacelle profile. (a) $M_\infty = 0.64$, $\alpha = 2°$, (b) $M_\infty = 0.62$, $\alpha = 2°$. Computations of Kraiko and P'yankov (2000). Reproduced by permission of the authors.

We notice that due to the transonic similarity rule and the similarity parameter

$$K_{\text{sim}} = \frac{1 - M_\infty^2}{\delta^{2/3}}$$

occurring in the theory of transonic flow past thin airfoils (Cole and Cook 1986), the smaller thickness of the airfoil is expected to affect the flow in the same way as a lower freestream Mach number does.

For comparison of the effect of the smaller δ, we simulated the flow over the airfoil (4.6) at $p = -0.10$ and $p = -0.12$ under the constant $M'_{\text{out}} = 0.675$. The corresponding heights $y_{\max}$ of the airfoil were 0.093 17 and 0.093 05; we recall that the height of the shockless airfoil $y_0(x)$ (at $p = 0$) was 0.093 79.

For $p = -0.10$ computations demonstrated that the reduced curvature of the airfoil (4.6) in the interval $1.35 < x < 1.50$ yields a strong compression wave which propagates from the airfoil towards the sonic line and, having been reflected from it, yields a valley in its shape (Plate 11). Further reflections of the compression wave from the airfoil and the sonic line create a wavy configuration of the shape of the supersonic region which is more pronounced as compared with that in Plate 5. However, after a few reflections, the compression wave finally decays due to the effect of expansion waves emanating from the airfoil. The shock wave terminating the local supersonic region appears near the rear upper part of the sonic line in accordance with the scheme discussed in Section 3.3. For $p = -0.12$ the curvature of the airfoil in the interval $1.35 < x < 1.50$ is further reduced; this results in the formation of a shock wave which splits the supersonic region into two separate ones (Plate 12), and the pattern is similar to that shown in Plate 6.

4.3.2 Flow control by airfoil shape modifications

The need for drag reduction as well as for the stability and safety of flight at transonic speeds has led to concepts of flow control aimed at the weakening of shock waves, the reduction of the viscous drag by maintaining laminar flow over the major portion of the wing surface, and the delay of the boundary-layer separation. The control can be performed in a passive or active way by means of devices which modify the shape of the airfoil or provide appropriate blowing/suction through a perforated surface of the wing. The first way refers to flow-control mechanisms not requiring any energy input, like the natural ventilation through porous sheets or a fixed bump. Conversely, the second way refers to tools, like the forced wall ventilation, the suction slots or flexible bumps, requiring some power consumption or the use of control systems. This technology can provide high aerodynamic efficiency over a range of Mach numbers and lift coefficients. Also, flow control makes it possible to maintain a fixed position of the shock wave on the wing, thus shifting the buffet onset at a higher freestream Mach number or lift.

A simple concept of flow control by local modifications of the airfoil shape, which can be obtained by mechanical devices, was explored by Tai *et al.* (1987). The authors conducted an experimental study aimed at improvement of the performance of supercritical airfoils at off-design conditions by reducing the wave drag. The research was based on the fact that the entropy change is the sole source of transonic wave drag, and the magnitude of the entropy change depends on the flow velocity ahead of the shock. Therefore, any attempt to reduce the wave drag should be directed to suppressing the local Mach number before the shock wave terminating the local supersonic region. Tai *et al.* used a 16% thick Korn–Garabedian airfoil at $M_\infty = 0.78$ and the lift coefficient $C_L = 0.4$ as the baseline airfoil. Modifications were introduced in the shock foot area by inserting a short bump of height 3×10^{-3} with the peak located at $x/c = 0.760$, where c is the airfoil length (see Figure 4.26). The front bump surface extended over the segment $0.725 \leqslant x/c \leqslant 0.760$. The curvature of the bump was tailored in such a way that the flow decelerated from a maximum local Mach number (about 1.35) to an ideal value of unity at the peak, and that an incipient separation condition was satisfied on the rear surface of the bump. Experiments were performed in the Mach number range from 0.40 to 0.85. For Mach numbers between 0.672 and 0.749, the shock wave was broken into two waves, one before the bump and the other at the

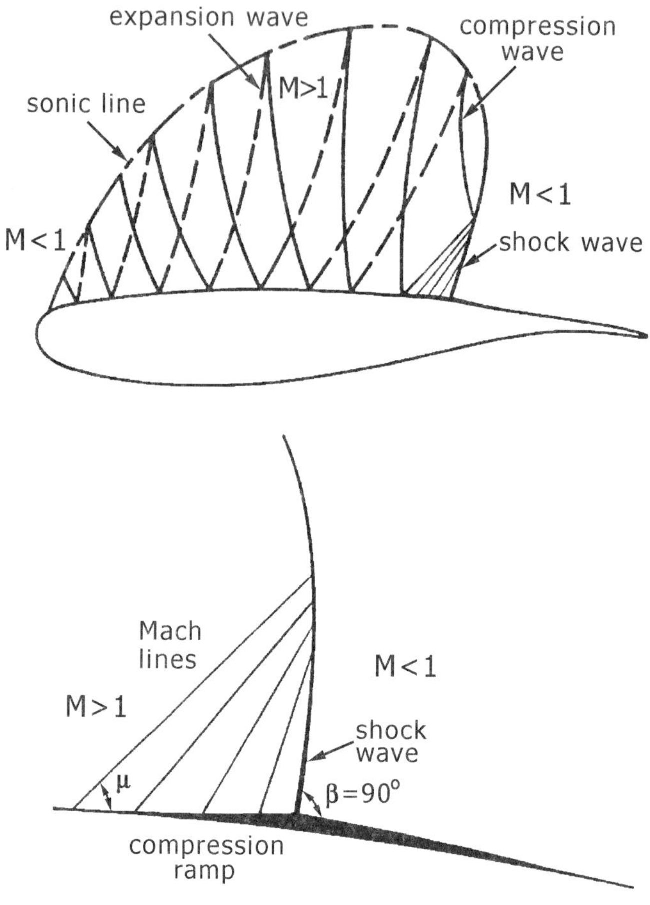

Figure 4.26 A scheme of transonic flow over a bumped airfoil. Reproduced from Tai *et al.* (1987).

bump. The breakage of the shock wave yielded immediate advantages to the airfoil moment and drag. The study demonstrated that a bumped airfoil generally had better aerodynamic characteristics at off-design high supercritical flow conditions than a regular airfoil. The principal conclusions were as follows.

1. The breakage of the shock wave by the bump led to a certain concentration of the lift near the bump that eventually stabilized the airfoil performance at off-design conditions.

2. The bumped airfoil led to less pressure drag for the high supercritical flow conditions that were tested.

3. The lift coefficient of the bumped airfoil, although generally reduced, was sufficient and stable for operation at high Mach numbers and angles of attack.

The exact shape of the airfoil modification, which provides shockless flow at a changed freestream Mach number, can be obtained by comparing the baseline profile with a new one designed for the changed Mach number. Sobieczky and Seebass (1984) demonstrated that a modification with two small peaks of the height 2×10^{-3} of the airfoil length, which is added on the upper surface of an airfoil, shifts the shock-free design conditions from $M_\infty = 0.73$ to $M_\infty = 0.75$ (or one can specify the transpiration velocity which would compensate for the difference between the given profile and the newly designed one if one employs blowing/suction through a perforated portion of the airfoil). Unfortunately, such a way of getting shockless control conditions requires implementation of control over a long portion of the airfoil, which makes it irrelevant for practical purposes.

Another example by Sobieczky *et al.* (1998) showed that the required modification of an original airfoil needed to provide a shock-free flow at $M_\infty = 0.71$ implies the reduction of the original airfoil thickness of the order 3×10^{-3} (Figure 4.27, curve 2). Instead of the impractical thickness reduction Sobieczky *et al.* suggested that similar curvatures can be obtained by the addition of two suitable bumps in the nose and rear area of the airfoil (Figure 4.27, curves 3–5).

In the next example considered the authors ignored the tail bump and concentrated on the modification of the nose shape, thus, admitting only partial improvement of the performance and an incomplete achievement of the shock-free conditions for the changed freestream Mach number. They undertook research of transonic flow control with short modifications to the airfoil shape which were located on the upper surface near the nose of the airfoil. The goal of the study was to show that certain local deformations should result in attractive improvements of the aerodynamic performance for varying transonic operating conditions. Both wave and viscous drag reductions were analysed. As a baseline airfoil Sobieczky *et al.* (1998) chose the transonic ONERA OA15T airfoil designed for the Mach number $M_\infty = 0.73$ and the lift coefficient $C_L = 0.65$. For the higher Mach number $M_\infty = 0.75$ and the higher lift of $C_L = 0.7$ at the Reynolds number of $Re = 6 \times 10^6$, the baseline airfoil exhibited a moderate shock.

The authors used a set of functions which provided a very smooth bump geometry. Basic parameters numbered the range and height, (anti-)symmetry, crest curvature and ramp exponents. The bump started at the nose of the airfoil, $x = 0$, and its maximum height of $2 \times 10^{-3}c$ was placed at $x = 0.05c$, where c is the airfoil length. A cubic ramp exponent ensured smooth connection of the bump with the baseline airfoil at $0.15c$. Crest curvature was an important parameter, as it influenced the local behaviour of the flow substantially.

Flow analysis using both a MSES code by M. Drela, in which the Euler equations are coupled with an integral boundary-layer method, and a Navier–Stokes code substantiated the employed concept of flow control. Both numerical methods showed good

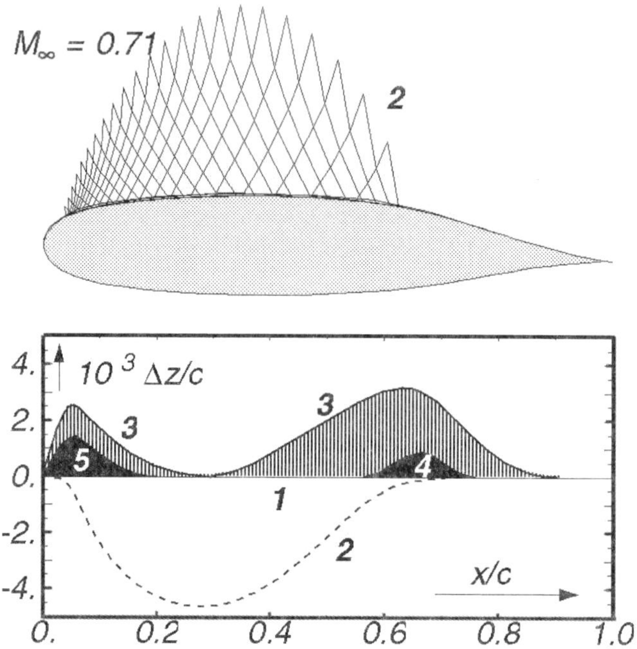

Figure 4.27 Airfoil in inviscid transonic flow. Bottom: shape changes of an initial airfoil (1) to become shockless (2), and a new airfoil (3) with surface bumps to model the curvature of the shockless airfoil (2). Reduced size bumps (4) and (5) to pick up part of the aerodynamic performance improvements. Top: characteristics patterns of shock-free flow (2). Reprinted from Sobieczky *et al.* (1998, Fig. 1) with kind permission from Kluwer Academic Publishers.

agreement in local surface pressure and gave improved ratios of lift over drag. The improvement resulting from the MSES code was $\Delta(L/D) = 15\%$, for the Navier–Stokes code it was 8.3%. The discrepancies were attributed to the imperfect computation of drag and insufficient grid resolution in both codes.

The problem of improving the aerodynamic efficiency at off-design conditions by the application of a shock control bump on a variable chamber airfoil was also investigated by Sommerer *et al.* (2000). The authors employed 20% chord length bumps in the area of the shock wave terminating the supersonic region, similarly to Figure 4.26. Four different type bumps were analysed; e.g. the shape of the fourth bump was set up by a polynomial of 11th order. To investigate the wave drag reduction potential of the bump, the authors specified the flow conditions $M_\infty = 0.77$, $C_\mathrm{L} = 0.524$, which were considerably out of the design region of the baseline airfoil. Since the bump has to be properly positioned and shaped to generate a favourable effect, critical geometrical parameters were investigated using numerical optimization tools. An optimization environment was developed consisting of a hybrid optimizer, a geometry module, and the MSES code coupling the Euler equations with the integral boundary-layer method for the aerodynamic analysis. Bump shapes were optimized for the specified off-design flight conditions. The influence of various geometric bump representations was investigated. All the bump shapes, being optimized, demonstrated impressive gains in wave drag of about 70%. Though the viscous drag was slightly increased, the reduction of the total drag for the considered off-design conditions for all bumps was found of about 16%. At the Mach numbers other than that specified for the control, the bumps caused a considerable increase in the drag coefficient as compared to that of the baseline airfoil.

Allison and Sewall (1995) studied airfoil modifications aimed at the increase of low-speed maximum lift for manoeuvring, while having a minimal effect on transonic performance. Changes of the section contour were confined to the leading-edge slat and trailing-edge flap regions of the wing. Experimental data were analysed from tests in the Langley 16-Foot Transonic Tunnel on the baseline and two modified wing–fuselage configurations with the slats and flaps in their retracted positions. The modified-wing configurations produced maximum lift coefficients which exceeded those of the baseline configuration at low-speed Mach numbers (0.300 and 0.400). This benefit was related to the behaviour of the wing upper surface leading-edge suction peak and the behaviour of the trailing-edge pressure. At transonic Mach numbers (0.725 to 0.900), the wing modifications produced a somewhat stronger nose-down pitching moment, a slightly higher drag at low-lift levels, and a lower drag at higher lift levels.

Makino and Mineo (1994) performed an experimental study with a deformable wing section and mentioned the great sensitivity of transonic flow pattern to slight deformation of the wing surface. In their experiment the wind tunnel model was a hollow wing, the outer plate of which was made of an aluminium alloy plate of suitable thickness, and the curvature of the wing surface could be changed smoothly and continuously by regulation of the air pressure inside the wing. The pressure distribution on the surface was measured for the Mach numbers 0.60–0.98. In addition, a colour video was taken by schlieren method in order to observe the shock wave patterns.

A problem related to the flow control is the effect of the accretion of structural ice on aircraft surfaces on aerodynamic performance in flight conditions. While deicing and anti-icing systems are available and work well in many applications, they add considerable weight to the aircraft and are expensive to maintain. Therefore, many aircraft surfaces remain unprotected. We mention here the impressive work of Hedde and Guffond (1995), who developed a three-dimensional model for computation of ice growth on airfoils and inlet guide vanes and also performed simulation of the flow over an airfoil with an Euler/boundary-layer solver in a range of freestream conditions.

4.3.3 Flow over a perforated airfoil

Passive control through a perforated surface provides blowing and suction in the area of the shock wave without externally supplied power, since the pressure behind the

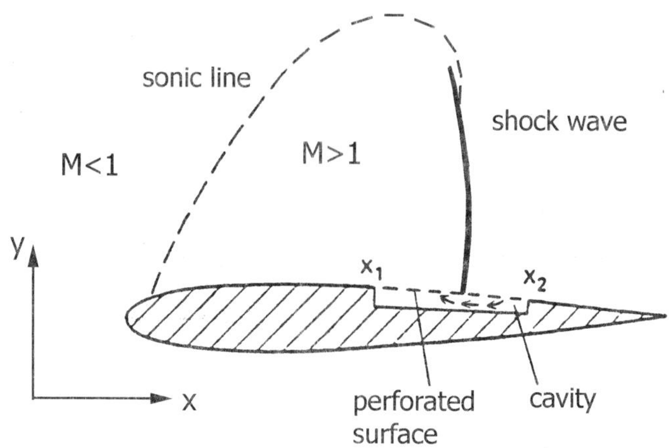

Figure 4.28 Sketch of transonic flow over a perforated airfoil.

shock wave is greater than that in the cavity beneath the perforation, while the pressure in front of the shock is less than that in the cavity (Figure 4.28). Early computations based on inviscid-flow models indicated that a porous surface can reduce the wave drag essentially and in some cases eliminate the shock completely (Savu and Trifu 1984; Chow *et al.* 1985, Fig. 11). However, when the shock-wave–boundary-layer interaction becomes strong, the inviscid models are not sufficient to describe the flow around the airfoil.

Bohning and Zierep (1985) calculated turbulent shock-wave–boundary-layer interaction at curved surfaces with active control using decomposition of the flow into three decks. They simulated ventilation through a perforated surface by prescribing a normal velocity distribution as a boundary condition on the wall and performed a thorough analysis clarifying the flow structure in the area of interaction.

Chen *et al.* (1989) studied numerically effects of porosity on the viscous drag of an airfoil. Both an interactive boundary-layer algorithm and a thin-layer Navier–Stokes algorithm in combination with a modified Baldwin–Lomax turbulence model were used in numerical simulation of the outer flow. The compressible flow in the shallow cavity was assumed to be inviscid and isentropic, with a total pressure different from that of the outer flow. A stream function formulation was used to model the inner flow in the cavity. The coupling procedure at the porous surface was based on Darcy's law (3.29), which states that the transpiration velocity through the perforated or porous surface is proportional to the pressure difference across it,

$$v_n = \frac{\sigma}{\rho_\infty v_\infty}(p_{\text{cavity}} - p_{\text{flow}}),$$

where $\sigma(x)$ is the porosity distribution function determined by the size and spacing of the holes in the porous surface as well as by the flow viscosity. Computations for a symmetric airfoil, that was generated by reflecting the upper surface of an RAE 2822 airfoil to the lower surface, were made with the thin-layer Navier–Stokes procedure. The flow parameters were $M_\infty = 0.85$, $Re = 6.74 \times 10^6$, and the transition was fixed at 0.03 chord. The porous surface was from 0.615 to 0.805 with $\sigma = 0.4$. The effect of porosity was described as follows. The shock was weakened near the airfoil surface, and the pressure on the porous airfoil was lower at the trailing edge. The original strong pressure jump on the airfoil turned into two consecutive weaker jumps, which represented the leading and rear legs of a lambda shock. Though the viscous loss was increased in the presence of the porosity, the total drag was reduced at $M_\infty = 0.85$, but was increased at $M_\infty = 0.82$ in agreement with experiments.

Also, Chen *et al.* (1989) studied the influence of either active blowing or suction in the shock foot area. It was shown that appropriate blowing ahead of a strong shock changes the shape of the wall streamline, which becomes concave and results in formation of compression waves and an oblique shock upstream of the injection region, thus resulting in a smaller pressure gradient and a smaller entropy rise. The additional kinetic energy supplied by blowing would increase the mixing rate in the boundary layer and act to prevent flow separation. However, strong blowing not only thickens the boundary layer, but also probably provokes an early separation as a side effect. On the other hand, active suction in the strong adverse-pressure gradient region would possibly delay separation but might produce a stronger shock and cause a higher wave drag as a side effect.

First experiments on an airfoil with a porous surface by Thiede and Krogmann (1989) demonstrated the potential of passive flow control. Bohning and Jungbluth (1989) and Braun (1990) undertook careful experimental investigations of passive control of flow over a specially designed wall which modelled the upper surface of an airfoil. The goal was to influence the flow field and the shock wave in the end of the supersonic region through a cavity which was placed in the wall and covered by a perforated metal sheet. The perforation of the sheet in the experiments was 2.8 22.7%. The emphasis was laid on flow control through ventilation aimed at (a) weakening the shock, (b) fixing the shock in its position, and (c) reducing the drag of the profile.

Figure 4.29 shows a Mach–Zehnder interferogram with a cavity of an extended length covered by a perforated 2.8% sheet. The dark and bright lines are lines of equal velocity. From the beginning of the cavity a weak oblique shock starts due to the pressure difference in the cavity and in the flow. This oblique shock is further reflected from the sonic line and the wall. A remarkable feature of this experiment is the damping of the multiply reflected shock by the perforation, so that the shock strength eventually vanishes at approximately 80% of the cavity length.

Figure 4.30 demonstrates interferograms of flow over cavities of different lengths and the 8.2% perforation of the sheet. We notice that the oblique shock, which originates at the beginning of the cavity, then reflects from the sonic line creating a pattern similar to the example of Mach reflection obtained in Section 4.2 and also to the results of numerical simulation of the reflection shown in Plates 4, 8 and 9. At the same time, the flow patterns in the vicinities of further reflections of the oblique shocks from the sonic line, which occur downstream, resemble the example of transonic flow with a cusp of the sonic line depicted in Figure 4.19c.

Figure 4.29 Passive control of transonic flow over a perforated wall. Experiments of Bohning and Jungbluth (1989, Fig. 7). Perforation of 2.8%. Reproduced by permission of Springer and the authors.

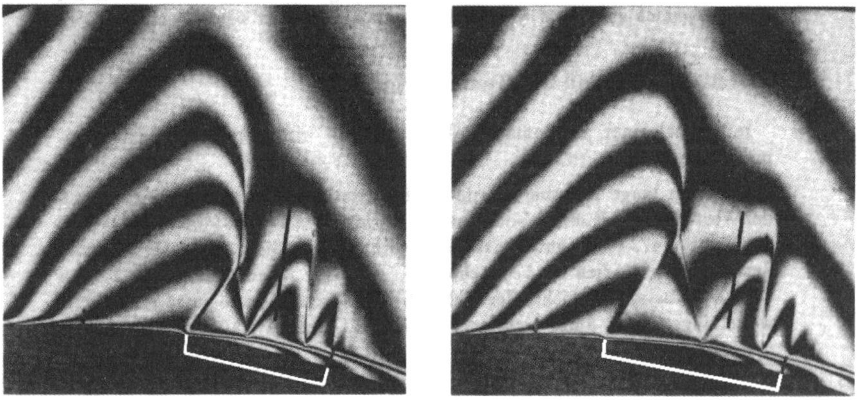

a) b)

Figure 4.30 Passive control of transonic flow over a perforated wall. Experiments of Braun (1990, p. 43). The length of the perforated 8.2% sheet is (a) 42 mm, (b) 52 mm. Reproduced by permission of Institut für Strömungslehre und Strömungsmaschinen, Universität Karlsruhe (TH).

Numerical simulations of Bohning and Jungbluth (1989) and Braun (1990) based on an interacting boundary-layer code yielded flow fields that corresponded in all details to the experimental observations. The normal velocity distribution on the wall which simulated the ventilation was derived from the calculated wall pressure distribution. The maximum normal velocity was 3.3 m s^{-1} in the front part of the cavity and 6.6 m s^{-1} in the back part. The conclusion was that if the cavity is placed properly, the shock strength and the pressure gradient accompanying it are weakened, and the shock position is fixed.

Studies of de Matteis and Dima (1998) were concerned with modifications to the airfoil shape and the flow transpiration through a porous sheet, either natural or forced, using different steady as well as time-accurate CFD techniques. These were based on both the solution of the Reynolds-averaged Navier–Stokes equations and viscous–inviscid coupling procedures. For two laminar airfoils analysed, it was confirmed that modifications with a short bump are effective in reducing the overall drag, mainly due to the reduction of the wave drag. The relative reduction of the total drag predicted numerically was found to be within the range 6–11% for both airfoils investigated, lower than the value of 15% resulting from experiments. Numerical simulation of shock control through wall ventilation showed that the wave drag reduction obtainable using such a control is overcompensated by a viscous drag penalty, due to the presence of flow separation and a significant excrescence drag. The application of suction in the cavity was not sufficient to reduce the overall drag, even though the drag increase was lower than in the case of natural ventilation.

An impressive numerical study of active control by suction of the boundary layer as a means of improving the drag characteristics of supercritical airfoils was performed by Merchant and Drela (1996). The developed MSES method was based on coupling the steady Euler equations and a two-equation integral boundary-layer formulation which covered both laminar and turbulent regimes and involved an e^9 method to predict transition. A parabolic distribution of mass flux in the suction region was used and a series of airfoils of varying thickness were designed to operate with boundary-layer suction. Significant gains in the lift coefficient, the overall drag reductions and delay in shock-induced separation were achieved.

A fundamental investigation of airfoil/wing drag reduction by passive shock control was carried out in 1993–95 by a few European research organizations within the project EUROSHOCK I. Regarding the shock control effectiveness for laminar airfoils, the study predicted a considerable reduction in the wave drag but also an increase in total drag due to control for all flow cases investigated. For a turbulent airfoil, computational results showed a reduced influence of control on viscous drag which also leads, at certain conditions, to a reduction in total drag, again in agreement with experiment. Concluding the research carried out, Stanewsky *et al.* (1997, p. 78) wrote:

> Passive control is not a viable method for reducing wing drag, but a distinct improvement in the buffet boundary can be achieved. Also other configurations, such as engine inlets, where passive control may be of merit, have been identified. The investigations have furthermore shed new light on the complex subject of shock wave boundary layer interaction and have highlighted new active control techniques which will constitute the next steps in the drag reduction efforts to be pursued in the follow-on project EUROSHOCK II. . . The numerical codes achieved generally good qualitative agreement with most of the experimental data; however, the absolute values of the total drag coefficients (lift coefficients were generally prescribed) showed in some instances considerable

> deviations between computed results and between calculations and the corresponding experimental data even w/o control. Considerable scatter also occurred in the ventilation velocities calculated by the various codes, which can only partly be attributed to the use of different control laws but is also dependent on the numerical method employed and on the implementation of the control law within a code. The scatter in the ventilation velocities seems to be the origin of strong deviations in the boundary layer properties within the control region which contributes to the differences observed in the local pressure distributions and in the subsequent flow development in the case of control. Concerning the control laws, it was found that of the laws employed (not including the new developments mentioned above) Poll's formula, derived from the calibration of laser-drilled plates, gives the most reliable results for the present control cases. Finally, the results of the unsteady calculations have shown considerable larger deviations than the steady flow cases.
>
> (the quotation is given by permission of Vieweg Publisher)

Promising experimental results on active and hybrid control of the shock-wave–turbulent-boundary-layer interaction and porous plate transpiration flow have been recently obtained by Bohning and Doerffer (1999) and Bur *et al.* (1997).

4.4 Unsteady Transonic Flow

4.4.1 Self-excited oscillations

As mentioned in Section 3.7, experimental studies of airfoils at zero incidence demonstrated that self-excited oscillations occur in a narrow range of freestream Mach numbers which depends on the airfoil geometry. In numerical simulations, important features of the shock buffet on a circular-arc airfoil have been computed with Navier–Stokes, thin-layer Navier–Stokes, and interacting boundary-layer codes (see Gillan 1995; Bartels and Rothmayer 1995; Rumsey *et al.* 1995). Those studies highlighted the role of the shock-induced separation for the onset of shock buffet, and the sensitivity of this problem to the type of turbulence model. Although details of the shock buffet depend on these factors, all computations have determined the onset Mach number for the circular-arc airfoil quite accurately.

Similar results were obtained for several more widely used airfoils (see Edwards 1993; Bartels 1995). For the NACA 0012 airfoil, experimental data on the buffet onset boundary (McDevitt and Okuno 1985) and computational results obtained with interacting boundary-layer codes (Edwards 1993; Le Balleur and Girodroux-Lavigne 1989) may be approximated by the formula,

$$\alpha = 82.9(0.815 - M_\infty) - 350.5(0.815 - M_\infty)^2 \quad \text{(deg)}$$

at $0.70 < M_\infty < 0.815$. Buffet boundaries for different types of airfoils are discussed in detail in a review paper of Lee (2001).

For symmetric airfoils at zero incidence, Gibb (1988) postulated that disturbances on the surface will cause an upward deflection of the wake similar to the deflection of a trailing-edge flap. The asymmetric wake causes the shock on the other surface to move toward the rear. The shock waves then move in antiphase, and the wake is displaced toward the surface on which the separation is taking place.

For an airfoil at nonzero incidence, when a shock wave exists only on the upper surface, there are different explanations of the self-sustained oscillations. Stanewsky and Basler (1990) performed a wind-tunnel investigation of the buffet characteristics of the CAST 7/DOA1 airfoil. They argued that the thickening of the boundary layer at the trailing edge and the corresponding drop in trailing-edge pressure drive the shock upstream since the shock strength must adjust according to the trailing-edge pressure. The thickening of the boundary layer at the trailing edge also causes a decambering of the airfoil. This reduces the circulation and results in a decrease of the supersonic region and hence of the shock strength. The generated pressure disturbances must travel via the lower surface where the flow is being accelerated due to the pressure drop at the trailing edge. On reaching the leading-edge region, these waves cause a change in the stagnation point location. A new flow field is developed on the upper surface.

Lee (1990) proposed a possible mechanism of self-sustained oscillations using a feedback loop which consists of downstream propagation of disturbances in the separated boundary layer and upstream propagation of waves in the inviscid flow outside the region of separated flow. This concept provides a method to estimate the frequency of oscillations of the shock wave in buffeting flows. The model assumed that the shock wave oscillates on the upper airfoil surface about a mean position, and the flow behind the shock boundary-layer interaction is fully separated. Pressure waves, formed as a result of the shock motion, propagate downstream in the separated flow region. On reaching the trailing edge, the disturbances generate upstream-moving waves as a result of satisfying the unsteady Kutta condition. The upstream-moving waves, termed Kutta waves, interact with the shock and impart energy to return it to its initial location. The loop is then completed. The period of shock wave oscillations was found by Lee to agree with the time it takes for a disturbance to propagate from the shock to the trailing edge plus the duration for an upstream wave to reach the shock from the trailing edge.

The proposed buffet mechanism involves velocities of upstream-moving waves outside the separated flow region. The velocities were estimated using an empirical formula given by Tijdeman (1977). For a better understanding of the phenomenon, Lee *et al.* (1994) studied numerically propagation of wavefronts in transonic flow using the small disturbance equation. An impulse source of perturbation was located at the trailing edge of the airfoil. The shock front was obtained using asymptotic expansions and solving numerically the equation governing characteristic lines.

Though a relation between the amplitude of the disturbance initiated at the trailing edge and the magnitude of shock displacement were not derived to study the inter-

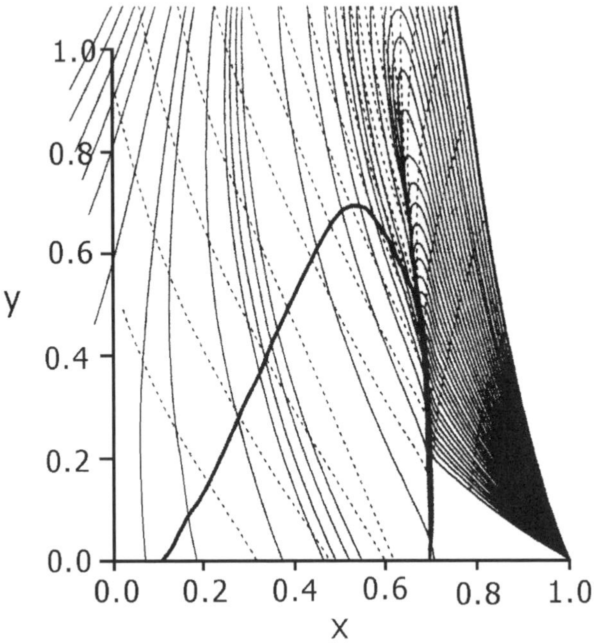

Figure 4.31 Wavefronts and rays on the upper surface of a NACA 64A006 airfoil at $M_\infty = 0.88$ generated by an impulse source at the trailing edge: - - - -, wavefront; —, ray; ▬, sonic line. Computations of Lee *et al.* (1994). Reproduced by permission of the authors.

action process, the results presented were sufficient to give a better insight into the feedback mechanism. For the NACA 64A006 airfoil and the freestream Mach number $M_\infty = 0.85$, Lee *et al.* computed the upstream wave propagation at different times after a disturbance was generated by the impulse source. The behaviour of the disturbances, initiated at an airfoil trailing edge, was determined by their convection in the inviscid flow region around the airfoil. The obtained results confirmed the sketch given in Figure 3.33. The range of initial ray angles at the trailing edge was 107–174°. The nonuniformity of the flow field caused the rays to bend toward the airfoil. At this freestream Mach number, a shock wave was formed at approximately midchord. The Mach number on the airfoil surface upstream of the shock was 1.07. Crowding or steepening of the wavefronts was most sharp in the area behind the shock wave, where the flow velocity was close to the sound speed. The perturbation propagating in the upstream direction slowed down and the impulses were crowded in this area because the propagation velocity decreased due to the convection as the wavefront approached that region. The interaction of a finite-amplitude disturbance with the shock wave and the resulting amplitude of shock motion could not be determined from the analysis.

For an increase in M_∞ from 0.85 to 0.88 the wavefronts and rays obtained by Lee *et al.* (1994) are represented in Figure 4.31. The supersonic region extends to

0.7 chord in the vertical direction, the Mach number on the airfoil upstream of the shock is 1.18. The subsonic velocity behind the shock wave is smaller than that in the previous case. Therefore, it is the area above the shock, not behind it, where the perturbation propagating in the upstream direction slows down abruptly and the impulses are crowded.

A similar pattern of wave propagation was observed in numerical simulation of transonic flow over airfoils in the channel pointed out in Section 4.1. For instance, in the case of the airfoil (4.4) with $n = 3$, an impulse change in $M'_{\text{out}} = 0.675$ led to upstream wave propagation in the subsonic region. Having reached the minimal cross-section of the subsonic region above the sonic line, the waves turned down and entered the supersonic region yielding an indentation in the shape of the sonic line (Figure 4.32). Then, having passed the supersonic region, the waves travelled upstream again in the subsonic region, thus performing a number of rotations in the counterclockwise direction. This was accompanied by changes in the configuration of the sonic line behind the triple point from the shape with a cusp to the one with an inflection point and vice versa. After a few hundred of cycles, the oscillations decayed, and the flow relaxed to a new steady state.

A curious example obtained for an RAE 5214 type airfoil is shown in Plate 14, where one can observe valleys at the tops of the two supersonic regions over the airfoil. The valleys look similar, whereas their physical origins are quite different. The first valley is due to a compression wave caused by a small concave arc of the airfoil shape in the nose area, which occurred because of a coarse prescription of the airfoil coordinates. The indentation in the second top is caused by the unsteady waves generated by the change $M'_{\text{out}} = 0.75$ for $M'_{\text{out}} = 0.76$ similarly to those in Figure 4.32. Therefore, the second indentation is transient and vanishes when the transonic flow achieves the steady state.

Self excited oscillations due to shock-induced separation on airfoils/wings can be amplified by oscillation of the flexible wing structure, as the pressure fluctuations due to the boundary-layer separation can excite a response of the structural modes, called buffeting. The aircraft structure acts as a selective filter for the excitation, so that the spectra of buffeting always contain pronounced peaks at structural frequencies.

> Wings with low angles of sweep are characterized beyond the buffet boundary at transonic speeds by leading-edge or trailing-edge separations. These separations form bubbles on the wing which usually excite heavy buffeting. At transonic speeds the presence of strong shock waves nearly parallel with the leading edge add to the difficulties of predicting the flow... Swept wings are characterized by a combination of mixed flows (Rogers and Hall 1980) that is difficult to predict. The separated flows on a swept wing at transonic speeds may include shock waves (which vary in intensity across the span), bubbles (from the leading edge or the trailing edge), and vortices. Thus, a small increase in Mach number may dramatically alter the position of a shock wave or the reattachment point of a bubble. Similarly, an increase in unit Reynolds number or

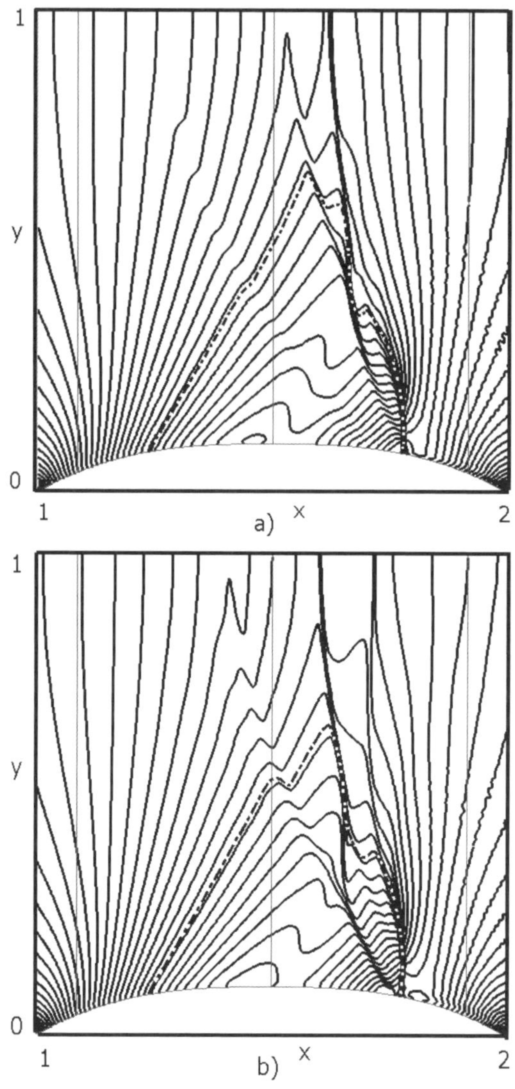

Figure 4.32 Unsteady flow over the airfoil (4.4) with $n = 3$ under an impulse change in $M'_{\text{out}} = 0.675$ applied at the outlet. The difference between instants (a) and (b) is $\Delta t = 3.57 \times 10^{-3}$, where the time is nondimensionalized by the channel height and the reference velocity 279.76 m s^{-1}.

> a change of the roughness band used to fix transition on the model in wind-tunnel tests may completely alter the character of the shock-wave–boundary-layer interaction...
>
> (Mabey 1989)

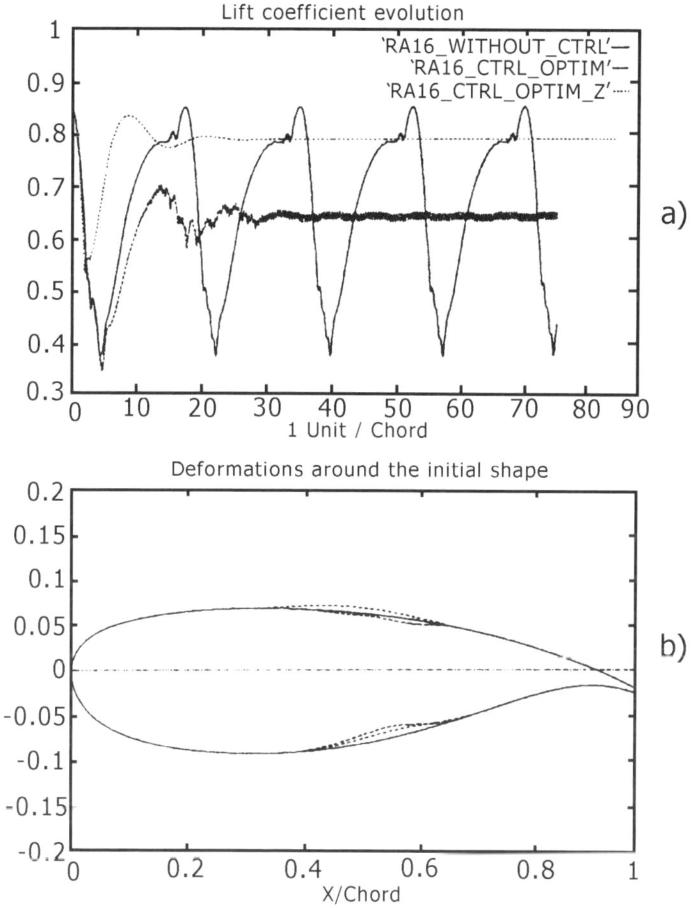

Figure 4.33 Buffeting control for an RA16 airfoil at 3.5° of incidence, $M_\infty = 0.723$ and $Re = 4.2 \times 10^6$: (a) lift coefficient histories without and with control for two simultaneous with the injection devices placed between $0.4 < (x/\text{chord}) < 0.6$ or between $0.35 < (x/\text{chord}) < 0.65$; (b) two different instantaneous deformations simulated by injection boundary conditions. Computations of Mohammadi (1998). © John Wiley & Sons, Ltd. Reproduced with permission.

The existing codes for numerical simulation of self-excited oscillations are extremely time consuming. Therefore, further theoretical studies, leading to a better understanding of the self-induced oscillations, their suppression and control, are also expected to contribute to the development of efficient codes for their computation. For further details regarding self-excited oscillations and buffeting phenomena we refer the interested reader to the review paper of Lee (2001).

4.4.2 Methods using optimization tools

Methods incorporating optimization techniques for the airfoil/wing design and unsteady flow control have been developed for both the Euler equations (see, for example, Baysal and Ghayour 2001) and the Navier–Stokes equations (Nielsen and Anderson 1999).

An application of a gradient-based optimization tool to a buffeting control problem over an airfoil in turbulent transonic flow was worked out by Mohammadi (1998). The problem was solved using the Navier–Stokes equations and two equation-type turbulence models. Computations of flow over the RA16 profile at the freestream Mach number 0.723, the angle of attack 3.5°, and the Reynolds number 4.2×10^6 showed a periodic motion of the upper-surface shock with the main frequency of about 15 Hz in the absence of control. The aim of the control was to reduce the shock displacement by mass injection/suction in the part of the profile between 0.4 and 0.6 and between 0.35 and 0.65 chord. The minimization problem and the cost function were chosen according to the above objective. The evolution of the computed lift coefficient is represented in Figure 4.33a. It demonstrates a remarkable stabilization of the shock on the upper surface of the airfoil under the control used. In the case of injection placed in the interval $0.4 < x/c < 0.6$, where c is the chord length, the stabilization was accompanied by an improvement of the lift. Figure 4.33b shows two different instantaneous deformations simulated by injection boundary conditions. We would like to emphasize that the goal of this study was the stabilization of the shock wave position, not its weakening. The resulting shock in the flow under control remained of a moderate strength, and the concave portions existing on the modified airfoil geometry indicate that the obtained flow is far from shock-free conditions.

Though methods using optimization techniques are a powerful tool for airfoil design and flow control, the high sensitivity of transonic flow to small details of the airfoil shape may cause difficulties in practical application of those methods. The use of just a few design variables may be insufficient for the accurate modelling of transonic flow. Hager *et al*. (1994) solved the Euler equations with the aim of improving the off-design performance of the airfoil. The authors reported that the single-point design demonstrated two fairly strong shocks, whereas the two-point design exposed one fairly strong and one very weak shock. This is in accordance with the discussion presented in the previous sections which showed that very small changes of the airfoil curvature may influence the flow structure crucially.

On the other hand, the employment of a large number of design variables leads to a highly multidimensional search space, extremely complicated objective function distribution, and inappropriately large computing time and cost. Therefore, the application of optimization algorithms to real-world problems may not be straightforward.

Oyama *et al*. (1999) discussed this problem as follows.

> A parametric study is often conducted by varying one parameter at a time or by trial and error for a limited number of parameters. However, such approaches only lead to incomplete knowledge for a large design space. An exhaustive search, in contrast, requires unacceptably large number

of experiments and thus they are not suitable for real-world problems. For instance, a complete study of a design space of 10 parameters with 3 levels requires $3^{10} = 59\,049$ experiments...

...The choice of design parameters should be based on the flow structure around an airfoil and therefore, aerodynamic performance. An airfoil shape is defined by basic geometric parameters instead of the coefficients of shape functions themselves: leading-edge radius, upper and lower crest location including curvatures, trailing-edge ordinate, thickness, direction and wedge angle.

4.4.3 Eigenfrequencies of acoustical oscillations in transonic flow

In this section, we describe a numerical method for finding eigenfrequencies of acoustical oscillations in a Laval nozzle. For simplicity, we consider axisymmetric oscillations in axisymmetric flow.

The problem reads as follows. Find such values of the complex frequency ω which provide the existence of a nonzero solution of the equation

$$\left[\frac{(\rho\lambda)_\lambda}{\rho}u_\varphi\right]_\varphi + (y^2\rho^2 u_\psi)_\psi \\ - \rho^{1-\gamma}[2\mathrm{i}\omega u_\varphi - \omega^2\lambda^{-2}u + \mathrm{i}\omega(\gamma-1)\lambda\rho^{1-\gamma}\lambda_\varphi u] = 0, \tag{4.28}$$

endowed with the homogeneous boundary conditions

$$\left.\begin{array}{lll} u = 0 & \text{at } \varphi = 0,\ 0 \leqslant \psi \leqslant \psi_2 & \text{(at the inlet)}, \\ u_\psi = 0 & \text{at } \psi = \psi_2,\ 0 \leqslant \varphi \leqslant \varphi_2 & \text{(on the wall)}, \\ u,\ u_\varphi,\ u_\psi & \text{are bounded as } \psi \to 0,\ 0 \leqslant \varphi \leqslant \varphi_2 & \text{(on the nozzle axis)}. \end{array}\right\} \tag{4.29}$$

Equation (4.28) with respect to the amplitude $u(\varphi, \psi)$ of oscillations of the acoustical potential is obtained from the linearized equations of unsteady, axisymmetric, potential flow in the same way as Equation (2.124) for oscillations in plane flow. The velocity $\lambda(\varphi, \psi)$, density $\rho(\varphi, \psi)$, and the radial coordinate $y = y(\varphi, \psi)$ involved in the coefficients of the equation are known functions of the streamline coordinates for a given transonic flow. The condition of the boundedness of u_φ is due to physical reasons, while the boundedness of u_ψ near the nozzle axis follows from the symmetry condition and the smoothness of the given flow which yield $u_y = O(y)$. This shows that $u_\varphi\varphi_y + u_\psi\psi_y = O(y)$. Hence, $\lambda\sin\theta u_\varphi + y\rho\lambda\cos\theta u_\psi = O(y)$, and owing to $\sin\theta = O(y)$, we obtain that u_ψ is bounded.

Multiplying Equation (4.28) by $\rho^{\gamma-1}\lambda^2$, we rewrite it in the form

$$A_\omega u + \omega^2 u = 0, \tag{4.30}$$

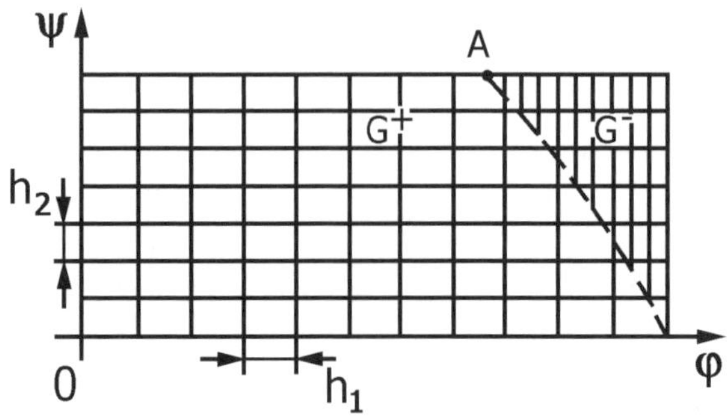

Figure 4.34 A mesh employed for solving problem (4.29), (4.32).

where

$$A_\omega u := \lambda^2\rho^{\gamma-1}\left[\frac{(\rho\lambda)_\lambda}{\rho}u_\varphi\right]_\varphi + \lambda^2\rho^{\gamma-1}(y^2\rho^2 u_\psi)_\psi$$
$$- 2\mathrm{i}\omega\lambda^2 u_\varphi - \mathrm{i}\omega(\gamma-1)\lambda^3\rho^{1-\gamma}\lambda_\varphi u. \tag{4.31}$$

To describe the numerical algorithm, we consider the equation

$$A_\omega u + \mu u = 0, \tag{4.32}$$

with a parameter μ. First we concentrate on Problem (4.29), (4.32) of finding spectral values of the parameter μ under a fixed frequency ω. Using the technique developed in Chapter 2, one can prove the Fredholm alternative for Problem (4.29), (4.32) with the complex-valued μ when the flow acceleration is positive on the sonic line. Consequently, there exists at most a numerable set of spectral μ-values. Once these values μ_j are found under a fixed ω, we generally get the inequality $\mu_j \neq \omega^2$. Hence, ω must be chosen in such a way that the difference $|\mu_j - \omega^2|$ would be sufficiently small; such a value of ω can be accepted as an approximate value of the eigenfrequency.

In order to solve numerically Problem (4.29), (4.32) with respect to μ_j, we introduce a Cartesian mesh $\{(\varphi_n, \psi_m) : \varphi_n = (n-1)h_1, \psi_m = (m-1)h_2\}$, $n = 1, 2, \ldots, N$, $m = 1, 2, \ldots, M$, in the domain $G = \{(\varphi, \psi) : 0 < \varphi < \varphi_2, 0 < \psi < \psi_2\}$ shown in Figure 4.34, where the dashed line is the sonic line $\lambda(\varphi, \psi) = 1$. We denote by G^+ the elliptic part of G and by G^- the hyperbolic one. Let $u^{(0)}$, $\mu^{(0)}$ be initial approximations for an eigenfunction and a spectral value, which can be obtained, e.g. by solving the problem on acoustical oscillations in the nozzle in the one-dimensional approximation. Using $u^{(0)}$ on the sonic line and $u_\psi = 0$ on the wall as boundary conditions, we can calculate the solution $u^{(1)}(\varphi, \psi)$ of Equation (4.32) in G^- with a marching scheme in the φ-direction. In order to perform the first step we notice that the derivative u_ψ vanishes at the beginning A of the sonic line due to

(4.29). Moreover, the derivative of u along the sonic line is known; that is why we can find u_φ at the point A.

Thus, we get the first approximation $u^{(1)}$ in G^- and, in particular, the derivative $u_\varphi^{(1)}$ on the sonic line. After that, using this derivative and the boundary conditions (4.29), we can calculate the solution of Equation (4.32) in the elliptic subdomain G^+, e.g. with the expression

$$u^{(k+1)} = u^{(k)} - \tau^{(k+1)}(A_\omega u^{(k)} + \mu^{(k)} u^{(k)}),$$

where the parameter $\tau^{(k+1)}$ is determined by the condition of minimal norm of the residual

$$r^{(k+1)} = A_\omega u^{(k+1)} + \mu^{(k)} u^{(k+1)}.$$

One can derive the expressions for the real and imaginary parts of $\tau^{(k+1)}$ in a standard way (Samarsky 1977) as follows,

$$\tau_1^{(k+1)} = \|Br^{(k)}\|_0^{-2} \int_{G^+} \mathrm{Re}\{Br^{(k)} \cdot \bar{r}^{(k)}\}\,\mathrm{d}G,$$

$$\tau_2^{(k+1)} = -\|Br^{(k)}\|_0^{-2} \int_{G^+} \mathrm{Im}\{Br^{(k)} \cdot \bar{r}^{(k)}\}\,\mathrm{d}G,$$

where $Br^{(k)} = A_\omega u^{(k)} + \mu^{(k)} u^{(k)}$ and $\|\cdot\|$ is the norm in $L^2(G^+)$. The derivatives involved in the operator A_ω are calculated with central differences in the subdomain G^+.

Once $u^{(1)}$ is found, the approximation $\mu^{(1)}$ for μ can be calculated by the formula

$$\mu^{(k+1)} = -\|u^{(k+1)}\|_0^{-2} \int_{G^+} A_\omega u^{(k+1)} \cdot \bar{u}^{(k+1)}\,\mathrm{d}G,$$

which easily follows from Equation (4.32). Thus, we get the first approximation $u^{(1)}$ in G^+ and on the sonic line. Then we can solve the boundary-value problem in G^- again and find a second approximation $u^{(2)}$ in G^- and, in particular, the derivative $u_\varphi^{(2)}$ on the sonic line, etc. The described procedure iterates until the difference between two sequential approximations becomes small enough.

An example of eigenfrequencies calculation was performed for the nozzle geometry shown in Figure 4.35. First, the spectral problem was solved with the one-dimensional approach in which the dependence of λ on x was chosen the same as on the axis of the given two-dimensional flow. Longitudinal oscillations are governed by the ordinary differential equation which follows from Equation (4.30) when $u_\psi \equiv 0$:

$$\lambda^2 \rho^{\gamma-1} \left[\frac{(\rho\lambda)_\lambda}{\rho} u_\varphi \right]_\varphi - 2\mathrm{i}\omega\lambda^2 u_\varphi - \mathrm{i}\omega(\gamma-1)\lambda^3 \rho^{1-\gamma} \lambda_\varphi u + \omega^2 u = 0.$$

The numerical solution of this equation was carried out in the direction from the critical section of the nozzle to the inlet. At the critical section, the condition

$$-[(\gamma+1)\lambda_\varphi + 2\mathrm{i}\omega]u_\varphi + [\omega^2 - \mathrm{i}\omega(\gamma-1)\lambda_\varphi]u = 0$$

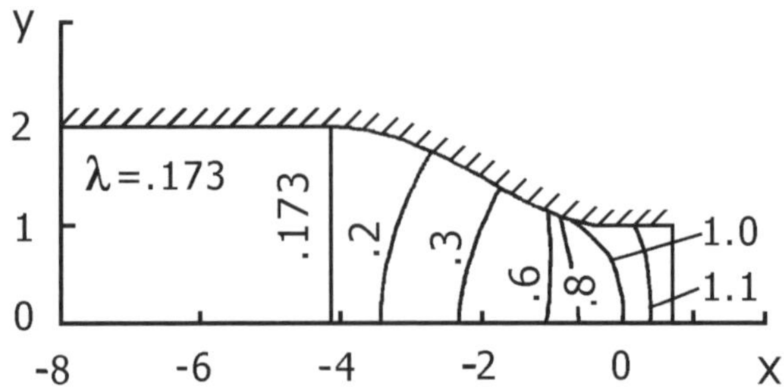

Figure 4.35 A nozzle geometry and velocity isolines $\lambda(x, y)$ = const. in transonic flow to undergo acoustical oscillations. Reproduced from Kuz'min (1992a) by permission of Birkhäuser.

was imposed; it follows from the requirement of the boundedness of $u_{\varphi\varphi}$ (Natanzon 1986). The value of $(\gamma + 1)\lambda_\varphi/2$ at the critical section was 0.31. The complex-valued parameter ω was fitted to provide the condition $u = 0$ at the inlet. In this way the following three discrete eigenfrequencies were found,

$$\left.\begin{aligned} \omega_{\mathrm{I}} r^* &= \pm 0.270 + 0.045\mathrm{i}, \\ \omega_{\mathrm{II}} r^* &= \pm 0.756 + 0.127\mathrm{i}, \\ \omega_{\mathrm{III}} r^* &= \pm 1.624 + 0.260\mathrm{i}, \end{aligned}\right\} \tag{4.33}$$

where r^* is the radius of the critical section. The imaginary parts are positive; thus, the oscillations are decaying in accordance with what was said in Section 2.7. The magnitudes of the acoustical potential $|u(\varphi)|$ and the pressure $|p_{\mathrm{a}}(\varphi)|$ (related by (2.132)) are shown in Figure 4.36.

Then the iterative method described above was applied to calculate eigenfrequencies and eigenfunctions in the 2D flow shown in Figure 4.35. In the streamline coordinates, the domain under consideration was $G = \{(\varphi, \psi) : 0 < \varphi < \varphi_2,\ 0 < \psi < \psi_2\}$ with $\varphi_2 = 2.419$ and $\psi_2 = 0.3417$. A mesh of 81×41 with $h_1 = 0.030\,24$ and $h_2 = 0.008\,54$ was employed in the region of subsonic flow G^+. The functions $u(\varphi)$ obtained with the one-dimensional approach (Figure 4.36a) were used as initial approximations. The convergence of $u^{(k)}(\varphi, \psi)$ was obtained in about 1500 iterations. The calculated eigenfrequencies,

$$\begin{aligned} \omega_{\mathrm{I}} r^* &= \pm 0.270 + 0.051\mathrm{i}, \\ \omega_{\mathrm{II}} r^* &= \pm 0.821 + 0.140\mathrm{i}, \\ \omega_{\mathrm{III}} r^* &= \pm 1.710 + 0.270\mathrm{i}, \end{aligned}$$

demonstrate notable deviations of ω_{II} and ω_{III} from the results (4.33) obtained with the one-dimensional approach. The computed acoustical pressure field corresponding to ω_{II} is shown in Figure 4.37.

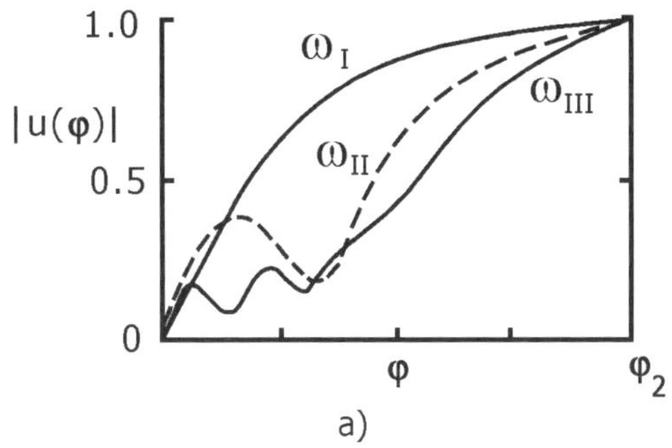

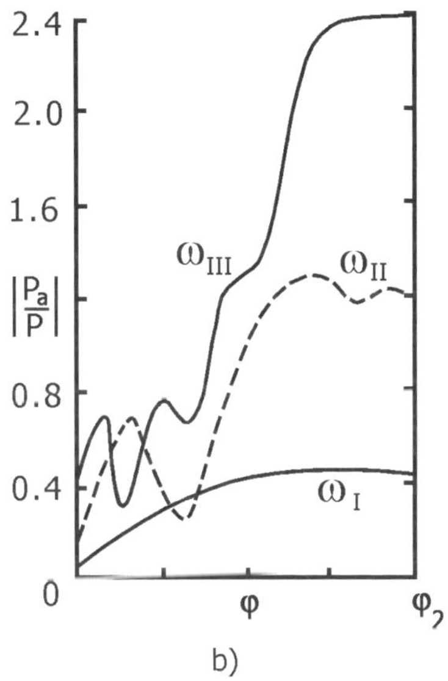

Figure 4.36 Magnitudes of the acoustical potential (a) and the pressure (b) of eigenoscillations in the one-dimensional approximation. Reproduced from Kuz'min (1992a) by permission of Birkhäuser.

4.5 Other Studies Concerned with the Flow Structure

4.5.1 Formation of oblique shocks in Laval nozzles

An experimental study of formation of shock waves in transonic flow through axisymmetric Laval nozzles with circular-arc throat was performed by Back and Cuffel (1966). An oblique shock was incipient near the upper wall, in the vicinity of the junction of the straight and circular-arc portions of the contour. The authors explained that the shock formation in the flow just downstream of the throat was due to the small radius of curvature of the nozzle contour in the throat area. They implied that, as the

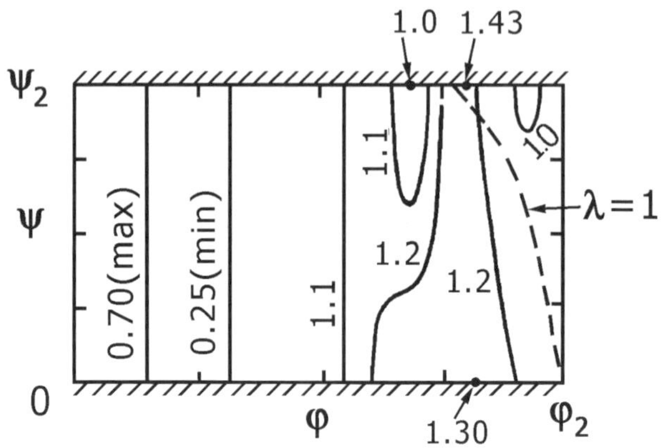

Figure 4.37 Isolines in the acoustical pressure field $|p_a/p|$ corresponding to the frequency ω_{II} in the 2D transonic flow. Reproduced from Kuz'min (1992a) by permission of Birkhäuser.

flow enters the conical section of the nozzle, the high angular momentum must be reduced to meet the tangential wall boundary conditions. This is accomplished by a compression wave, which occurs near the discontinuous change in wall curvature. The compression wave propagates and intensifies to a shock which finally intersects the axis of symmetry and reflects back.

Loth *et al.* (1992) studied the formation of shock waves in Laval nozzles numerically. They employed conservative nonlinear shock-capturing schemes and adaptive unstructured grids in order to get fine grid resolution near high flow gradients. The developed axisymmetric finite-element method based on the flux-corrected transport procedure was used to investigate the flow for two nozzle configurations. Adaptive remeshing was employed to optimize the distribution of grid points by refining areas of high density gradients and coarsening areas of low density gradients. For a first nozzle geometry, with the conical aft section, compression waves were initiated along the wall surface just downstream of the circular-arc throat and propagated toward the axis of symmetry, as shown in Figure 4.38a. These waves coalesced to form an oblique shock that was further strengthened as it approached the axis, due to the axisymmetric reduction in the cross-sectional flow area. Figure 4.38b demonstrates flow field predictions for the second nozzle. This particular nozzle contour (fitted with a spline) employed a conical (constant slope) section between a circular-arc throat and a parabolic-arc aft section. The two slope changes resulted in the formation of two sets of compression waves near the wall, which eventually coalesced to form two oblique shocks. Axisymmetric effects increased the strength of shocks as they approached the centreline. Formation of shocks finally reduced the thrust efficiency by 7% as compared to an ideal shockless nozzle.

Mazaheri and Ayasoufi (1998) studied transonic flow in the conical nozzle shown in Figure 4.38a with an efficient adaptive method using a first-order conservative

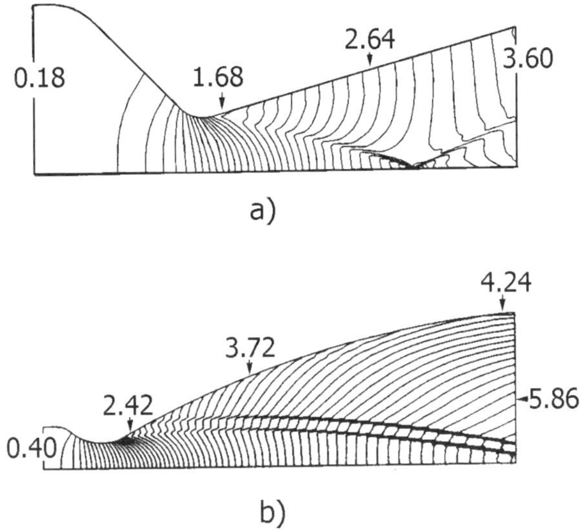

Figure 4.38 Mach number isolines in transonic flow through Laval nozzles. (a) A conical aft section and a circular-arc wall contour in the area of nozzle throat. (b) A conical section between a circular-arc throat and a parabolic-arc aft section. Computations of Loth *et al.* (1992).

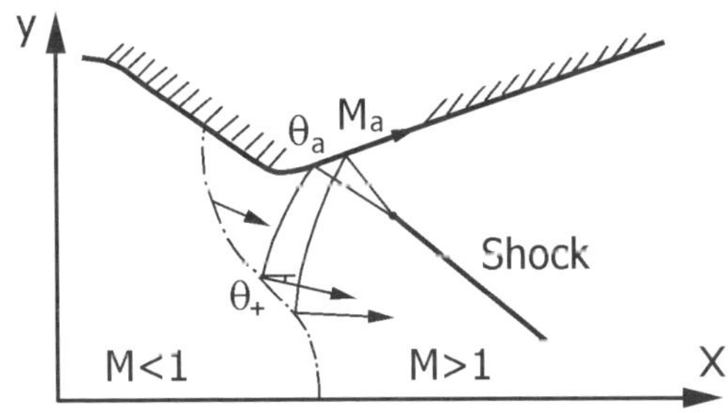

Figure 4.39 Scheme of formation of an oblique shock in a Laval nozzle.

finite-volume scheme based on Roe's flux difference splitting method and successive mesh refinement. A three-step adaptation procedure was developed in order to resolve delicate flow features. Computations confirmed the arising of a weak shock which originated near the abrupt changes in the curvature of the nozzle contour.

Theoretical considerations of Section 3.3 make it possible to clarify the scheme of the formation of shock waves near the nozzle throat. For simplicity, consider 2D steady inviscid flow instead of the axisymmetric one in which relations (3.15) and (3.16) along the characteristic lines are to be replaced by differential relations. Due to Nikolsky–Taganov's theorem on the monotonous behaviour of the deflection angle θ_+

on the sonic line (see (3.17)), this angle increases as the observation point runs along the sonic line toward the x-axis (Figure 4.39). Consequently, the relation (3.15) along the left-running characteristics shows that the difference $\theta_a - \zeta(M_a)$ on the wall of the nozzle must increase with increasing x. This inevitably results in the decrease of M_a if θ_a is constant along a supersonic portion of the wall, i.e. if this portion has a constant slope. Thus, the flow decelerates along the straight segment of the supersonic portion of the wall. This deceleration initiates compression waves which may coalesce and form a shock wave.

4.5.2 Transonic flow past an expansion corner

A wedge of a finite thickness can be considered as a limit of a symmetric airfoil whose middle part flattens, while the nose part sharpens. On the other hand, the flow near the shoulder of the wedge is similar to the flow near a corner point of the nozzle contour at its minimal cross-section. Such an expansion corner may be used in a nozzle design method for reducing the length of the nozzle (Pirumov and Roslyakov 1990). The sharp flow gradients in the vicinity of the wedge shoulder imply that viscosity effects play an important role in this region. However, the inviscid analysis appears to capture the principal features of the flow fairly well.

As known, the wall configuration downstream of the corner point can be chosen so as to provide a shock-free flow having just a weak singularity that propagates along the characteristic line emanated from the corner point (see Guderley 1957, Section 6.3). On the other hand, if the downstream portion of the wall has a constant slope, then a shock wave arises due to arguments similar to those illustrated in Figure 4.39. A simple model of the flow over a wedge was suggested by Cole (1951), who assumed that the sonic line is locally vertical and contended that the Prandtl–Meyer expansion at the wedge shoulder be terminated by an oblique shock wave.

Using the theory of self-similar solutions, Vaglio-Laurin (1960) obtained a principal term in the asymptotic expansion of the solution which described the flow in the region of its acceleration over the corner. Shifrin (1974) made a qualitative analysis and suggested a scheme in which a shock wave of vanishing strength originates at the corner point A and contacts the last characteristic of the expansion fan if the contour of the wall behind the corner point is a straight segment and no separation occurs (Figure 4.40a). As the shock develops, it becomes stronger due to the compression waves emanated from the wall behind the corner point. If there is separation beyond the corner point, then the flow is shock free. The asymptotic technique was used for construction of exact solutions within the above model by Lifshitz and Boichenko (1976) and Esin and Chernov (1977) for both 2D and the axisymmetric flows.

Another scheme of transonic flow over the corner point implies that the strength of the shock originated at the corner point is nonvanishing, and the flow is supersonic behind the shock (Figure 4.40b). Esin *et al.* (1976) constructed examples of exact solutions illustrating this scheme. Numerical verification was done by Liang and Fung (1987), who applied a subgrid refinement procedure to obtain a superior solution of the

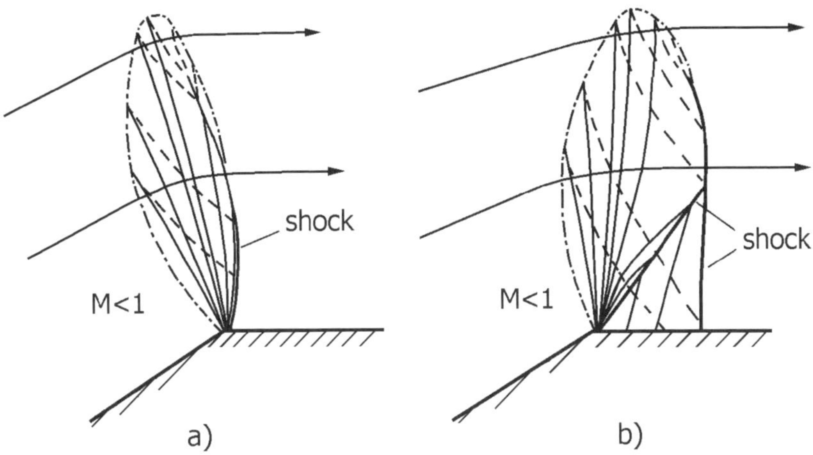

Figure 4.40 Scheme of formation of oblique shock waves in transonic flow over an expansion corner. (a) Flow is subsonic behind the shock originated at the corner point. (b) Flow is supersonic behind the shock originated at the corner point.

transonic small disturbance equation. The base mesh was a uniform one of spacing $\Delta x = \Delta y = 0.1$ over a computational domain of size 6×6 wedge chords. Two refined solutions were computed on meshes of spacings 0.05 and 0.025. Computations indicated the location of an oblique shock at an angle of about 41° with the x-axis due to the compression immediately after the Prandtl–Meyer expansion. The obtained flow deflection angle testified that the flow overexpanded as much as $-3.1°$ at a point 0.3 chord above and 0.6 chord downstream of the shoulder.

A local asymptotic analysis (see, for example, Sobieczky 1985) shows that the flow is to expand to a maximum velocity $u_{\max} = K_{\text{sim}} + (3/2)^{2/3} = 1.810$ right after the shoulder and decelerate along the wall at the rate

$$u(x, 0) = (3/2)^{2/3}[1 - 2^{-8/3}3x^{2/3}] + K_{\text{sim}},$$

where K_{sim} is the transonic similarity parameter, and x is the distance along the wall behind the corner point. Because of the compression, the characteristics coalesce into a weak shock wave starting with a slope of $(2/3)^{1/3} \simeq 41°$, a zero curvature, and a negative third derivative.

The maximum velocity $u_{\max}$ obtained by Liang and Fung was linearly extrapolated from the region downstream of the shoulder. It assumed the value 1.878 on the base grid and the values 1.864 and 1.819 after the first and second refinement cycles, respectively, approaching the theoretical value as the solution on the base mesh was improved.

A similar problem of flow over a corner point of a contour occurs in analysis of unsteady supersonic flow in a channel with a forward-facing step placed on the lower wall at a distance d from the inlet. A test problem with $M_{\text{inlet}} = 3$, $d = 0.6$ and

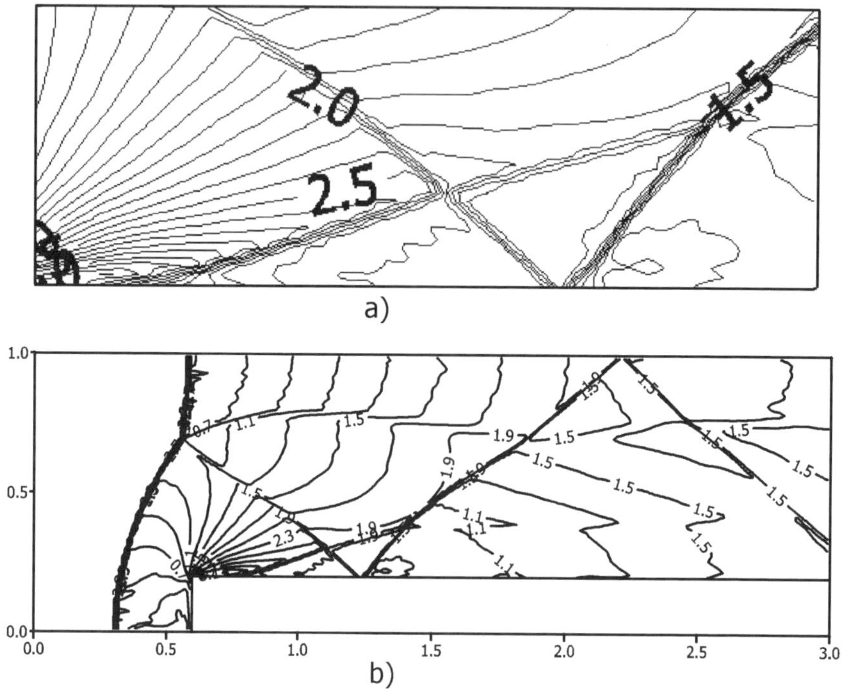

Figure 4.41 Mach number isolines in a channel with a step at $M_{\text{inlet}} = 3$: (a) fragment of the flow field in the vicinity of the corner; (b) flow pattern at $t = 8$. Computations of B. N. Azarenok (personal communication, 2001). Reproduced by permission of the author.

the height 0.2 of the step was analysed numerically by a number of authors (see, for example, Woodward and Colella 1984; Jiang and Shu 1996; Sonar and Süli 1998). In a paper of Azarenok and Ivanenko (2000), numerical simulation of the flow over the step was performed with a finite-difference method for solving the Euler equations on moving grids that condense toward solution singularities, substantially improving the computational accuracy. The mesh-generation algorithm was based on the theory of harmonic mappings. The computed flow field for $t = 8$ is represented in Figure 4.41. It can be seen that a weak shock wave is formed by compression waves emanating from a short segment of the wall located beyond the corner at $0.7 \leqslant x \leqslant 0.8$.

4.5.3 Flow emerging from an orifice

This problem is also of practical interest in connection with the flow pattern in the vicinity of a corner point of the nozzle contour. Kosolapov (1991, 1988a,b) performed a numerical study of flow emerging from the orifice of a reservoir in the hodograph plane. In order to solve the Tricomi problem for Equation (2.72) governing the stream function, Kosolapov used a marching finite-difference scheme in the

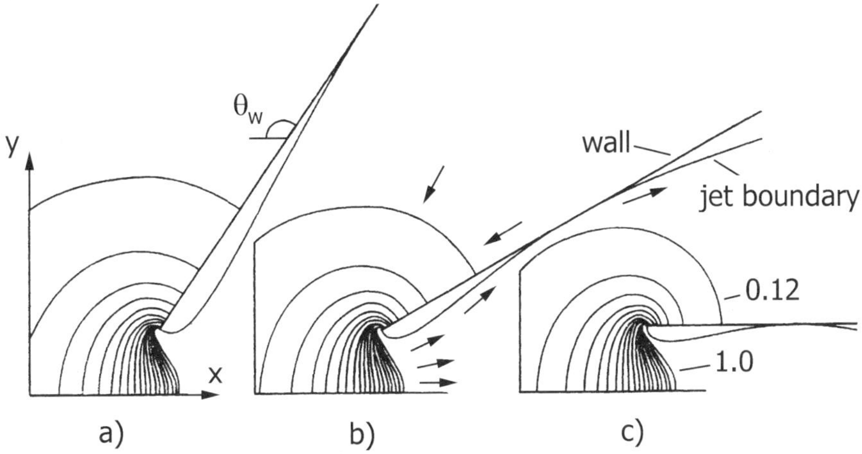

Figure 4.42 Velocity magnitude isolines $\lambda(x, y) = \text{const.}$ in transonic flow emerging from an orifice. Computations of Kosolapov (1991). Reproduced by permission of the author.

region of supersonic velocities, while the system of algebraic equations obtained in the subsonic region was solved with a successive line over-relaxation method. Details of the numerical method are given in Kosolapov (1988a,b). Examples of computations on meshes of 31×51 and 61×101 involved choked flow regimes at various angles θ_w made by the reservoir wall with the x-axis, including large ones. Figure 4.42 shows the jet boundary and the lines of constant velocity $\lambda = \text{const.}$ with $\Delta\lambda = 0.05$ in the range $0.12 \leqslant \lambda \leqslant 1.0$ for the pressure ratios $\varepsilon = p_{\text{outer}}/p_0$ and the angles θ_w listed below:

(a) $\theta_w = 124.5°$, $\varepsilon_* = 0.0093$ (Figure 4.42a);

(b) $\theta_w = 150°$, $\varepsilon_* = 0.0582$ (Figure 4.42b);

(c) $\theta_w = 180°$, $\varepsilon_* = 0.2711$ (Figure 4.42c).

4.5.4 Dynamic stall over an oscillating airfoil

Chandrasekhara *et al.* (1998) conducted a fundamental experimental investigation of the onset of dynamic stall from a leading-edge shock wave over oscillating NACA 0012 airfoils, and suggested new explanation of the onset mechanisms. The work focused on details of the development of the local supersonic region and its complex interactions with the laminar separation bubble or the attached boundary layer and demonstrated that this interaction dictates further events in the flow. With increasing freestream Mach number, the flow attained supersonic velocities after the laminar separation bubble had already formed. The growth of the supersonic region and the tendency of the bubble to burst with increasing angle of attack presented a situation when complex interactions occurred, leading to a new source of the dynamic stall.

Interferograms obtained for the 3 in (7.6 cm) NACA 0012 airfoil at $M_\infty = 0.45$ and the frequency of oscillations $k = \pi f c / U_\infty = 0.05$ demonstrated the dynamic stall onset from interactions of supersonic flow and the laminar bubble. Tiny disturbance waves arose in the supersonic region. These were interpreted as expansion and compression waves reflecting off the sonic line above the leading edge of the airfoil and the shear layer.

> The unsteady transonic flow waves can cause dramatic variations in pressure along the weak shear layer enclosing the bubble, which could either alter the bubble dynamics by influencing transition and reattachment or simply force separation by preventing reattachment, causing dynamic stall, depending on their instantaneous strength. If the bubble continues to exist, its bursting is dictated by the pressure distribution in the supersonic flow unlike that seen for the lower freestream Mach number case where the flow was subsonic everywhere. Whereas dynamic stall eventually ensues, these are fine-scales events that compete with each other and play a major role in dynamic stall onset.
>
> (Chandrasekhara *et al.* 1998)

The interferograms presented in Figure 4.43 for the 6 in (15.2 cm) NACA 0012 airfoil at $M_\infty = 0.45$ and $k = 0.05$ show several interesting features. By the time the angle of attack increases to $\alpha = 8.98°$, the laminar separation bubble is enclosed by the supersonic flow (Figure 4.43a). The bubble must respond to this change in the external flow, and further developments in the bubble occur under this supersonic external boundary condition. As before, some expansion and compression waves arise in the local supersonic region, and the largest wave, located close to the end of the bubble, can be interpreted as a shock. At $\alpha = 9.49°$, a multiple shock pattern forms (Figure 4.43b). The shear layer thickens significantly and lifts off the surface. A dramatic change in the boundary-layer thickness can be seen across the shock as well. This signals of the onset of dynamic stall for this flow condition. The Mach number at the foot of the shock exceeds 1.2, and thus the pressure rise across the shock is large enough to induce boundary-layer separation. The dynamic stall onset point has moved farther upstream as the shear layer enveloping it moves outward.

4.5.5 Wavy patterns occurring in design procedures

Patterns of multiple oblique shocks in the local supersonic region may occur in intermediate modes of transonic airfoil/wing design. Of significant interest is the work of Hirose *et al.* (1987), who developed an iterative method for airfoil design utilizing a Navier–Stokes analysis code to attain specified pressure distributions. The code was based on an implicit approximate factorization scheme for solving the time-averaged compressible Navier–Stokes equations. A design procedure of the residual-correction type was used as follows. A target pressure distribution over the airfoil at a specified freestream Mach number, angle of attack, and Reynolds number was fixed. The flow

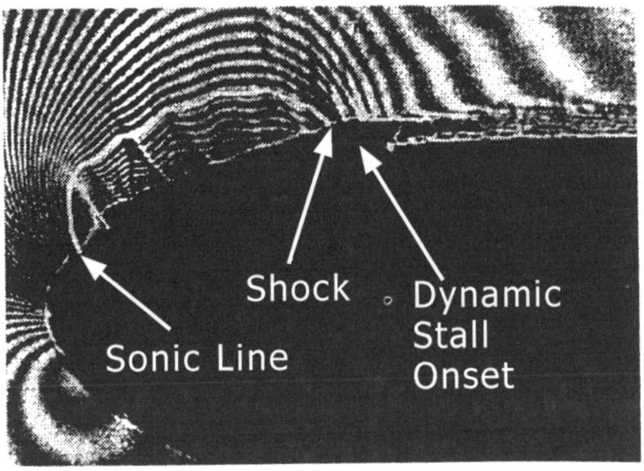

a)

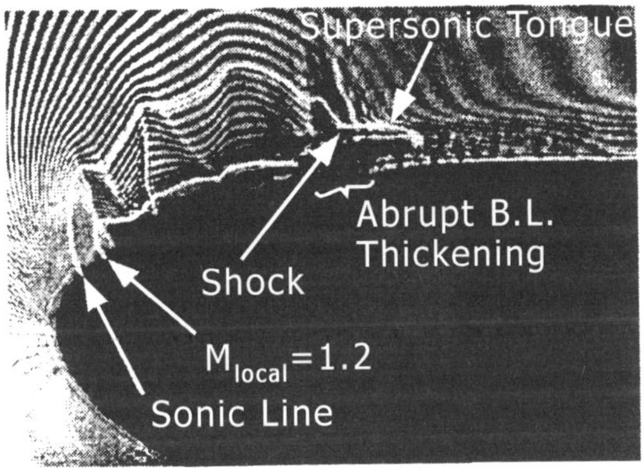

b)

Figure 4.43 Interferograms of transonic over an oscillating NACA 0012 airfoil at $M_\infty = 0.45$ and the reduced frequency of oscillations $k = 0.05$: (a) $\alpha = 8.98°$, (b) $\alpha = 9.49°$. Reproduced from Chandrasekhara *et al*. (1998).

field at the specified freestream conditions was computed for an arbitrarily chosen initial geometry. Then the pressure difference from the target distribution was checked and the necessary geometry adjustment was calculated with a geometry correction code. At the next iteration, the Navier–Stokes code was used again to obtain the new flow field and pressure distribution for the corrected geometry, and the pressure difference was checked. The iterative loop encompassing the analysis mode and the design mode was repeated until the pressure difference became small. The geometry design code was based on the integral equation formulation of the geometry correction problem for airfoils and wings proposed by Takanashi (1985).

In an example of the shockless airfoil design presented by Hirose *et al*. the target pressure distribution was chosen to correspond to the Garabedian–Korn airfoil 75-06-12 at its design conditions: $M_\infty = 0.75$, the angle of attack $\alpha = 0.8°$, and Reynolds

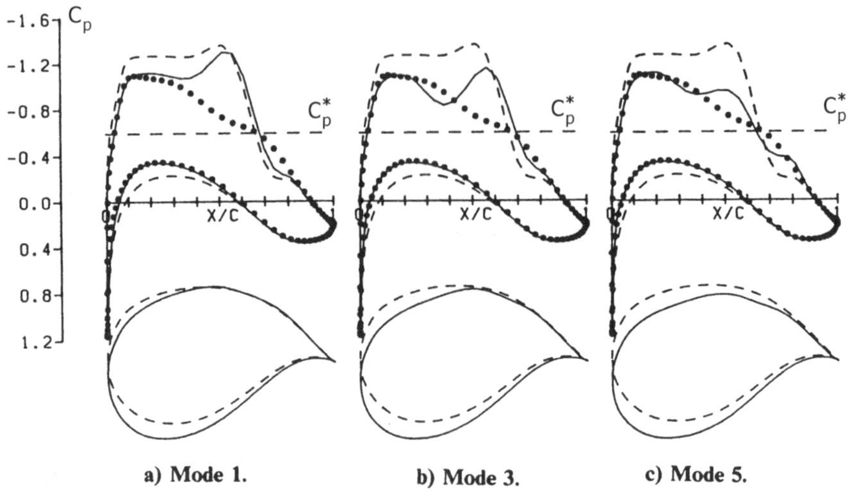

Figure 4.44 Pressure coefficients over intermediate airfoil configurations (top): · · · · · ·, target; - - - -, initial; ——, obtained at an iteration mode. Airfoil configurations (bottom): ——, designed at the iteration mode; - - - -, initial geometry. Computations of Hirose *et al.* (1987). Copyright © 1987 AIAA. Reproduced with permission.

number $Re = 1.3 \times 10^7$. A 125×51 C-type mesh with 93 points on the airfoil surface was used. The far-field boundary was placed 12 chord lengths away from the airfoil.

The same airfoil at a different angle of attack $\alpha = 2.0°$ was selected for an initial geometry. The initial pressure distribution contained a strong shock wave, and the lift coefficient was 0.87 as compared to the on-design value 0.6. During the iterative design procedure, the angle of attack in the analysis mode was fixed to the same value as in the initial iteration. Only the coordinates of the airfoil were corrected by the design mode output and the computational mesh was reconstructed for the next analysis mode. To understand how the correction proceeds, the geometry and pressure distributions at each iterative mode were examined. Plots of the pressure coefficient $C_p = 2(p - p_\infty)/(\rho_\infty v_\infty^2)$ obtained for three iterations are shown in Figure 4.44, where C_p^* denotes the critical pressure coefficient corresponding to $M = 1$. The designed geometry (solid line) is compared with the initial geometry (dashed line) in the bottom plots. The scale is zoomed in the normal direction to illustrate the correction in the geometry. It is seen that the geometry designed at each iteration gradually rotates counterclockwise retaining the shape of the initial airfoil. This is an expected result, since the initial geometry differs from the target geometry only in the angle of attack by 1.2° while its shape is the same.

The pressure distribution obtained at the mode 3 reveals the existence of a strong deceleration along the airfoil at $0.25 < x/c < 0.40$. A supersonic minimum of the flow velocity is reached at $x/c \approx 0.40$. This implies the existence of either a compression wave or an oblique shock which is significantly smeared because of the coarse mesh employed. After reaching the minimum, the flow accelerates again and

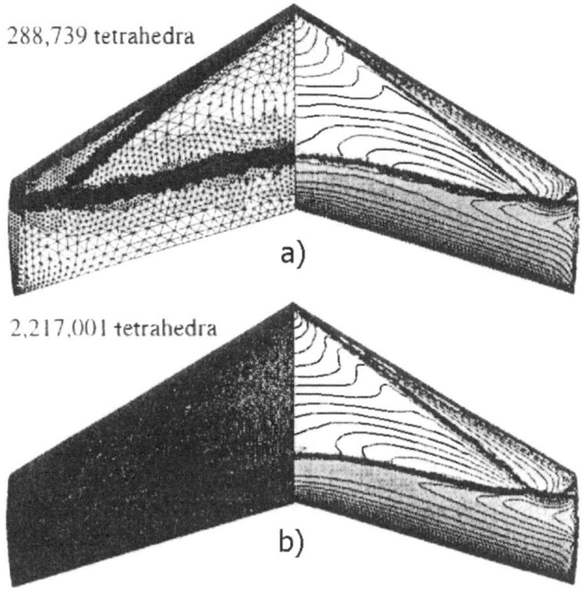

Figure 4.45 Grids and surface pressure contours on the ONERA M6 wing, $M_\infty = 0.84$, $\alpha = 3.06°$, adapted mesh and solution. Computations of Pirzadeh (1999).

attains a second maximum at $x/c \approx 0.55$ which is followed by a shock terminating the supersonic region. The convergence history of the pressure distribution was oscillatory in the intermediate modes until the shock waves finally disappeared. In the above example, a total of eight iterations produced the desired result of attaining the target pressure distribution.

4.5.6 Three-dimensional flow over a wing

A well-known example of a 3D transonic flow is the one over ONERA M6 wing, which has a leading-edge sweep angle of 30°, an aspect ratio of 3.8, and a taper ratio of 0.562. The airfoil section of the wing is the ONERA ‘D’ airfoil, which is a 10% maximum thickness-to-chord ratio conventional section. The test freestream conditions are the Mach number of 0.84 and the angle of attack 3.06. To resolve the flow field near the wing with reasonable accuracy, one needs good quality meshes and an efficient method for adaptation to the essential features of the flow field.

Pirzadeh (1999) used an upwind, cell-centred, finite-volume, unstructured mesh solver for calculation of inviscid flow over the ONERA M6 wing. The far field was placed 20 chords away from the wing. An adaptive method by mesh subdivision, local remeshing, and mesh movement was employed. After the solutions fully converged on coarser meshes, three levels of solution-based grid adaptation were performed. Figure 4.45 shows the surface grids and the corresponding pressure contours at the

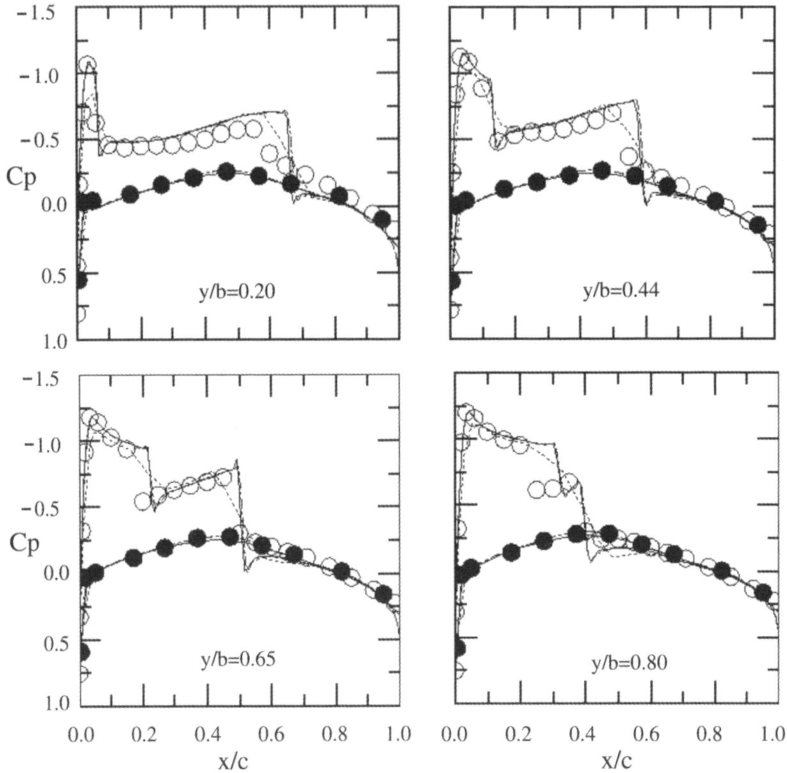

Figure 4.46 Pressure coefficient on the ONERA M6 wing at four different span locations: - - - -, coarse mesh; ——, adapted mesh; — — · — — · — —, fine grid; •, ○, experiment. Reproduced from Pirzadeh (1999).

second and third levels of adaptation. Figure 4.46 demonstrates the obtained pressure coefficient on the wing and experimental data at four different span locations, where y is the Cartesian coordinate in the span direction. It can be seen that two shock waves form on the upper surface of the wing, one near the leading edge and the other at 40–70% of the chord length. At 80% span the double-shock structure is still well captured. The double shocks merge into one and move forward near the wing tip.

Similar patterns of flow over ONERA M6 wing were obtained in a number of other studies (see Pandya and Hafez 1999; Wang *et al.* 1999; Nakahashi *et al.* 1999; Hirsh 1990). There are only minor differences between the flow fields obtained in these papers, and between the adapted and fine grid results.

Though three-dimensional phenomena play a significant part in the case at hand, shocks arise in a similar way to that discussed in Section 3.3 of this book. It is determined by an interplay between the rate of decrease of the flow deflection angle on the sonic surface and the curvature distribution over the wing, which governs the arising of compression waves. Qualitative patterns of the double shock formation at spanwise stations $y/b = \text{const.} \leqslant 0.80$ are the same as shown in Plates 6 and 13 for two-dimensional flow over airfoils.

Generally, the computed pressure coefficient distributions shown in Figure 4.46 agree quite well with experimental data except at the root station where viscous effects are important. Addition of viscous effects to inviscid solutions usually weakens shock waves and moves them upstream to the correct locations.

4.6 Nonunique Numerical Solutions

4.6.1 Transonic small-disturbance and full-potential equations

In the beginning of the 1980s, it was found that numerical approximations to the differential equations governing steady flow over an airfoil admit nonunique solutions in a range of freestream Mach numbers and angles of attack. Steinhoff and Jameson (1982) revealed both symmetric and nonsymmetric solutions for the Joukowski airfoil at zero incidence. They used a conservative scheme for the full-potential equation. Verifications included the analysis of convergence of the solution with grid refinement. The convergence observed with several 'O'-type meshes of 96×24 to 384×192 served as an evidence of the existence of the nonsymmetric solution at zero angle of attack.

Salas *et al.* (1984) and Salas and Gumbert (1985) studied flows over five different airfoils and found nonunique numerical solutions with negative lift coefficients at positive angles of incidence. For certain freestream conditions, there were three solutions, none of them having a physical significance. It was demonstrated that the nonuniqueness occurred due to the employed scheme of approximation of the differential equation, i.e. the conservative scheme for the full-potential equation. The difference between the solutions of the full-potential equation and the system of the Euler equations became more pronounced with increasing Mach numbers but remained relatively limited for the nonlifting airfoils. For lifting airfoils, however, these discrepancies could become very strong and lead to very different flow patterns at the same freestream conditions. The employment of a nonconservative scheme or the Euler equations led to shock jumps of lower strength and eliminated the nonuniqueness in the examples considered.

Bristeau *et al.* (1985) obtained nonunique solutions of the full-potential equation with a finite-element method and a conjugate-gradient-type algorithm. For the airfoil NACA 0012 at zero angle of attack and $M_\infty = 0.82$, in agreement with Salas's results, three solutions were obtained: (i) a symmetric solution, (ii) an asymmetric solution with positive lift, and (iii) an asymmetric solution with negative lift. For a lower Mach number $M_\infty = 0.81$, Bristeau *et al.* obtained five numerical solutions.

Though the question of the nonuniqueness of the solutions to the boundary-value problem in the differential form remained open, the obtained results demonstrated the sensitivity of numerical results to the choice of numerical schemes and small discretization errors. On the other hand, they showed that the nonuniqueness may arise due to an inadequate mathematical model employed for description of transonic flow, in particular, due to inaccurate jump conditions across the shock waves.

The jump conditions (3.62), (3.63) typically used in the potential model conserve mass and energy but not momentum, thus, introducing a deviation from the correct Rankine–Hugoniot jump relations which follow from the physical conservation laws. In order to reduce these errors and increase the validity range of the potential flow model, it was proposed to introduce nonisentropic corrections into the jump conditions across the shock discontinuities. Hafez and Lovell (1981), Chen (1983) and Klopfer and Nixon (1983) discussed different options available within the potential model with respect to the conservation laws over a shock discontinuity. A comprehensive analysis of various models including the entropy corrections was carried out by Whitlow (1988) and Hafez and Lovell (1988), who also mentioned that in some cases the use of artificial viscosity in the subsonic region, as well as in the supersonic one, made it possible to obtain unique solutions for the same conditions where standard codes (lacking artificial viscosity in the subsonic region) produced nonunique solutions. Nixon (1987) suggested a nonconservative/conservative algorithm which may preclude the appearance of nonunique solutions.

Williams *et al.* (1985) studied unsteady flow over the NACA 0012 airfoil using the time-dependent transonic small-disturbance equation (TSD). They examined the flow response to a pitch pulse and found that for some range of freestream Mach numbers the lift coefficient did not decay to zero, thus indicating the presence of nonunique solutions. Fuglsang and Williams (1985) developed nonisentropic modifications to the TSD equation, which involved a modified flux term, a correction to the pressure coefficient equation, and a correction to the wake jump condition to allow for entropy convection. These modifications refined the multiple solutions and enabled the use of the TSD equation for a wider range of problems. Gibbons *et al.* (1986) extended this method to three dimensions and also found evidence of similar multiple solutions. However, Murty (1993) computed the same problem of flow response to a pitch pulse using the time-dependent full-potential equation and showed that the results of the full-potential code and the small disturbance code are different. Moreover, he could not verify the nonunique solutions obtained in certain cases by his predecessors.

Thus, the numerical studies have demonstrated that the inconsistencies of the potential model with respect to the Euler model may be circumvented through nonisentropic corrections, though there is no guarantee that this works in all cases. The potential model with nonisentropic corrections was recognized as a useful tool for simulation of transonic flow problems as long as the maximum Mach number remains below $M \simeq 1.3$.

We recall that the uniqueness and solvability issues for transonic flow problems were rigorously studied in this book only for shock-free flows. Moreover, the uniqueness 'on the whole' was only proved for the flow with positive acceleration in a nozzle (Theorem 2.1). The existence of eigenvalues and eigensolutions appears to be quite plausible for nonlinear equations and solutions with shock waves, and one cannot rule out the situation when an eigenvalue vanishes. Then the problem in the differential form would have nonunique solutions. Nixon (1987) discussed eigensolutions of the von Kármán equation using an integral equation technique. Under some special assumptions he demonstrated that the von Kármán equation can admit eigensolutions that satisfy all of the boundary conditions generally found in such problems; there were an infinite number of these eigensolutions. He showed that for the existence of the eigensolutions, certain consistency requirements must be satisfied. These requirements are met if the average of the velocity on the upper and lower surfaces of the airfoil is supersonic. The eigensolutions provide lift and act like an additional asymmetric source term in the flow field. If real lift is present, the fictitious component provided by the nonuniqueness appears as a simple additive term to the real lifting component. There is no way to distinguish or uncouple the two components in a given numerical solution. The stability terms used in algorithms tend to initiate the appearance of eigensolutions thus making the computations of a real solution different.

Mathematically, the problem of finding eigenvalues and eigensolutions is extremely difficult even for finite domains, smooth solutions, and linear equations of mixed type. In Section 1.6 we mentioned papers where the existence of an eigenvalue and, hence, the nonuniqueness of solutions were rigorously proved. The difficulties become more severe in the case of infinite domains, nonlinear equations, nonsmooth solutions representing flows with shock waves, and the Kutta condition at the trailing edge.

The aforementioned numerical studies, thus, showed that if rigorous results on the uniqueness of solutions to problems for the Kármán equation were obtained for flow with shock waves under the jump conditions (3.62), (3.63) they would be of little practical use, because they would not be able to give an answer to the question whether the nonuniqueness can occur due to real physical phenomena. To get an answer, one must study the full-potential equation with the nonisentropic jump conditions across the shock waves, or the system of the Euler equations.

4.6.2 The Euler and Navier–Stokes equations

Baruzzi *et al.* (1989) employed the Euler equations in order to analyse the two-dimensional flow in a symmetric convergent–divergent nozzle with a smooth area variation under supersonic inlet conditions. Using a fully implicit technique, both stable and unstable nonunique solutions were obtained unlike the case of time-marching procedures, which produced only the stable solution. The stability of the solutions was investigated by solving the unsteady Euler equations with a backward Euler finite-difference discretization of the time-dependent terms: shock waves were perturbed about the steady-state position and the shock motion examined for each case. The shock in the divergent part was proved to be stable, i.e. having been misplaced to the right or left of the steady-state position it returned back to it. The shock wave in the convergent part, however, was unstable and depended on the initial displacement. If the shock was initially placed to the right side of the steady state, it was eventually swallowed by the throat, where the flow was sonic, and disappeared. On the other hand, if the shock was placed on the left side of the steady-state position in the convergent part of the nozzle, it continuously moved towards the inlet.

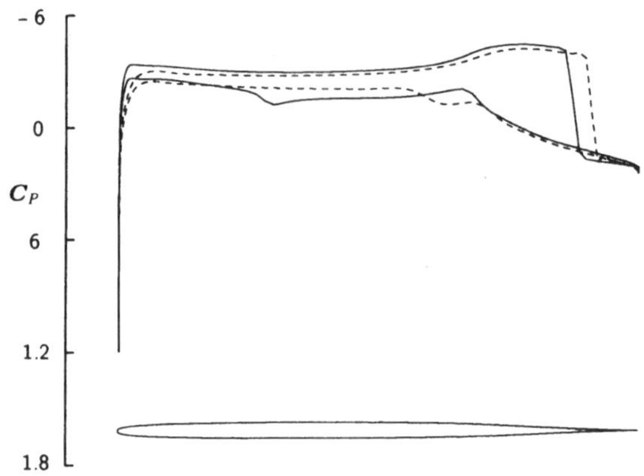

Figure 4.47 Nonunique solutions to both the full-potential and the Euler equations for a 3% thick symmetric airfoil, $\alpha = 0$, an O-type mesh of 160×40: ——, full-potential solver, $M_\infty = 0.924$, $C_L = 0.168$, $C_D = 0.0039$; - - - - -, Euler solver $M_\infty = 0.936$, $C_L = 0.161$, $C_D = 0.0035$. Computations of McGrattan (1992). Copyright © 1992 AIAA. Reproduced with permission.

Mathematical considerations related to this problem were presented in Section 3.5, though we treated decelerating flow in a channel with a perforated wall. The problem on the stability of the decelerating flow with weak shock waves in a channel with impermeable walls can probably be studied with the technique developed in Chapter 3 if one employs the third-order governing Equation (3.84)

$$\varepsilon(\varphi_{xx} + \varphi_{yy})_x + (\gamma + 1)(1 - \varphi_x)\varphi_{xx} + \varphi_{yy} = 0$$

involving the streamline viscosity term.

McGrattan (1992) presented a study of transonic flow over an airfoil which yielded both a nonunique potential solution and a nonunique solution to the Euler equations. A special thin supercritical airfoil was designed using a method of complex characteristics. The goal of the design was to find an airfoil that gives rise to a large supersonic zone and relatively weak shocks when there is little difference between the Euler and the potential model. Remarkably, the potential solver revealed a nonunique solution at the design Mach number $M_\infty = 0.924$, and this nonunique solution was used as an initialization of the Euler computation. The Euler solver did not yield nonunique solutions as easy as the potential one, though it eventually demonstrated a nonunique solution at $M_\infty = 0.936$ and zero incidence (Figure 4.47).

The behaviour of the nonunique solutions was identical to that of the real solution: the question arose as to whether these solutions are physically realizable or not. McGrattan argued that the obtained nonuniqueness had been possibly caused by implementation of Kutta–Joukowski condition at the trailing edge:

> The Euler solver yields a nonunique solution for a Mach number slightly above design. The Euler residual, the root mean square of the quantity $\partial\rho/\partial t$, was reduced to the order of 10^{-6}. The wave drag in both cases is low, suggesting that the nonuniqueness has more to do with implementation of the Kutta–Joukowski condition than the isentropic assumption. The potential model explicitly restricts the flow from turning around the trailing edge, and the velocity potential throughout the entire flow field is readjusted at each iteration to account for the new value of circulation about the airfoil. The Euler solver does not have a similar mechanism of enforcing the Kutta–Joukowski condition. Instead, the Euler solver relies on higher-order artificial viscosity terms to restrict the flow from turning around the trailing edge... The explicit mechanism by which the potential solver enforces the Kutta–Joukowski condition yields, as an unwanted side effect, nonunique solutions for a certain range of Mach number. The fact that the Euler solver does not have a similar mechanism means that it will not as easily generate nonunique solutions. However, because the fully conservative potential and the Euler solutions are very similar in the presence of weak shocks, there is no reason why a nonunique potential solution with weak shocks should not have an Euler counterpart. Indeed, such a solution exists for the constructed thin airfoil. The exact cause of this phenomenon is yet to be determined, but it is unlikely that it is due to the isentropic assumption of the potential model.

A series of special airfoils was designed by Jameson (1991), who obtained nonunique solutions of the Euler equations for these airfoils.

We would like to emphasize that it is the high sensitivity of transonic flow to small approximation errors which may cause spurious nonuniqueness in some cases, so that the rapid development of very small approximation errors may invoke multiple numerical solutions, while the original problem for the differential full-potential equation or the Euler equations may have a unique solution. We recall the results of Pfenninger *et al.* (1986) which demonstrated the great sensitivity of transonic flow to small perturbation in M_∞, and numerical examples presented in Plates 5 and 6 for $M_{\text{out}} = 0.673$ and $M_{\text{out}} = 0.671$, correspondingly. Also, the change of the flow pattern shown in Plate 4 under the very small modification ($h = 1.5 \times 10^{-4}$) of the airfoil shape is very instructive.

The sensitivity of solutions to the numerical viscosity was discussed by Hänel and Breuer (1989), who showed that the accuracy of the numerical solution at high Reynolds numbers may be essentially influenced by the numerical damping, caused by the discretization of the inviscid terms of the Navier–Stokes equations. Obviously, the exact inviscid solution itself is nondissipative, but numerical dissipation is always present and is used to suppress high-frequency errors and to capture weak solutions. In the Navier–Stokes solution, where viscous layers with strong gradients and curvature occur, this dissipation becomes large and is added to the physical viscous effects. The amount of the numerical dissipation depends on the type of Euler solver used. Hänel and Breuer (1989) demonstrated the influence of the discretization and TVD flux limiters on the skin friction coefficient. Also, they showed results of an experimental study of flow over the CAST-7 airfoil which represents a class of supercritical airfoils of high sensitivity. The interferograms clearly demonstrated that near the design point ($\alpha = 0$, $M = 0.76$) a small change of the angle of attack α from 0 to $0^\circ\ 10'$ altered the pattern from nearly shock-free flow to flow with a remarkable shock.

The necessity of using very fine meshes in transonic computations was emphasized by Sobieczky and Seebass (1984), who pointed out again the great sensitivity of transonic flow to changes in airfoil or wing geometry. They mentioned that very small changes in the airfoil configuration may have rather dramatic effects on the pressure distribution, and the supersonic expansion near the nose of the airfoil appears to be an especially delicate one. A typical computational grid used in the airfoil design with the fictitious gas method was very dense in the beginning and end of the local supersonic region. The method of characteristics clustered the grid points on the upper surface of the shockless airfoil near the supersonic expansion. If the shock-free flow is to be tested with a Euler code, then the employed mesh must resolve the local upper-surface curvature quite accurately.

Provided the influence of discretization errors was minimized, numerical simulations did supply convincing evidence of the existence of nonunique solutions.

Hafez and Dimanlig (1996) obtained a numerical solution of the unsteady Euler equations for compressible flow over a circular cylinder. The equations, written in cylindrical coordinates, were discretized on an orthogonal grid via central differences for spatial derivatives, using a simple second-order artificial viscosity form and a special treatment of the boundary conditions. Backward differences in time were employed, resulting in a large system of nonlinear difference equations at each step. A direct solver, based on an efficient Gaussian elimination procedure for banded matrices, was used to solve the linearized system of equations. The stability of the nonunique solutions of the steady Euler equations was investigated. It was demonstrated that the symmetric solutions, with zero circulation, are not stable. For a certain Mach number range, a periodic solution was obtained where the shock oscillation persisted. If a periodic circulation (within a certain frequency range) was enforced in the far field, an irregular solution emerged with unpredictable shock motions. For such a solution, the Lyapunov exponent was shown to be greater than zero, indicating the appearance of chaos. The obtained results, therefore, testified that the nonuniqueness of the Euler equations is not restricted to particular airfoil shapes, but even for very basic shapes, such as a cylinder, one can find multiple steady-state solutions.

A fundamental numerical study was performed by Hafez and Guo (1999a), who analysed the nonuniqueness of solutions to two-dimensional Euler equations in the case of special wavy shapes of airfoils. An implicit central difference scheme was used with fourth-order and second-order artificial dissipations in order to establish the nonuniqueness. It goes without saying that the existence of concave arcs of the supersonic portion of the airfoil results in the arising of shock waves. Consequently, several concave arcs in the Hafez and Guo's airfoils yielded the formation of complex flow patterns involving interaction of multiple shocks. The symmetric solution was

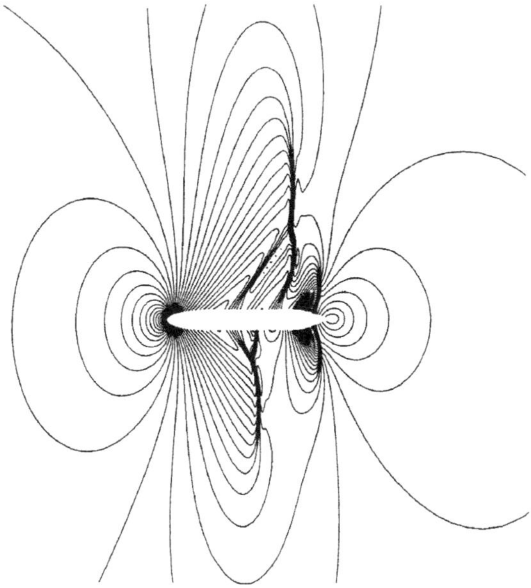

Figure 4.48 Pressure contours over an airfoil with a wavy surface. The asymmetric solution obtained for $M_\infty = 0.84$, $\alpha = 0$. Computations of Hafez and Guo (1999a). Copyright © 1999 Elsevier Science. Reproduced with permission.

shown to be unstable, so that small perturbations led to the asymmetric solution (Figure 4.48). Computations were carried out on a C-grid of 501×61 points. A different code developed by the authors and based on the Cauchy–Riemann equations and Crocco's relations yielded the same results on the nonuniqueness on the same grid. Also, an example for a three-dimensional wing admitting nonunique solutions of the potential and Euler equations was demonstrated. In another paper, Hafez and Guo (1999b) established nonunique solutions of the Navier–Stokes equations and showed that a steady state (independent of time) may cease to exist for a certain range of Reynolds number. It was concluded that though some of the numerical solutions may have no physical relevance, for example, exhibit negative density or be physically unstable, other nonunique solutions may be stable and hence of interest for practical applications.

Therefore, it is likely that the nonuniqueness can appear at certain conditions as a feature of the boundary-value problem for the Euler equations in the differential form and as a real physical phenomenon. This conclusion is supported by the fact that there are other kinds of transonic flows for which the development of self-excited oscillations under the steady boundary conditions is well documented. Moreover, the viscosity plays little role in the mechanism of the oscillations. An example of this kind is given by the transonic cavity oscillations. Here, the supersonic flow or jet interacts with the mainly subsonic flow inside a cavity in such a way that the unsteady behaviour is dominated by the acoustic resonances of the cavity. A typical case is the Hartmann-oscillator, where a supersonic jet is injected into a cavity. The heavy oscillations of such a system can be explained and modelled by the shock motion and related entropy changes caused by the varying shock strength during the periodic change of the flow field in the cycle of oscillations. Models are described in Jungowski and Meier (1984) and Jungowski and Grabitz (1987). These models are able to predict the unsteady behaviour with sufficient accuracy without resorting to viscosity effects. A related problem is that of self-induced oscillations of the supersonic underexpanded jet impinging on a rigid plate normal to the jet axis. In particular, for large plates, the acoustic feedback through the near field to the edges of the orifice is of primary importance for the mechanism explaining the flow instability, the arising of oscillations, and the radiation of intense sound of discrete frequencies which are independent of the cellular structure of the jet (Powell 1988).

Appendix A

Boundary-Value Problem with a Damping Condition

Below we prove the Fredholm solvability of a boundary-value problem for the linear equation

$$Lu - \lambda u := k(x,t)u_{tt} + [a(x,t)u_x]_x - \alpha(x,t)u_t - \lambda u = f(x,t), \qquad \text{(A.1)}$$

where $a(x,t) \geqslant \delta > 0$. The coefficient $k(x,t)$ changes the sign in the rectangle $D = \{(x,t) \in \mathbb{R}^2;\ -1 < x < 1,\ 0 < t < T\}$, while $k(x,0)$ and $k(x,T)$ are supposed to be positive. Therefore, the type of the equation is elliptic on the lower and upper sides of D. Let the coefficients of Equation (A.1) be smooth enough: $k, a \in C^3(\bar{D})$; $\alpha \in C^2(\bar{D})$, and λ be a real-valued parameter. The boundary conditions are

$$\left.\begin{aligned} u &= 0 && \text{at } t = 0,\ -1 < x < 1,\\ u_t &= 0 && \text{at } t = T,\ -1 < x < 1,\\ u_x + R_1(t)u_t &= 0 && \text{at } x = -1,\ 0 < t < T,\\ u_x + R_2(t)u_t &= 0 && \text{at } x = 1,\ 0 < t < T. \end{aligned}\right\} \qquad \text{(A.2)}$$

The Fredholm solvability of Problem (A.1), (A.2) with $R_1 \equiv R_2 \equiv 0$ was established in Section 1.4 under the condition $2\alpha \pm k_t > 0$ at the characteristic points and endpoints of the type degeneracy line $k(x,t) = 0$. For the transonic flow with a local supersonic region, this condition is violated at the endpoint B of the sonic line where the flow decelerates from supersonic to subsonic velocities, i.e. the acceleration is negative. That is why the aim of this appendix is to study Problem (A.1), (A.2) without the condition $2\alpha \pm k_t > 0$ at point B. Instead, we employ the oblique derivative condition $u_x + R_2u_t = 0$ on a segment embracing point B.

Let $\|\cdot\|_n$ again denote $\|\cdot\|_{W^{n,2}(D)}$ for brevity.

Lemma A.1 (Kuz'min 2001). *Let the line of type degeneracy defined by the equation $k(x,t) = 0$ be a simple curve with endpoints* A *and* B *on the side $x = 1, 0 \leqslant t \leqslant T$, of D, and $k_x \leqslant -\delta < 0$ on* AB, *while $-k_t/k_x$ monotonously increases along* AB *(see Figure A.1). If $2\alpha \pm k_t > 0$ on a portion* AQ *of the type degeneracy line, and the*

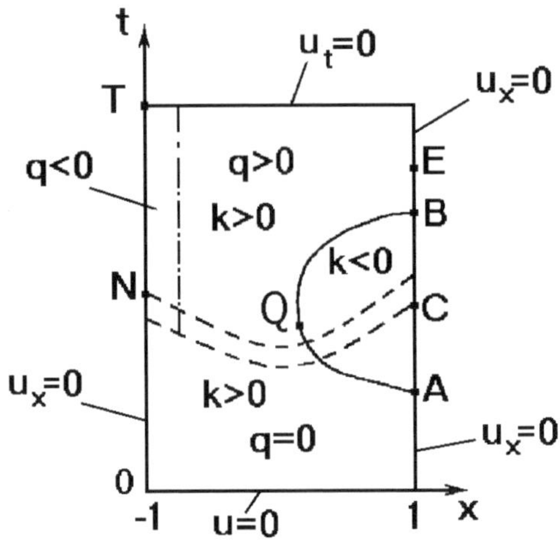

Figure A.1 A domain in which problem (A.1), (A.2) is studied. Reproduced from Kuz'min (2001) by permission of Birkhäuser.

coefficients $R_1(t)$, $R_2(t)$ satisfy the conditions

$$\begin{aligned} &R_1(t) = 0 && \text{at } 0 \leqslant t \leqslant t_N, \\ &R_1(t) < 0 && \text{at } t_N < t \leqslant T, \\ &R_2(t) = 0 && \text{at } 0 \leqslant t \leqslant t_C \text{ and } t_E \leqslant t \leqslant T, \\ &R_2(t) > 0 && \text{at } t_C < t < t_E, \end{aligned}$$

$$R_2(t_B) < 2 \min_{QB} \left| \frac{k_x}{2\alpha \pm k_t} \right|, \tag{A.3}$$

then the estimates

$$\|u\|_n \leqslant m\|Lu - \lambda u\|_{n-1}, \qquad n = 1, 2, \quad m > 0, \tag{A.4}$$

are true for sufficiently large λ and each function $u \in W^{2,2}(D)$ satisfying boundary conditions (A.2) and such that $Lu \in W^{1,2}(D)$. Moreover, Problem (A.1), (A.2) is Fredholm solvable in the space $W^{2,2}(D)\ \forall f \in W^{1,2}(D)$, i.e. the problem is uniquely solvable for any real λ with the exception of at most a sequence of eigenvalues λ_i, $i = 1, 2, \ldots$. The only $\lambda = -\infty$ can be a limit for this sequence.

If, in addition, $Lu - \lambda u = f \in W^{3,2}(D)$ vanishes near $t = 0$, $t = T$, and we have $2\alpha - 3k_t > 0$, $2\alpha - 5k_t > 0$ on AQ,

$$R_2(t_B) < 2 \min_{QB} \left\{ \left| \frac{k_x}{2\alpha - 3k_t} \right|, \left| \frac{k_x}{2\alpha - 5k_t} \right| \right\}, \tag{A.5}$$

then (A.4) is true at $n = 3, 4$, and the solution u belongs to $W^{4,2}(D)$.

Proof. As in Section 1.4, we shall analyse the identity obtained with Green's formula,

$$
\begin{aligned}
&-\int_D (Lu-\lambda u)\cdot(2u_t+2qu_x+rku)\mathrm{e}^{-\mu t}\,\mathrm{d}D \\
&\quad=\left(\int_{D^-}+\int_{D^+}\right)[(2\alpha+k_t-qk_x-kq_x-\mu k+rk^2)u_t^2 \\
&\qquad\qquad+2(\alpha q-\mu kq+qk_t+kq_t)u_tu_x \\
&\qquad\qquad+(\mu a-a_t-a_xq+aq_x+rak)u_x^2 \\
&\qquad\qquad+rak_xu_xu+rk(\alpha-\mu k+2k_t)u_tu \\
&\qquad\qquad+\lambda(\mu-q_x+rk)u^2]\mathrm{e}^{-\mu t}\,\mathrm{d}D \\
&\qquad+\int_\Gamma[(qku_t^2-2au_tu_x-qau_x^2+q\lambda u^2-raku_xu)\mathrm{e}^{-\mu t}\,\mathrm{d}t \\
&\qquad\qquad+(ku_t^2+2qku_tu_x-au_x^2-\lambda u^2+rk^2u_tu)\mathrm{e}^{-\mu t}\,\mathrm{d}x], \qquad \text{(A.6)}
\end{aligned}
$$

where D^- and D^+ are the hyperbolic and elliptic parts of D in which $k<0$ and $k>0$, respectively. The coefficient $r(x,t)$ vanishes in $\bar{D}^-$, while $r(x,t)=\text{const.}>0$ in D^+. The function $q(x,t)$ is defined by the expression

$$q(x,t)=q_0\cdot\psi(x)\cdot\varphi(x,t)\exp[-\varepsilon\mu k^2(x,t)], \qquad \text{(A.7)}$$

which is the crucial point for the proof. The positive parameters q_0, ε, μ, and the cut-off functions $\varphi\in C^\infty(\bar{D})$, $\psi\in C^\infty[-1,1]$ will be chosen so as to provide sufficiently large values of q on a portion QB of the type degeneracy line and sufficiently small $|q|$ outside a vicinity of QB. Then identity (A.6) will yield estimate (A.4) with $n=1$.

We fix $\psi(x)<0$ at $-1\leqslant x\leqslant -1+\sigma$ and $\psi(x)=1$ at $-1+2\sigma\leqslant x\leqslant 1$, where $\sigma>0$, so that q is negative in a narrow strip adjacent to the segment $x=-1$, $0\leqslant t\leqslant T$, while $q>0$ at $x>-1+\sigma$ (see Figure A.1). A proof of the first estimate $\|u\|_1\leqslant m\|Lu-\lambda u\|_0$ is split into four steps.

Step 1. Let us construct two nonintersecting lines located below point Q so that the upper line begins at point N, while the lower one ends at point C, $t_A<t_C<t_B$, of the boundary $\Gamma=\partial D$ (see the dashed lines in Figure A.1). We choose such a function $\varphi(x,t)\geqslant 0$ that $\varphi\equiv 1$ in the subdomain above the upper dashed line, and $\varphi\equiv q\equiv 0$ below the lower one. Obviously, $\varphi=\psi=1, q(x,t)=q_0$ along the arc QB of the type degeneracy line. Therefore, one can fix q_0 large enough, $q_0=\max_{\mathrm{QB}}|(2\alpha\pm k_t)/k_x|$, so that $2\alpha\pm k_t-q_0k_x\geqslant\delta>0$ on QB. Also, $2\alpha\pm k_t\geqslant\delta>0$ along the arc AQ due to the conditions of Lemma A.1. Then the coefficient in front of u_t^2 in the second line of identity (A.6) is positive on the whole AQB.

Denote by I_{D^-}, I_{D^+}, I_Γ the three integrals on the right-hand side of (A.6).

By choosing a sufficiently large μ and small $\varepsilon>0$, due to the term μa in front of u_x^2 and the term $-\mu k$ in front of u_t^2, where $-k>0$, we can obtain an estimate for the integral I_{D^-}. Indeed, let us represent the coefficient in front of u_t^2 in the form

$$(2\alpha+k_t-qk_x+\mu|k|/2)+|k|(\mu/2+q_x), \qquad \text{(A.8)}$$

and choose ε so small that the sum $-q(\mu)\cdot k_x+\mu|k|/2$ becomes an increasing function of μ, $\forall(x, y) \in D^-$. At sufficiently large μ we obtain $(2\alpha + k_t - qk_x + \mu|k|/2) \geqslant \delta > 0$ in D^-. After that, we choose ε so small that the inequality

$$\tfrac{1}{2}\mu \pm q_x(\mu) = \mu(\tfrac{1}{2} \mp 2\varepsilon k k_x q) + O(1) \geqslant \tfrac{1}{3}\mu + O(1)$$

is true in D in addition to the previous one. Thus, (A.8) becomes positive in $\bar{D}^-$. The coefficient in front of $u_t u_x$ in (A.6) is bounded from above by a constant independent of μ, because the functions $-\mu k q$ and $k q_t$ are bounded at $0 \leqslant \mu < \infty$. In the meanwhile, the coefficient in front of u_x^2 unlimitedly increases with increasing μ due to μa. Therefore, under sufficiently large μ, the functions u_t and u_x constitute a positive definite form:

$$I_{D^-} \geqslant \int_{D^-} [m(u_t^2 + u_x^2) + \lambda(\mu - q_x)u^2]\mathrm{e}^{-\mu t}\,\mathrm{d}D \geqslant m\|u\|_1^2.$$

Step 2. The integration over the boundary Γ in (A.6) is carried out in the counter-clockwise direction. Using the Dirichlet condition at $t = 0$ and the Neumann one at $t = T$, we find

$$I_\Gamma = \int_{S_1 \cup S_2} (qku_t^2 - 2au_t u_x - qau_x^2 + q\lambda u^2 - raku_x u)\mathrm{e}^{-\mu t}\,\mathrm{d}t$$
$$+ \int_{-1}^{1} ku_t^2\big|_{t=0}\,\mathrm{d}x + \int_{-1}^{1} (au_x^2 + \lambda u^2)\big|_{t=T}\mathrm{e}^{-\mu T}\,\mathrm{d}x,$$

where $S_1 = \{(x,t) : x = -1, 0 \leqslant t \leqslant T\}$ is the left-hand side of Γ and $S_2 = \{(x,t) : x = 1, 0 \leqslant t \leqslant T\}$ is the right-hand side. The second and third integrals are evidently nonnegative at $\lambda \geqslant 0$. Therefore,

$$I_\Gamma \geqslant I_{S_1} + I_{S_2} = \int_{S_1 \cup S_2} (qku_t^2 - 2au_t u_x - qau_x^2 + q\lambda u^2 - raku_x u)\mathrm{e}^{-\mu t}\,\mathrm{d}t.$$

We assume $t_C < t_N$ for simplicity. Due to the boundary condition $u_x(-1, t) = 0$ (that is $R_1(t) = 0$) at $t \leqslant t_N$, and $q \equiv 0$ at $t \leqslant t_C$, $q(0, t) < 0$ at $t_C < t \leqslant T$, we have

$$I_{S_1} = -\int_{t_N}^{T} (qku_t^2 - 2au_t u_x - qau_x^2 + q\lambda u^2 - raku_x u)\big|_{x=-1}\mathrm{e}^{-\mu t}\,\mathrm{d}t$$
$$- \int_{t_C}^{t_N} (qku_t^2 + q\lambda u^2)\big|_{x=-1}\mathrm{e}^{-\mu t}\,\mathrm{d}t$$
$$\geqslant \int_{t_N}^{T} (2au_t u_x + qau_x^2 - q\lambda u^2 + raku_x u)\big|_{x=-1}\mathrm{e}^{-\mu t}\,\mathrm{d}t.$$

By substituting the boundary condition $u_x = -R_1 u_t$ at $t_N < t \leqslant T$, where $R_1(t) < 0$, we obtain

$$I_{S_1} \geqslant \int_{t_N}^{T} [(2|R_1| + qR_1^2)au_t^2 - q\lambda u^2 + rakR_1 u_t u]\big|_{x=-1}\mathrm{e}^{-\mu t}\,\mathrm{d}t.$$

The coefficient $2|R_1| + qR_1^2$ is positive as $|q(-1,t)|$ is small enough at sufficiently large μ owing to (A.7). That is why the integrand is nonnegative if λ is large enough as compared to r^2, therefore $I_{S_1} \geqslant 0$. Thus,

$$I_\Gamma \geqslant I_{S_2} = \int_{S_2} (-2au_tu_x + qku_t^2 - qau_x^2 + q\lambda u^2 - raku_xu)\big|_{x=1} e^{-\mu t}\, dt. \quad (A.9)$$

Step 3. On the segments $0 \leqslant t \leqslant t_C$ and $t_E \leqslant t \leqslant T$ of S_2, we prescribe the boundary condition $u_x = 0$ (that is $R_2(t) = 0$). Since q vanishes below the lower dashed line in Figure A.1 and $q(1,t) > 0$ above the upper one, integral (A.9) reduces to

$$\begin{aligned} I_{S_2} &= \int_{t_C}^{t_E} (-2au_tu_x + qku_t^2 - qau_x^2 + q\lambda u^2 - raku_xu)\big|_{x=1} e^{-\mu t}\, dt \\ &\quad + \int_{t_E}^{T} (qku_t^2 + q\lambda u^2)\big|_{x=1} e^{-\mu t}\, dt \\ &\geqslant \int_{t_C}^{t_E} (-2au_tu_x + qku_t^2 - qau_x^2 + q\lambda u^2 - raku_xu)\big|_{x=1} e^{-\mu t}\, dt \end{aligned}$$

at $\lambda \geqslant 0$. By substituting the boundary condition $u_x = -R_2u_t$ at $t_C < t < t_E$, where $R_2(t) > 0$, we obtain

$$I_{S_2} \geqslant \int_{t_C}^{t_E} [(2aR_2 + qk - qaR_2^2)u_t^2 + q\lambda u^2 + rakR_2u_tu]\big|_{x=1} e^{-\mu t}\, dt. \quad (A.10)$$

Let us show that

$$2aR_2 + qk - qaR_2^2 \geqslant \delta_1 > 0 \quad (A.11)$$

on the segment CE. At point B, inequality (A.11) reduces to $2 - q_0R_2(t_B) \geqslant \delta_1 > 0$. The latter is true owing to the choice of q_0 at Step 1 and condition (A.3). Consequently, (A.11) holds in a vicinity of point B due to the continuity of the functions on the left-hand side. Outside the vicinity, (A.11) can be provided by sufficiently large μ which makes q small enough owing to (A.7). At point E, where $R_2 = 0$, inequality (A.11) holds due to the term $qk > 0$. Thus, (A.10) and (A.11) yield

$$\begin{aligned} I_{S_2} &\geqslant \int_{t_C}^{t_E} [\delta_1u_t^2 + q\lambda u^2 + rakR_2u_tu]\big|_{x=1} e^{-\mu t}\, dt \\ &\geqslant \int_{t_B}^{t_E} [\delta_1u_t^2 + q\lambda u^2 + rakR_2u_tu]\big|_{x=1} e^{-\mu t}\, dt \end{aligned}$$

since $r = 0$ at $t_C \leqslant t \leqslant t_B$. The latter integral is nonnegative due to the term $q\lambda u^2$ if λ is large enough as compared to r^2. Then (A.9) shows that $I_\Gamma \geqslant 0$.

Step 4. For the integral I_{D^+} in (A.6), choosing sufficiently large r_0 under fixed μ, we obtain

$$\begin{aligned} I_{D^+} \geqslant \int_{D^+} [m(u_t^2 + u_x^2) &+ rak_xu_xu + rk(\alpha - \mu k + 2k_t)u_tu \\ &+ \lambda(\mu - q_x + rk)u^2] e^{-\mu t}\, dD. \end{aligned}$$

The coefficients in front of $u_x u$ and $u_t u$ do not involve λ, while the coefficient in front of u^2 unlimitedly increases with increasing λ because $\mu - q_x \geqslant \mu/2 + O(1) > 0$ (see Step 1). Therefore, at sufficiently large λ, we obtain $I_{D^+} \geqslant m\|u\|_1^2$. Also, $I_\Gamma \geqslant 0$ owing to the conclusion of the previous paragraph. That is why all the integrals on the right-hand side of (A.6) are nonnegative at sufficiently large μ, r_0, λ. Moreover,

$$-\int_D (Lu - \lambda u) \cdot (2u_t + 2qu_x + rku)\mathrm{e}^{-\mu t}\,\mathrm{d}D \geqslant m\|u\|_1^2. \tag{A.12}$$

Now, using Young's inequality, we obtain the estimate $\|u\|_1 \leqslant m\|Lu - \lambda u\|_0$.

The remainder of the proof of Lemma A.1 consists of two steps.

Step 5. The *a priori* estimate (A.4) with $n = 2$ can be obtained from the inequality

$$-2\int_D (Lv - \lambda v)_t \cdot (v_{tt} + qv_{tx})\mathrm{e}^{-\mu t}\,\mathrm{d}D \geqslant m\|v_t\|_1^2 - \|v_t\|_1\|v\|_1 \tag{A.13}$$

similar to (A.12), where $v = \zeta u$, the cut-off function $\zeta(t)$ vanishes in the strips $0 \leqslant t \leqslant \eta/2$ and $T - \eta/2 \leqslant t \leqslant T$, while it equals unity at $\eta \leqslant t \leqslant T - \eta$. The function ζ is employed in order to provide the appropriate boundary conditions $v_t(x, 0) = v_{tt}(x, T) = 0$. Estimate (A.13) yields $\|v_t\|_1 \leqslant m\|f_t\|_0$ and $\|u_t\|_{W^{1,2}(D_\eta)} \leqslant m\|f\|_1$, where $D_\eta = D \cap (\eta < t < T - \eta)$. Since Equation (A.1) is elliptic in $D\backslash D_\eta$, the norm $\|u_t\|_{W^{1,2}(D\backslash D_\eta)}$ can be estimated by $\|u\|_1$ and $\|f\|_0$. By summing the inequalities for $\|u_t\|_1$ in D_η and $D\backslash D_\eta$, we obtain $m\|u_t\|_1 \leqslant \|f\|_1$. It can be seen from Equation (A.1) that the norm $\|u_{xx}\|_0$ can be estimated by $\|u_{tt}\|_0$, $\|Lu - \lambda u\|_0$, and $\|u\|_1$. Therefore, we finally obtain $\|u\|_2 \leqslant m\|Lu - \lambda u\|_1$.

The latter estimate permits one to prove the unique solvability of Problem (A.1), (A.2) in $W^{2,2}(D)$ under sufficiently large λ with the method of prolongation by a parameter γ as follows. The equation

$$L^0 u := ku_{tt} + (au_x)_x + \alpha_0 u_t - \lambda u = f \tag{A.14}$$

endowed with conditions (A.2) is uniquely solvable in $W^{2,2}(D)$ if α_0 is sufficiently large. This is proved in Section 1.4 in the case of the Neumann condition at $x = \pm 1$, and the proof needs only minor modifications if the oblique derivative condition is prescribed with R_1 and R_2 pointed out in Lemma A.1. The inverse operator $(L^0)^{-1}$ from the whole $W^{1,2}(D)$ into $W^{2,2}(D)$ is bounded due to (A.4). That is why the family of equations $L^\gamma u := L^0 u + \gamma(\alpha - \alpha_0)u_t = f$, where $0 \leqslant \gamma \leqslant 1$, under boundary conditions (A.2) can be represented in the equivalent form $u + \gamma(L^0)^{-1}[(\alpha - \alpha_0)u_t] = (L^0)^{-1} f$. This equation is solvable in $W^{2,2}(D)$ for $0 \leqslant \gamma \leqslant \gamma_1$ if γ_1 is small enough, since $\|(L^0)^{-1}[(\alpha - \alpha_0)u_t]\|_2 \leqslant m\|(\alpha - \alpha_0)u_t\|_1 \leqslant m\|u\|_2$. In the same way, using the estimate $\|u\|_2 \leqslant m\|L^\gamma u\|_1$, which is uniform with respect to γ, one can prove the solvability of the equation $L^\gamma u = f$ for $\gamma_1 \leqslant \gamma \leqslant 2\gamma_1$. Thus, in a finite number of steps, one gets the solvability of the equation $L^\gamma u = f$ at $\gamma = 1$, i.e. the solvability of Problem (A.1), (A.2).

In order to prove the Fredholm alternative at any real λ, let us represent Equation (A.1) in the form $L_{\lambda_0} u := Lu - \lambda_0 u = f + (\lambda - \lambda_0)u$. Under sufficiently large λ_0, this equation is equivalent to $u = (\lambda - \lambda_0)L_{\lambda_0}^{-1}u + L_{\lambda_0}^{-1}f$ due to the conclusion of the previous paragraph. Taking into consideration the compactness of the embedding operator from $W^{2,2}(D)$ into $W^{1,2}(D)$, we obtain that $L_{\lambda_0}^{-1}$ is a compact operator in $W^{2,2}(D)$, which means the Fredholm solvability of Problem (A.1), (A.2).

Step 6. From an inequality similar to (A.13), one can derive the estimate $\|u_{tt}\|_1 \leqslant m\|f\|_2$ valid under conditions of Lemma A.1 for functions f vanishing in the vicinities of $t = 0$ and $t = T$. The vanishing of f guarantees that $u_{tt} \in W^{1,2}$ in the vicinities of the corner points $t = 0$, $t = T$, $x = \pm 1$ of the boundary Γ. The fact that u_{tt} does belong to $W^{1,2}$ can be proved by extending the equation across the segments $t = 0$ and $t = T$, $-1 \leqslant x \leqslant 1$, and then by treating boundary-value problems in the extended domains. Equation (A.1) makes it possible to estimate the norm $\|u_{xx}\|_1$ by $\|u_{tt}\|_1$, $\|f\|_1$, and $\|u\|_2$, therefore, we arrive at the estimate $\|u\|_3 \leqslant m\|f\|_2$. In the same way, we can obtain $\|u\|_4 \leqslant m\|f\|_3$. Then using the method of prolongation by a parameter and the solvability of (A.14) in $W^{4,2}(D)$ under sufficiently large α_0, one can prove that the solution u of Problem (A.1), (A.2) belongs to $W^{4,2}(D)$. □

Remark A.2. If the Neumann condition is prescribed at $x = 1$, $t_C < t < t_E$, i.e. $R_2(t) \equiv 0$, then the proof of Lemma A.1 fails, because inequality (A.11) is violated, therefore the estimate $I_{S_2} \geqslant 0$ is not true.

References

Adams RA 1975 *Sobolev Spaces*. Academic Press.

Adamson TC 1972 Unsteady transonic flows in two-dimensional channels. *J. Fluid Mech.* **52**, 437–449.

Adamson TC and Feo A 1975 Interaction between a shock-wave and a turbulent boundary layer in transonic flow. *SIAM J. Appl. Math.* **29**, 121–145.

Adamson TC and Messiter AF 1980 Analysis of two-dimensional interactions between shock-waves and boundary layers. *A. Rev. Fluid Mech.* **12**, 103–138.

Agrawal S and Messiter AF 1984 Turbulent boundary-layer interaction with a shock wave at a compression corner. *J. Fluid Mech.* **143**, 23–46.

Allison DO and Sewall WG 1995 Airfoil modification effects on subsonic and transonic pressure distributions and performance for the EA-6B airplane. NASA TP-3516.

An C-F and Barron RM 1995 Transonic Euler computation in streamfunction coordinates. *Int. J. Numer. Meth. Fluids* **20**, 75–94.

Azarenok BN and Ivanenko SA 2000 Application of adaptive grids in numerical analysis of time-dependent problems in gas dynamics (in Russian). *Zh. Vychisl. Matem. Mat. Fiz.* **40**, 1330–1349.

Aziz AK, Bell JB and Schneider M 1982 Existence and uniqueness for an equation of mixed type in a rectangle. *Math. Meth. Appl. Sci.* **4**, 307–316.

Aziz AK, Lemmert R and Schneider M 1990 A finite element method for the nonlinear Tricomi problem. *Numer. Math.* **58**, 95–108.

Back LH and Cuffel RF 1966 Detection of oblique shock in a conical nozzle with a circular-arc throat. *AIAA J.* **4**, 2219–2221.

Ballhaus WF and Goorjian PM 1977 Implicit finite difference computation of unsteady transonic flow about airfoils including the treatment of irregular shock wave motions. AIAA Paper 77-205.

Bartelheimer W 1995 An improved integral equation method for the design of transonic airfoils and wings. AIAA Paper 95-1688, pp. 463–473.

Bartels RE 1995 Interactive boundary layer computations using the improved k–ε turbulence model. In *Numerical Methods in Laminar and Turbulent Flows* (ed. C Taylor and P Durbetaki), vol. 9, pt 1, pp. 561–538.

Bartels RE and Rothmayer AP 1994 An efficient interacting boundary layer method for supersonic flow in the transonic regime. *Comput. Fluids* **23**, 55–62.

Bartels RE and Rothmayer AP 1995 An IBL approach to multiscaled shock induced oscillations. AIAA Paper 95-2157.

Baruzzi G, Habashi WG and Hafez MM 1989 Non-unique solutions of the Euler equations. In *Advances in Fluid Dynamics. Proc. Symp. in Honor of Maurice Holt on His 70th Birthday* (ed. WF Ballhaus Jr and MY Hussaini), pp. 1–10. Williamsburg.

Bauer F, Garabedian P, Korn D and Jameson A 1977 *Supercritical wing sections*. III. Lecture Notes in Economics and Mathematical Sciences. Springer.

Bayliss A and Turkel E 1982 Far field boundary conditions for compressible flow. *J. Computat. Phys.* **48**, 182–199.

Baysal O and Ghayour K 2001 Continuous adjoint sensitivities for optimization with general cost functionals on unstructures meshes. *AIAA J.* **39**, 48–55.

Berger H 1989 A convergent finite element formulation for transonic flow. *Numer. Math.* **56**, 425–447.

Berger H and Feistauer M 1993 Analysis of the finite element variational crimes in the numerical approximation of the transonic flow. *Math. Computat.* **61**, 493–521.

Berger H, Warnecke G and Wendland WL 1997 Analysis of a FEM/BEM coupling method for transonic flow computations. *Math. Computat.* **66**, 1407–1440.

Bers L 1954 Function-theoretical properties of solutions of partial differential equations of elliptic type. Annals of Mathematics Studies, no. 33, pp. 69–94. Contributions to the Theory of Partial Differential Equations. Princeton University Press.

Bers L 1958 Mathematical aspects of subsonic and transonic gas dynamics. *Surv. Appl. Math.* **3**, 1–208.

Bodonyi RJ and Kluwick A 1982 Supercritical transonic trailing-edge flow. *Q. J. Mech. Appl. Math.* **35**, 265–277.

Bodonyi RJ and Kluwick A 1998 Transonic trailing-edge flow. *Q. J. Mech. Appl. Math.* **51**, 297–310.

Bodonyi RJ and Smith FT 1986 Shock-wave laminar boundary layer interaction in supercritical transonic flow. *Comput. Fluids* **14**, 97–108.

Boerstoel JW and Huizing GH 1974 Transonic airfoil design using Cartesian coordinates. AIAA Paper 74-539.

Bohning R and Doerffer P 1999 Hybrid and active control of the shock wave–turbulent boundary layer interaction and porous plate transpiration flow. Final Technical Report. EUROSHOCK II. Contract BRPR-CT95-0076. Karlsruhe University.

Bohning R and Jungbluth H 1989 Turbulent shock–boundary-layer interaction with control. Theory and experiment. In *Symp. Transsonicum III*, pp. 389–398. Springer.

Bohning R and Zierep J 1985 Calculation of 2D-turbulent shock/boundary layer interaction at curved surfaces with suction and blowing. In *IUTAM Symp. on Turbulent Shear Layer/Shock Wave Interactions*, pp. 105–112. Palaiseau.

Bowles RI and Smith FT 1989 On boundary-layer transition in transonic flow. UTRC Report 89-26.

Bowles RI and Smith FT 1993 On boundary-layer transition in transonic flow. *J. Engng Math.* **27**, 309–342.

Braun W 1990 Experimentelle Untersuchung der turbulenten Stoß-Grenzschicht-Wechselwirkung mit passiver Beeinflussung: Dis. Karlsruhe University. Fakultaet für Maschinenbau.

Brilliant HM and Adamson TC 1973 Shock-wave boundary-layer interactions in laminar transonic flow. AIAA Paper 73-239.

Bristeau MO, Pironneau O, Glowinski R, Periaux J, Perrier P and Poirier G 1979 On the numerical solution of nonlinear problems in fluid dynamics by least squares and finite element method. I. Least squares formulations and conjugate gradient solution of the continuous problem. *Computat. Meth. Appl. Mech. Engng* **17/18**, 619–657.

Bristeau MO, Pironneau O, Glowinski R, Periaux J, Perrier P and Poirier G 1985 On the numerical solution of nonlinear problems in fluid dynamics by least squares and finite element method. II. Application to transonic flow simulations. *Computat. Meth. Appl. Mech. Engng* **51**, 363–394.

Bur R, Délery J and Corbel B 1997 Basic study of passive control applied to a two-dimensional transonic interaction. In *EUROSHOCK – Drag Reduction by Passive Shock Control*. Notes on Numerical Fluid Mechanics, vol. 56, 89–112.

Busemann A 1949 The drag problem at high subsonic speeds. *J. Aeronaut. Sci.* **16**, 337–344.

Carpenter MH and Casper JH 1997 The accuracy of shock capturing in two spatial dimensions. AIAA Paper 97-2107, pp. 488–498.

Chandrasekhara MS, Wilder MC and Carr LW 1998 Competing mechanisms of compressible dynanmic stall. *AIAA J.* **36**, 387–393.

Chen C and Garabedian P 1998 Complex analysis of transonic flow. In *Frontiers of computational fluid dynamics* (ed. DA Caughey and MM Hafez), pp. 107–124. World Scientific.

Chen C-L, Chow C-Y, Van Dalsem WR and Holst TL 1989 Computation of viscous transonic flow over porous airfoils. *J. Aircraft* **26**, 1067–1075.

Chen T 1983 Transonic flowfield computation using a modified shock-point operator. In *Second Symp. on Numerical and Physical Aspects of Aerodynamic Flows*. California State University, Long Beach.

Cherry TM 1949 Flow of compressible fluid about a cylinder. *Proc. R. Soc. Lond.* A **196**, 1–31.

Cherry TM 1959 Some nozzle flows found by the hodograph method. *J. Austral. Math. Soc.* **1**, 80–94.

Cherry TM 1960 Nozzle flows found by the hodograph method. II. *J. Austral. Math. Soc.* **1**, 357–367.

Chow C-Y, Chen C-L, Holst TL and Van Dalsem WR 1985 Numerical simulation of transonic flow over porous airfoils. AIAA Paper 85-5022.

Ciarlet PG 1978 *The finite element method for elliptic problems*. Amsterdam: North-Holland.

Ciavaldini J-F, Pogu M and Tournemine G 1986 Existence and regularity of stream functions for subsonic flows past profiles with a sharp trailing edge. *Arch. Ration. Mech. Analysis* **93**, 1–14.

Cole JD 1951 Drag of a finite wedge at high subsonic speed. *J. Math. Phys.* **30**, 79–93.

Cole JD 1983 Note on the axisymmetric sonic jet. *SIAM J. Appl. Math.* **43**, 944–948.

Cole JD and Cook LP 1986 *Transonic aerodynamics*. Amsterdam: North-Holland.

Cook LP 1978 A uniqueness proof for a transonic flow problem. *Indiana Univ. Math. J.* **27**, 51–71.

Cook LP, Newman E, Rimbey S and Schleiniger G 1999 Sonic and subsonic axisymmetric nozzle flows. *SIAM J. Appl. Math.* **59**, 1812–1824.

Cosentino GB 1983 Modifying a general aviation airfoil for supercritical flight. *J. Aircraft* **20**, 377–379.

Courant R and Hilbert D 1962 *Methods of mathematical physics*. II. *Partial differential equations*. Interscience.

Crocco L and Chen S-I 1956 *Theory of combustion instability in liquid propellant rocket motors*. Butterworth Scientific.

Culick FEC 1994 Some recent results for nonlinear acoustics in combustion chambers. *AIAA J.* **32**, 146–169.

Dachev GD 1982 Boundary value problems for a class of equations of mixed type. *Diff. Eqns* **18**, 1356–1363.

Dachev GD 1987 Smoothness of the solutions of the generalized Tricomi problem with the Neumann condition on the boundary (in Russian). *Annuaire Univ. Sofia Fac. Math. Inform.* **81**, 103–121.

Dachev GD 1991 On the existence of smooth solutions of boundary value problems for a class of equations of mixed type. *Diff. Eqns* **27**, 435–441.

Daripa P 1991 Solvability condition and its application to fast numerical solution of overposed problems in compressible flows. *J. Computat. Phys.* **95**, 436–449.

Delanaye M, Geuzaine Ph and Essers JA 1997 Development and application of quadratic reconstruction schemes for compressible flows on unstructured adaptive grids. AIAA Paper 97–2120, pp. 250–260.

Délery J 1989 Recent basic studies on transonic shock-wave/turbulent boundary-layer interactions. In *Symp. Transsonicum III*, pp. 357–378. Springer.

de Matteis PP and Dima C 1998 Numerical prediction of shock wave and boundary layer control effects on airfoils at transonic speed. In *Proc. ECCOMAS 98. Computational Fluid Dynamics, 2* (ed. KD Papailiou *et al.*), pp. 604–612.

Dokolin AYu, Osipov IL and Shipilin AV 1992 *Computation of transonic gas flow past plane turbine cascades.* Modern Problems in Computational Aerohydrodynamics. Advances in Science and Technology in the USSR, pp. 19–33. Boca Raton, FL.

Edwards JW 1993 Transonic shock oscillations calculated with a new interactive boundary layer coupling. AIAA Paper 93-0777.

Egorov IE 1987 Solvability of a boundary value problem for a high-order equation of mixed type. *Diff. Eqns* **23**, 1075–1081.

Egorov IE and Fedorov VE 1995 *Nonclassical higher-order equations in mathematical physics* (in Russian). Novosibirsk.

Egorov IE, Pyatkov SG and Popov SV 2000 *Nonclassical operator—differential equations* (in Russian). Novosibirsk: Nauka.

Eidelman S, Colella P and Shreeve RP 1983 Application of the Godunov method and its second-order extension to cascade flow modeling. AIAA Paper, no. 83-1941.

Emmons HW (ed.) 1958 *Fundamentals of gas dynamics.* Princeton University Press. High Speed Aerodynamics and Jet Propulsion (3).

Engquist B and Osher S 1980 Stable and entropy satisfying approximations for transonic flow calculations. *Math. Computat.* **34**, 45–75.

Engquist B and Sjogreen B 1998 The convergence rate of finite difference schemes in the presence of shocks. *SIAM J. Numer. Analysis* **35**, 2464–2485.

Eriksson K and Johnson C 1988 An adaptive finite element method for linear elliptic problems. *Math. Computat.* **50**, 361–383.

Esin AI and Chernov IA 1977 A shock in flow over a convex corner (in Russian). *Prikl. Mat. Mech.* **41**, 292–297.

Esin AI, Fal'kovitch SV and Chernov IA 1976 Transonic flow of an ideal gas in the vicinity of a corner point of a profile (in Russian). In *Aerodinamika* (Saratov University) **5**(8), 3–10.

Feistauer M 1984 On irrotational flows through cascades of profiles in a layer of variable thickness. *Aplikace Matematiky* **29**, 423–458.

Feistauer M and Nečas J 1985 On the solvability of transonic potential flow problems. *Z. Analysis Anwend.* **4**, 305–329.

Feistauer M and Nečas J 1988 Viscosity method in a transonic flow. *Commun. PDEs* **13**, 775–812.

Ferrand P, Aubert S, Smati L and Bron O 1998 Nonlinear interaction of upstream propagating sound with transonic flow in a nozzle. AIAA Paper 98-2213.

Ferrari C and Tricomi FG 1968 *Transonic aerodynamics*. Academic Press.

Ferziger JH and Peric M 1995 *Computational methods for fluid dynamics*. Springer.

Fichera G 1960 *On a unified theory of boundary value problems for elliptic–parabolic equations*, pp. 97–120. University of Wisconsin Press, Madison.

Fishelov D 1988 The spectrum and the stability of the Chebyshev collocation operator for transonic flow. *Math. Computat.* **51**, 559–579.

Fletcher CAJ 1991 *Computational techniques for fluid dynamics.* 1, 2. Springer Series in Computational Physics.

Frankl FI 1933 Über zweidimensionale Luftströmungen in Kanälen bei Geschwindigkeiten nahe der Schallgeschwindigfeit. *Mat. Sb.* **40**, 59–72.

Frankl FI 1945 On the theory of Laval nozzles (in Russian). *Izv. Akad. Nauk SSSR Matem.* **9**, 387–422.

Frankl FI 1947 On the formation of shock waves in subsonic flows with local supersonic velocities (in Russian). *Prikl. Mat. Mech.* **11**, 199–202. (Transl. into English: NACA Tech. Mem. 1251, 1950.)

Frankl FI 1951 On a new boundary problem for the equation $y(\partial^2 z/\partial x^2) + (\partial^2 z/\partial y^2) = 0$ (in Russian). *Moskovsky Gosud Univ. Učenye Zapiski Mekh. 152*, 99–116.

Frankl FI 1955 An example of transonic gas flow with the region of supersonic velocities bounded downstream by a shock terminating within the flow (in Russian). *Prikl. Mat. Mech.* **19**, 385–392.

Frankl FI 1959 An existence theorem on a weak solution and first approximation of the direct problem in the theory of the plane-parallel Laval nozzle (in Russian). *Izv. Vysš. Učebn. Zaved. Matem.* **6**, 192–201.

Frankl FI 1961 Studies in the area of transonic flow (in Russian). *Inzhenernii Zh.* (*Engng J.*) **1**, 29–34.

Fuglsang DF and Williams MH 1985 Non-isentropic unsteady transonic small disturbance theory. AIAA Paper 85-0600.

Gadd GE 1960 The possibility of normal shock waves on a body with convex surfaces in inviscid transonic flow. *Z. Angew. Math. Phys.* **11**, 51–55.

Gamba IM and Morawetz CS 1996 A viscous approximation for a 2-D steady semiconductor or transonic gas dynamic flow: existence theorem for potential flow. *Commun. Pure Appl. Math.* **49**, 999–1049.

Gaponov SA 1977 Interaction of supersonic boundary layer with acoustical disturbances. *Izv. Akad. Nauk SSSR Mekh. Zhid. Gaza* **6**, 51–56.

Garabedian PR and Korn DG 1971 Numerical design of transonic aorfoils. In *Numerical solution of partial differential equations* (ed. B Hubbard), vol. 2, pp. 253–271.

Gibb J 1988 The cause and cure of periodic flows at transonic speeds. In *Proc. 16th Congress of the Int. Council of Aeronautical Science*, pp. 1522–1530. AIAA.

Gibbons MD, Whitlow Jr W and Williams MH 1986 Nonisentropic unsteady three dimensional small disturbance potential theory. AIAA Paper 86-0863.

Giles MB and Drela M 1987 Two-dimensional transonic aerodynamic design method. *AIAA J.* **25**, 1199–1206.

Gillan MA 1995 Computational analysis of buffet alleviation in viscous transonic flow over a porous airfoil. *AIAA J.* **33**, 769–772.

Gittel HP 1995 A variational approach to boundary value problems for a class of nonlinear systems of mixed type. *Z. Analysis Anwend.* **14**, 575–591.

Glazatov 1997 On periodic transonic flow of viscous gas (in Russian). *Sibirsky Mat. Zh.* **38**, 69–77.

Glowinski R 1984 *Numerical methods for nonlinear variational problems*. Springer Series in Computational Physics.

Glowinski R and Pironneau O 1978 On the computation of transonic flows. In *Functional analysis and numerical analysis* (ed. H Fujita), pp. 143–173. Japan Society for the Promotion of Science.

Göhner U and Warnecke G 1994 A shock indicator for adaptive transonic flow computations. *Numer. Math.* **66**, 423–448.

Göhner U and Warnecke G 1995 A second order finite difference error indicator for adaptive transonic flow computations. *Numer. Math.* **70**, 129–161.

Goldstein S 1948 On laminar boundary layer flow near a point of separation. *Q. J. Mech. Appl. Math.* **1**, 45–69.

Goldstein S, Lighthill MJ and Craggs JW 1948 On the hodograph transformation for high-speed flow. Part I. A flow without circulation. *Q. J. Mech. Appl. Math.* **1**, 442–450.

Gottlieb D 1972 Strang-type difference schemes for multidymensional problems. *SIAM J. Numer. Analysis* **9**, 650–661.

Gottlieb D and Orszag S 1977 *Numerical analysis of spectral methods: theory and applications*. CBMS-NSF Regional Conference Series in Applied Mathematics, no. 26. Philadelphia, PA: SIAM.

Goursat E 1933 *Cours d'analyse mathematique*, 1. Paris: Gauthier-Villars.

Guderley KG 1947 New aspects intransonic flow theory. ATI Technical Data Digest, Wright Field.

Guderley KG 1953 On the presence of shocks in mixed subsonic–supersonic flow patterns. *Adv. Appl. Mech.* **3**, 145–184.

Guderley KG 1957 *Theorie Schallnaher Strömungen*. Springer.

Habashi W and Hafez MM 1995 *Computational fluid dynamics technique*. Gordon & Breach.

Hafez MM and Dimanlig A 1996 Some anomalies of the numerical solutions of the Euler equations. *Acta Mechanica* **119**, 131–140.

Hafez MM and Guo WH 1999a Some anomalies of numerical simulation of shock waves. Part I. Inviscid flows. *Comput. Fluids* **28**, 701–719.

Hafez MM and Guo WH 1999b Some anomalies of numerical simulation of shock waves. Part II. Effect of artificial and real viscosity. *Comput. Fluids* **28**, 721–739.

Hafez MM and Lovell D 1981 Transonic small disturbance calculations using entropy corrections. In *Symp. on Numerical and Physical Aspects of Aerodynamic Flows*. California State University, Long Beach.

Hafez MM and Lovell D 1988 Entropy and vorticity corrections for transonic flows. *Int. J. Numer. Meth. Fluids* **8**, 31–53.

Hafez M, Whitlow Jr W and Osher S 1984 Improved finite difference schemes for transonic potential calculations. AIAA Paper 84-0092.

Hager JO, Eyi S and Lee KD 1994 Two-point transonic airfoil design using optimization for improved off-design performance. *J. Aircraft* **31**, 1143–1147.

Halt M 1962 The design of plane and axisymmetric nozzles by the method of integral relations. In *Symposium Transsonicum* (ed. K Oswatitsch), pp. 310–324. Springer.

Hänel D and Breuer M 1989 On the computation of transonic viscous flows. In *Symp. Transsonicum III*, pp. 75–86. Springer.

Harten A, Engquist B, Osher S and Chakravarthy SR 1987 Uniformly high-order accurate essentially non-oscillatory schemes. III. *J. Computat. Phys.* **71**, 231–303.

Hedde T and Guffond D 1995 ONERA three-dimensional icing model. *AIAA J.* **23**, 1038–1045.

Hirose N, Takanashi S and Kawai N 1987 Transonic airfoil design procedure utilizing a Navier–Stokes analysis code. *AIAA J.* **25**, 353–359.

Hirsh Ch 1990 *Numerical computation of internal and external flows. 2. Computational methods for inviscid and viscous flows*. Wiley.

Holst TL 1999 Fast transonic aerodynamic analysis suitable for the preliminary design environment. AIAA Paper 99-3133, pp. 246–258.

Holten RP 1962 Generalized Goursat problem. *Pac. J. Math.* **12**, 207–224.

Ilinca F, Pelletier D and Ignat L 1998 Adaptive finite element solution of compressible turbulent flows. *AIAA J.* **36**, 2187–2193.

Jameson A 1976 Numerical solutions of nonlinear partial differential equations of mixed type. In *Numerical Solutions of Partial Differential Equations III*, pp. 275–320. Academic Press.

Jameson A 1991 Nonunique solutions to the Euler equations. AIAA Paper 91-1625.

Jameson A 1995 The present status, challenges, and future developments in computational fluid dynamics. In *Proc. 7th AGARD Fluid Dynamics Panel Symp.* (http://aerocomlab.stanford.edu/publications.html).

Jiang GS and Shu C-W 1996 Efficient implementation of weighted ENO schemes. *J. Computat. Phys.* **126**, 202–228.

Johnson DA and Bachalo WD 1980 Transonic flow about a two-dimensional airfoil – inviscid and turbulent flow properties. *AIAA J.* **18**, 16–24.

Jungowski WM and Grabitz G 1987 Self-sustained oscillation of a jet impinging upon a Helmholtz resonator. *J. Fluid Mech.* **179**, 77–103.

Jungowski WM and Meier GEA 1984 Planar jets impinging on various obstacles and some models of flow oscillations. Mitt. Max-Planck-Institut für Strömungsforschung, Göttingen, 78.

Kal'menov TSh 1977 On the spectrum of the Tricomi problem for the Lavrent'ev–Bitsadze equation (in Russian). *Differentsial'nye Uravneniya* **13**, 1418–1425

Kal'menov TSh 1983 On strong solutions of the Darboux and Tricomi problems. *Sov. Math. Dokl.* **28**, 672–674.

Karabutov AA and Sapozhnikov OA 1986 On the reflection of acoustical disturbances from the nearsonic region of Laval nozzle (in Russian). *Acoust. J.* **34**, 538–542.

Karatoprakliev GD 1976 A class of mixed type equations in multidimensional domains. *Sov. Math. Dokl.* **17**, 1379–1383.

Karatoprakliev GD 1977 Theory of boundary value problems for mixed equations in multidimensional regions. I. *Diff. Eqns* **13**, 43–51.

Kellogg B and Liu B 2000 The analysis of a finite element method for the Navier–Stokes equations with compressibility. *Numer. Math.* **87**, 153–170.

King LS and Karamcheti K 1973 Propagation of plane waves in the flow through variable area duct. AIAA Paper 73-1009.

Klopfer GH and Nixon D 1983 Non-isentropic potential formulation for transonic flows. AIAA Paper 83-0375.

Klouček P 1994 On the existence of the entropic solution for the transonic flow problem. *Nonlinear Analysis* **22**, 467–480.

Klouček P and Nečas J 1990 The solution of transonic flow problems by the method of stabilization. *Applicable Analysis* **37**, 143–167.

Kohn JJ and Nirenberg L 1967 Degenerate elliptic–parabolic equations of second order. *Commun. Pure Appl. Math.* **20**, 797–872.

Korolev GL 1983 On the theory of transonic viscous flow over a corner (in Russian). *Uchyonie Zapiski TsAGI* (Central Aerohydrodynamics Institute, Moscow) **14**, 105–109.

Kosolapov YuS 1988a Numerical solution of the problem on supersonic flow from a plane slot in the hodograph variables. *Computat. Math. Math. Phys.* **28**, 93–96.

Kosolapov YuS 1988b Numerical solution of the problem on gas flow from a vessel with plane walls (in Russian). *Izv. Akad. Nauk SSSR Mekh. Zhid. Gaza* **4**, 177–181.

Kosolapov YuS 1991 On the gas flow from plane vessels with big inclination of walls (in Russian). *Izv. Akad. Nauk SSSR Mekh. Zhid. Gaza* **1**, 144–151.

Kotov AI and Kuz'min AG 2000 The structure of transonic flow with weak shocks in a local supersonic region (in Russian). *Vestnik St Petersburg University*, Series 1, no. 3, pp. 87–91.

Kraft EM and Lo CF 1977 Analytical determination of blockage effects in a perforated-wall transonic wind tunnel. *AIAA J.* **15**, 511–517.

Kraiko AN and P'yankov KS 2000 Construction of airfoils and engine nacelles that are supercritical in a transonic perfect-gas flow. *Zh. Vychisl. Matem. Mat. Fiz.* **40**, 1890–1904.

Krupp JA and Cole JD 1976 Studies in transonic flow. IV. Unsteady transonic flow. UCLA Engng Dept. Rep. 76/04.

Kuz'min AG 1975 Boundary value problems for mixed type equations with characteristics twice intersecting the degeneration line. I. *Diff. Eqns* **11**, 1385–1391.

Kuz'min AG 1984 The existence a solution of the direct Lavalle nozzle problem in variations. *USSR Computat. Math. Math. Phys.* **24**, 171–174.

Kuz'min AG 1985 Some boundary value problems for equations of mixed type. *Diff. Eqns* **21**, 62–67.

Kuz'min AG 1986a On an equation of mixed type associated with the direct problem of the Laval nozzle theory (in Russian). *Vestnik Leningrad Univ. Mat. Mekh. Astronom.* Series 1, no. 4, pp. 65–70.

Kuz'min AG 1986b Boundary-value problems for equations of mixed type containing the mixed derivatives. *Diff. Eqns* **22**, 52–59.

Kuz'min AG 1986c Asymptotic types of transonic vortex flows (in Russian). *Vestnik Leningrad Univ. Mat. Mekh. Astronom.* Series 1, no. 2, pp. 61–65.

Kuz'min AG 1987 Second-order equations with elliptic operators with respect to the space variables. *Diff. Eqns* **23**, 74–82.

Kuz'min AG 1990 *Non-classical equations of mixed type and their applications in gas dynamics* (in Russian). Izdat Leningrad Univ., Leningrad.

Kuz'min AG 1992a *Non-classical equations of mixed type and their applications in gas dynamics*. International Series of Numerical Mathematics, vol. 109. Birkhäuser.

Kuz'min AG 1992b The stability of transonic decelerating flow in a channel. *Computat. Math. Math. Phys.* **32**, 1461–1470.

Kuz'min AG 1994a Cumulative and nonlinear phenomena in the local supersonic region. *Computat. Math. Math. Phys.* **34**, 1091–1101.

Kuz'min AG 1994b Formation of shock waves in transonic flow with a local supersonic region. *Vestnik St Petersburg University* Series 1, no. 3, pp. 76–81.

Kuz'min AG 1996 The interaction of a shock wave with a sonic line. *Computat. Math. Math. Phys.* **36**, 1145–1152.

Kuz'min AG 2000a Solvability of a boundary value problem for transonic flow in a nozzle. *Z. Analysis Anwen.* **19**, 729–746.

Kuz'min AG 2000b Convergence of finite elements for shock-free transonic flows. In *Analysis and approximation of boundary value problems* (ed. P Neittaanmäki and L Rivkind), pp. 97–113. University of Jyväskylä, Finland.

Kuz'min AG 2001 Solvability of a problem for transonic flow with a local supersonic region. *J. Nonlin. Diff. Eqns Applic.* **8**, 299–321.

Kweon JR and Kellogg RB 1997 Compressible Navier–Stokes equations in a bounded domain with inflow boundary condition. *SIAM J. Math. Analysis* **28**, 94–108.

Ladyzhenskaya OA 1985 *The boundary value problems of mathematical physics*. Applied Mathematical Sciences, vol. 49. Springer.

Landahl MT 1964 Linearized theory for unsteady transonic flow. In *Symposium Transsonicum*, pp. 414–439. Springer.

Landau LD and Lifshits EM 1987 *Fluid mechanics*, 2nd edn. Oxford: Pergamon.

Larkin NA 1983 Galerkin's method and stabilization method for variable type equations. Technische Hochschule Darmstadt, Fachbereich Mathematik, Preprint 789, pp. 1–18.

Larkin NA 1991 *Smooth solutions of the equations of transonic gas dynamics* (in Russian). Novosibirsk: Nauka, Sibirsk Otdel.

Larkin NA and Schneider M 1990 A stabilization method for the Tricomi problem. *Z. Analysis Anwen.* **9**, 193–202.

Larkin NA and Schneider M 1994 A finite element method and stabilization method for a non-linear Tricomi problem. *Math. Meth. Appl. Sci.* **17**, 681–695.

Larkin NA and Schneider M 1995 Uniqueness theorems for a nonlinear Tricomi problem and the related evolution problem. *Math. Meth. Appl. Sci.* **18**, 591–601.

Larkin NA, Novikov VA and Yanenko NN 1983 *Nonlinear equations of variable type* (in Russian). Novosibirsk: Nauka, Sibirsk Otdel.

Latypov AM 1993 Numerical solution of steady Euler equations in streamline-aligned orthogonal coordinares. (http://www.ima.umn.edu/preprints/November1993/1182.ps).

Lavrent'ev MM 1957 On the Cauchy problem for linear elliptic equations of the second order (in Russian). *Dokl. Akad. Nauk SSSR* **112**, 195–197.

Lax PD 1957 Hyperbolic systems of conservation laws. II. *Commun. Pure Appl. Math.* **10**, 537–566.

Lax PD 1973 Hyperbolic systems of conservation laws and the mathematical theory of shock waves. Conference Board of the Mathematical Sciences Regional Conference Series in Applied Mathematics, no. 11. Philadelphia: SIAM.

Le Balleur JC and Girodroux-Lavigne P 1989 Viscous–inviscid strategy and computation of transonic buffet. In *Symp. Transsonicum III*, pp. 49–63. Springer.

Lebedinsky EV, Natanzon MS and Yarlikova NI 1982 Acoustical, entropy and vortex disturbances in a channel of variable cross section (in Russian). *Izv. Akad. Nauk SSSR Mekh. Zhid. Gaza* **1**, 91–98.

Lee BHK 1990 Oscillatory shock motion caused by transonic shock boundary-layer interaction. *AIAA J.* **28**, 942–944.

Lee BHK 2001 Self-sustained shock oscillations on airfoils at transonic speeds. *Prog. Aerospace Sci.* **37**, 147–196.

Lee BHK, Murty H and Jiang H 1994 Role of Kutta waves on oscillatory shock motion on an airfoil. *AIAA J.* **32**, 789–796.

Liang S-M and Fung K-Y 1987 Refined numerical solution of the transonic flow past a wedge. *AIAA J.* **25**, 1171–1175.

Lifshitz YuB 1965 On flow in the vicinity of the meeting point of the sonic line and a shock wave (in Russian). *Inzhenernii Zh.* **5**, 29–34.

Lifshitz YuB 2000 Instability of steady transonic flow over airfoil. In *Analysis and approximation of boundary value problems* (ed. P Neittaanmäki and L Rivkind), pp. 135–146. University of Jyväskylä, Finland.

Lifshitz YuB and Boichenko VS 1976 Transonic flow over a convex corner (in Russian). *Trans. Zhukovskii Central Aerohydrodynamics Institute* **7**, 8–16.

Lifshitz YuB and Ryzhov OS 1966 On the variation of the gas flux at the on-design operation of a Laval nozzle. *Zh. Vychisl. Matem. Mat. Fiz.* **6**, 276–287.

Lifshitz YuB, Solntsev IA and Velichko SA 1996 Analysis of transonic flow by method of viscous–inviscid interaction: correction for wind tunnel wall interference. AIAA Paper 96-1981.

Lighthill MJ 1945 A new method of two-dimensional aerodynamic design. Aeronautical Research Council, London, R&M 2112.

Lighthill MJ 1947 The hodograph transformation in transsonic flow. III. Flow round a body. *Proc. R. Soc. Lond.* A **191**, 352–369.

Lighthill MJ 1953 On boundary layers and upstream influence. II. Supersonic flow with separation. *Proc. R. Soc. Lond.* A **217**, 478–494.

Lin CC, Reissner E and Tsien HS 1948 On two-dimensional non-steady motion of a slender body in a compressible fluid. *J. Math. Phys.* **27**, 220–231.

Lions P-L 1998 *Mathematical topics in fluid mechanics. 2. Compressible models.* Oxford Lecture Series in Mathematics and Its Applications, vol. 2, no. 10. Oxford University Press.

Liu B 2000 The analysis of a finite element method with streamline diffusion for the compressible Navier–Stokes equations. *SIAM J. Numer. Analysis* **38**, 1–16.

Liu X-D, Osher S and Chan T 1994 Weighted essentially non-oscillatory schemes. *J. Computat. Phys.* **115**, 200–212.

Loth E, Baum J and Lohner R 1992 Formation of shocks within asisymmetric nozzles. *AIAA J.* **30**, 268–272.

Lupo D and Payne KR 2000a On the maximum principle for generalized solutions to the Tricomi problem. *Commun. Contemp. Math.* **2**, 535–557.

Lupo D and Payne KR 2000b Existence of a principal eigenvalue for the Tricomi problem. *Electron. J. Diff. Eqns Conf.* **5**, 173–180.

Mabey DG 1989 Physical phenomena associated with unsteady transonic flows. In *Unsteady transonic aerodynamics* (ed. D Nixon). Progress in Astronautics and Aeronautics, vol. 120, pp. 1–55. AIAA.

McDevitt JB and Okuno AF 1985 Static and dynamic pressure measuremants on a NACA 0012 airfoil in the Ames high Reynolds number facility. NASA Technical Paper 2485.

McDevitt JB, Levy LL and Deiwert GS 1976 Transonic flow about a thick circular-arc airfoil. *AIAA J.* **14**, 606–613.

McGrattan K 1992 Comparison of transonic flow models. *AIAA J.* **30**, 2340–2343.

Makino M and Mineo A 1994 An experimental study of transonic flow behavior over a deformable wing section. In *Proc. 1st Asian–Pacific Conf. on Aerospace Technology and Science, Hangzhou, China*, pp. 463–467. Beijing: International Academic Publishers.

Mandel J and Nečas J 1987 Convergence of finite elements for transonic potential flow. *SIAM J. Numer. Analysis* **24**, 985–996.

Maršik F and Daněk V 1993 Stability of the numerical solution of transonic potential flow. *Z. Angew. Math. Mech.* **73**, 173–178.

Mazaheri K and Ayasoufi A 1998 Adaptive solution of the Euler equations for internal axisymmetric compressible flows. In *Proc. ECCOMAS 98. Computational Fluid Dynamics. Grid Generation in CFD*, vol. 2, pp. 119–124. Wiley.

Melnik RE and Grossman B 1976 Further development in an analysis of the interaction of a weak normal shock-wave with a turbulent boundary layer. In *Symp. Transsonicun II* (ed. K Oswatitsch and D Rues), pp. 262–272. Springer.

Melnik ZO 1981 A non-classical boundary value problem for a hyperbolic system of first order with two independent variables. *Diff. Eqns* **17**, no. 6.

Merchant A and Drela M 1996 Design and analysis of supercritical suction airfoils. AIAA Paper 96-2397-CP, pp. 206–215.

Messiter AF 1970 Boundary-layer flow near the trailing edge of a flat plate. *SIAM J. Appl. Math.* **18**, 241–257.

Meyer HD 1996 Effect of inlet reflections on fan noise radiation. *AIAA J.* **34**, 1771–1777.

Meyer Th 1908 Über zweidimensionale Bewegungvorgänge in einem Gas das mit Überschallgeschwindigkeit strömt. Forschungshefte, 62.

Mises R 1958 *Mathematical theory of compressible fluid flow*. Academic Press.

Mock MS 1980 Systems of conservation laws of mixed type. *J. Diff. Eqns* **37**, 70–88.

Mohammadi B 1998 Flow control using shape optimization for unsteady flows. In *Proc. ECCOMAS 98. Computational Fluid Dynamics. Tools of Numerical Analysis* (ed. KD Papailiou *et al.*), pp. 338–341. Wiley.

Morawetz CS 1954 A uniqueness theorem for Frankl's problem. *Commun. Pure Appl. Math.* **7**, 697–703.

Morawetz CS 1956 On the non-existence of continuous transonic flow past profiles. I. *Commun. Pure Appl. Math.* **9**, 45–68.

Morawetz CS 1957 On the non-existence of continuous transonic flow past profiles. II. *Commun. Pure Appl. Math.* **10**, 107–131.

Morawetz CS 1964 Non-existence of transonic flow past a profile. *Commun. Pure Appl. Math.* **17**, 357–367.

Morawetz CS 1966 Mixed equations and transonic flow. *Rendi. Mat. Applic. 'La Sapienza'* (5) **25**, 482–509.

Morawetz CS 1970a Profile problems for transonic flows with shocks. *Atti Accad. Naz. Lincei. Rendi. Classe Sci. Fis. Mat. Natur.* (8) **49**, 347–356.

Morawetz CS 1970b The Dirichlet problem for the Tricomi equation. *Commun. Pure Appl. Math.* **23**, 587–601.

Morawetz CS 1995 On steady transonic flow by compensated compactness. *Meth. Appl. Analysis* **2**, 257–268.

Morfey CL 1971 Acoustic energy in nonuniform flows. *J. Sound Vib.* **14**, 159–170.

Morice Ph and Viviand H 1980 Équations de conservation et condition d'irréversibilité pour les écoulements transsoniques potentiels. *C. R. Acad Sci. Paris* Sér. A–B **291**, B235–B238.

Mostrel MM 1989 On some numerical schemes for transonic flow problems. *Math. Computat.* **52**, 587–613.

Moulden TH 1984 *Fundamentals of transonic flow*. Wiley.

Murman EM 1974 Analysis of embedded shock waves calculated by relaxation methods. *AIAA J.* **12**, 626–632.

Murman EM and Cole JD 1971 Calculation of plane steady transonic flows. *AIAA J.* **9**, 114–121.

Murty HS 1993 Nonunique solutions in unsteady transonic flow. *J. Aircraft* **30**, 129–131.

Nakahashi K, Togashi F and Sharov D 1999 An intergrid–boundary definition method for overset unstructured grid approach. AIAA Paper 99-3304, pp. 496–506.

Natanzon MS 1986 *Instability of the burning* (in Russian). Moscow.

Nebeck HE, Seebas AR and Sobieczky H 1980 Inviscid–viscous interactions in the nearly direct analysis of shock-free supercritical airfoils. AGARD CP-291. In *Conf. Comput. Viscous–Inviscid Interactions, Colorado Springs*, Paper 3.

Neiland VYa 1969 Towards a theory of separation of the laminar boundary layer in a supersonic stream (in Russian). *Izv. Akad. Nauk SSSR Mekh. Zhid. Gaza* **4**, 53–57.

Ni RH 1982 A multiple-grid scheme for solving the Euler equations. *AIAA J.* **20**, 1565–1571.

Nielsen EJ and Anderson WK 1999 Aerodynamic design optimization on unstructured meshes using the Navier–Stokes equations. *AIAA J.* **37**, 1411–1419.

Niewland GY and Spee RB 1973 Transonic airfoils: recent developments in theory, experiment, and design. In *Annual reviews of fluid mechanics* (ed. MD Van Dyke and WG Vencenti), vol. 5, pp. 119–150. Palo Alto, CA.

Nikolsky AA 1948 *Equations with respect to variations of plane adiabatic gas flows*. Trudi TsAGI, Central Aerohydrodynamics Institute, Moscow. (Republished 1981 Trudi TsAGI, no. 2122, pp. 30–34.)

Nikolsky AA and Taganov GI 1946 Gas motion in a local supersonic region and conditions of breakdown of potential-flow (in Russian). *Prikl. Mat. Mech.* **10**, 481–502. (Transl. into English: (a) Tech. Memos Nat. Adv. Commun. Aeronaut. 1949, no. 1213, (b) Translation A9-T-17. The Graduate Division of Applied Mathematics, Brown University, 1948.)

Nixon D 1987 Observations on the occurrence of multiple solutions in transonic potential theory. *Acta Mechanica* **68**, 43–56.

Nocilla S 1957/58a Flussi transonici attorno a profili alari simmetrici con onda d'urto attaccata ($M_\infty < 1$). *Atti Accad. Sci. Torino: Classe Sci. Fis. Mat. Nat.* **92**, 282–307.

Nocilla S 1957/58b Flussi transonici attorno a profili alari simmetrici con onda d'urto attaccata ($M_\infty < 1$). II. *Atti Accad. Sci. Torino: Classe Sci. Fis. Mat. Nat.* **93**, 124–148.

Nocilla S 1986 Applications and developments of the Tricomi equation in the transonic aerodynamics. Mixed type equations. Teubner-Texte Math., vol. 90, pp. 216–242. Leipzig: Teubner.

Norton DJ 1972 Subsonic, transonic, and supersonic nozzle flow by the inverse technique. *J. Spacecraft Rockets* **9**, 457–459.

Novotny A 1998 Compactness of steady compressible isentropic Navier–Stokes equations via the decomposition method (the whole 3-D space). In *Theory of the Navier–Stokes equations* (ed. JG Heywood, K Masuda, R Rautmann and VA Solonnikov), vol. 47, pp. 106–120. Series on Advances in Mathematics for Applied Sciences. World Scientific.

Obayashi S 1995 Genetic optimization of target pressure distributions for inverse design methods. AIAA Paper 95-1649-CP. In *12th AIAA CFD Conf. A Collection of Technical Papers*, Part 1, pp. 33–42.

Oryol AA 1971 On the theory of the direct problem for a Laval nozzle (in Russian). In *Transonic flows* (ed. SV Fal'kovitch), no. 3, pp. 12–18. Saratov University.

Osher S 1986 Shock capturing algorithms for equations of mixed type. In *Numerical methods for partial differential equations* (ed. R Hariharan and TH Moulden), pp. 103–139. Pitman.

Osher S, Hafez M and Whitlow Jr W 1985 Entropy condition satisfying approximations for the full potential equation of transonic flow. *Math. Computat.* **44**, 1–29.

Osipov AA and Shirkovsky IA 1988 Using the finite element method for calculation of acoustical fields in nonuniform gas flows (in Russian). *Zh. Vychisl. Matem. Mat. Fiz.* **28**, 362–374.

Ostapenko VV 2000 Construction of high order accurate shock-capturing finite-difference schemes. *Zh. Vychisl. Matem. Mat. Fiz.* **40**, 1857–1874.

Oswatitsch K 1976 *Grundlagen der Gasdynamic*. Springer.

Oswatitsch K and Zierep J 1960 Das Problem des Senkrechten Stosses an einer gekrümmten Wand. *Z. Angew. Math. Mech.* **40**, 143–144.

Oyama A, Obayashi S, Nakahashi K and Hirose N 1999 Fractional factorial design of genetic coding for aerodynamic optimization. AIAA Paper 99-3298, pp. 429–439.

Özyörük Y and Long LN 1995 A new efficient algorithm for computational aeroacoustics on massively parallel computers. AIAA Paper 95-1693, pp. 527–539.

Pandya SA and Hafez M 1999 A finite-element approach for modeling inviscid and viscous compressible flows using prismatic grids. AIAA Paper 99-3310, pp. 537–546.

Payne KR 1997 Solvability theorems for linear equations of Tricomi type. *J. Math. Analysis Appl.* **215**, 262–273.

Payne KR 1998 Propagation of singularities phenomena for equations of Tricomi type. *Applicable Analysis* **68**, 195–206.

Pearcey HH 1962 The aerodynamic design of section shapes for swept wings. *Adv. Aeronaut. Sci.* **3**, 277–237.

Peyser G 1961 The Goursat problem for hyperbolic equations. *J. Math. Mech.* **10**, 91–109.

Pfenninger W, Viken J, Vemuru CS and Volpe G 1986 All laminar supercritical LFC airfoils with natural laminar flow in the region of the main wing structure. AIAA Paper 86-2625, pp. 1–45.

Pirumov UG and Roslyakov GS 1990 *Gas dynamics of nozzles* (in Russian). Moscow: Nauka.

Pirzadeh SZ 1999 An adaptive unstructured grid method by grid subdivision, local remeshing, and grid movement. AIAA Paper 99-3255.

Pivovarchik VN 1987 On the discrete spectrum of a problem associated with wave propagation through nonuniform media with viscous friction (in Russian). *Differentsial'nie Uravneniya* **23**, 1533–1538.

Powell A 1988 The sound-producing oscillations of round underexpanded jets impinging on normal plates. *J. Acoust. Soc. Am.* **83**, 515–533.

Prasad P 1973 Nonlinear wave propagation on an arbitrary steady transonic flow. *J. Fluid Mech.* **57**, 721–737.

Raushenbakh BV 1961 *The vibrational burning* (in Russian). Moscow.

Ringleb F 1940 Exakte Lösungen der Differentialgleichungen einer adiabatischen Gasströmung. *Z. Angew. Math. Mech.* **20**, 185–198.

Rogers EWE and Hall IM 1980 An introduction to the flow about plane swept back wings at transonic speeds. *Aeronaut. J.* **64**, 449–464.

Romanova NA 1994 On a finite difference scheme for an equation of mixed type (in Russian). *Mat. Zametki Yakutskogo Univ.* **1**, 35–40.

Romanova NA 1995 On the convergence of a finite difference scheme for a boundary value problem for a mixed type equation (in Russian). *Mat. Zametki Yakutskogo Univ.* **2**, 60–66.

Romine GL 1998 Nozzle flow separation. *AIAA J.* **36**, 1618–1625.

Roy ID and Eversman W 1993 Improved finite element modeling of the turbofan engine inlet radiation problem. American Society of Mechanical Engineers. ASME Paper 93-WA/NCA-13.

Ruban AI and Turkyilmaz 2000 On laminar separation at a corner point in transonic flow. *J. Fluid Mech.* **423**, 345–380.

Rumsey CL, Sanetrick MD, Biedron RT, Melson ND and Parlette EB 1995 Efficiency and accuracy of time-accurate turbulent Navier–Stokes computations. AIAA Paper 95-1835.

Rusak Z 1993 Transonic flow around the leading edge of a thin airfoil with a parabolic nose. *J. Fluid Mech.* **248**, 1–26.

Ryzhov OS 1978 Viscous transonic flows. *A. Rev. Fluid Mech.* **10**, 65–92.

Ryzhov OS 1989 Asymptotic methods in transonic flow theory. In *Symp. Transsonicum III*, pp. 143–155. Springer.

Ryzhov OS 1993 An asymptotic approach to separation and stability problems of a transonic boundary layer. In *Transonic aerodynamics. Problems in asymptotic theory* (ed. LP Cook). Frontiers in Applied Mathematics, vol. 12, pp. 29–53.

Sabitov KB 1992 On the existence of a solution to the Tricomi problem. *Diff. Eqns* **28**, 1737–1745.

Salas MD and Gumbert CR 1985 Breakdown of the conservative potential equation. AIAA Paper 85-0367.

Salas MD, Gumbert CR and Turkel E 1984 Nonunique solutions to the potential flow equations. *AIAA J.* **22**, 145–146.

Samarsky AA 1977 *Theory of difference schemes* (in Russian). Moscow: Nauka.

Savenkov IV 1999 On the instability of axisymmetric boundary layer at transonic speeds of the outer flow (in Russian). *Zh. Vychisl. Matem. Mat. Fiz.* **39**, 1405–1414.

Savu G and Trifu O 1984 Porous airfoils in transonic flow. *AIAA J.* **22**, 989–991.

Schneider M 1992 Regularity statements for solutions of the Tricomi problem (in German). *Results Math.* **22**, 749–760.

Sevost'yanov GD 1973 Equations of nearsonic flows of viscous gas in the natural coordinates (in Russian). In *Aerodinamika* (ed. SV Fal'kovitch), no. 2, pp. 42–48. Saratov University.

Shifrin EG 1966 Plane vortex flow in the vicinity of the orthogonality point of a sonic line to the velocity vector (in Russian). *Izv. Akad. Nauk SSSR Mekh. Zhid. Gaza* **6**, 144–146.

Shifrin EG 1974 On the shock wave in transonic flow over a convex corner (in Russian). *Izv. Akad. Nauk SSSR Mekh. Zhid. Gaza* **5**, 43–48.

Shifrin EG 1978 On the uniqueness 'on the whole' of the solution to the direct problem of Laval nozzle (in Russian). *Zh. Vychisl. Matem. Mat. Fiz.* **18**, 509–512.

Shu C-W 1990 Numerical experiments on the accuracy of ENO and modified ENO schemes. *J. Scient. Comput.* **5**, 127–149.

Shu C-W and Osher S 1989 Efficient implementation of essentially nonoscillatory shock-capturing schemes. II. *J. Comput. Phys.* **83**, 32–78.

Shu C-W and Osher S 1988 Efficient implementation of essentially nonoscillatory shock-capturing schemes. *J. Comput. Phys.* **77**, 439–471.

Smirnov MM 1966 *Degenerate elliptic and hyperbolic equations* (in Russian). Moscow: Nauka.

Smirnov MM 1978 *Equations of mixed type*. Translation Mathematical Monographs, vol. 51. Providence, RI: American Mathematical Society.

Smith FT 1973 Laminar flow over a small hump on a flat plate. *J. Fluid Mech.* **57**, 803–829.

Smith FT 1982 On the high Reynolds number theory of laminar flows. *IMA J. Appl. Math.* **28**, 207–281.

Smith FT 1985 A structure for laminar flow past a bluff body at high Reynolds number. *J. Fluid Mech.* **155**, 175–191.

Smith FT, Degani AT and Walker JDA 1998a Unsteady separation past moving surfaces. *J. Fluid Mech.* **375**, 1–38.

Smith FT, Li L, Walker JDA and Bowles RI 1998b Short-scale break-up in unsteady interactive layers: local development of normal pressure gradients and vortex wind-up. *J. Fluid Mech.* **374**, 335–378.

Sobieczky H 1971 Rheoelektrische Analogie zur Darstellung transsonischer Strömungen. DLR FB 71-26.

Sobieczky H 1978 Die Berechung lokaler räumlicher Überschallfelder. *Z. Angew. Math. Mech.* **58T**, 331–333.

Sobieczky H 1985 Transonic wedge shoulder flow singularity. University of Arizona Rep. CFML 85-04.

Sobieczky H 1989 Progress in inverse design and optimization in aerodynamics. In *AGARD Fluid Dynamics Panel Specialists Meeting on Computational Methods for Aerodynamic Design (Inverse) and Optimization*, Norway, pp. 1–10.

Sobieczky H, Geissler W and Hannemann M 1998 Expansion shoulder bump for wing section viscous/wave drag control. In *Proc. IUTAM 'FLOWCON' Symp. on Mechanics of Passive and Active Flow Control* (ed. GEA Meier and PR Viswanath). Fluid Mechanics and its Applications, vol. 53, pp. 29–34. Kluwer.

Sobieczky H and Seebass A 1984 Supercritical airfoil and wing design. *A. Rev. Fluid Mech.* **16**, 337–363.

Sobieczky H and Stanewsky E 1976 The design of transonic airfoils under consideration of shock wave boundary layer interaction. ICAS Papers no. 76-14, pp. 1–15.

Sobieczky H, Yu NJ, Fung K-Y and Seebass AR 1979 New method for designing shock-free transonic configurations. *AIAA J.* **17**, 722–729.

Sokolovsky GA and Gnesin VI 1986 *Unsteady transonic and viscous flows in turbomachines* (in Russian). Kiev.

Soldatov AP 1994a Problems of Dirichlet type for the Lavrent'ev–Bitsadze equation. I. Uniqueness theorems. *Russ. Acad. Sci. Dokl. Math.* **48**, 410–414.

Soldatov AP 1994b Dirichlet-type problems for the Lavrent'ev–Bitsadze equation. II. Existence theorems. *Russ. Acad. Sci. Dokl. Math.* **48**, 433–437.

Soldatov AP 1994c Dirichlet-type problems for the Lavrent'ev–Bitsadze equation. *Diff. Eqns* **30**, 1846–1853.

Sommerer A, Lutz Th and Wagner S 2000 Design of adaptive transonic airfoils by means of numerical optimisation. In *Proc. European Congress on Computational Methods in Applied Science Engineering, ECCOMAS-2000, Barcelona*.

Sonar T and Süli E 1998 A dual graph-norm refinement indicator for finite volume approximations of the Euler equations. *Numer. Math.* **78**, 619–658.

Spee BM 1969 Investigations on the transonic flow around airfoils. National Aerospace Laboratory NLR-TR 69122, University of Amsterdam.

Stanewsky E and Basler D 1990 Experimental investigation of buffet onset and penetration on a supercritical airfoil at transonic speeds. Aircraft Dynamic Loads Due to Flow Separation, AGARD-CP-483, pp. 4.1–4.11.

Stanewsky E, Délery J, Fulker J and Geißler W 1997 Synopsis of the project EUROSHOCK. In *EUROSHOCK – Drag Reduction by Passive Shock Control. Notes Numer. Fluid Mech.* **56**, 1–80.

Steinhoff J and Jameson A 1982 Multiple solutions of the transonic potential flow equations. *AIAA J.* **20**, 1521–1525.

Stepanov GYu 1962 *Hydrodynamics of turbomachinery cascades* (in Russian). Moscow: Fizmatgiz.

Stewartson K 1970a Is the singularity at separation removable? *J. Fluid Mech.* **44**, 347–364.

Stewartson K 1970b On supersonic boundary layers near convex corners. *Proc. R. Soc. Lond.* A **319**, 289–305.

Stewartson K 1971 On laminar boundary layers near corners. Corrections and an addition. *Q. J. Mech. Appl. Math.* **24**, 387–389.

Stewartson K and Williams PG 1969 Self-induced separation. *Proc. R. Soc. Lond.* A **312**, 181–206.

Stolyarov EP 1983 Sound generation by small disturbances of entropy and vorticity in space—nonhomogeneous flows of compressible gas (in Russian). In *Acoustics of nonuniform flows*, pp. 3–14. Acoustical Institute of the USSR Academy of Sciences, Moscow.

Sun HS 1985 Generalized Tricomi problem for the equation of mixed type of second order. *Kexue Tongbao* (English Edn) **30**, 1441–1446.

Sun HS and Xiao Z 1989 Generalized Tricomi problem for the nonlinear equation of mixed type. *J. Part. Diff. Eqns* **2**, 87–96.

Swinney HL and Gollub JP (eds) 1981 *Hydrodynamic instabilities and the transition to turbulence* (with contribution by FH Busse, RA Davies *et al.*). Springer.

Takami H 1956 A study of the compressible flow past a Cherry profile. *J. Phys. Soc. Jpn* **11**, 145–154.

Takanashi S 1985 Iterative three-dimensional transonic wing design using integral equations. *J. Aircraft* **22**, 655–660.

Tai TC, Huson GG, Taylor DW, Hicks RM and Gregorek GM 1987 Transonic characteristics of a humped airfoil. AIAA Paper 87-1239.

Tam CKW and Webb JC 1993 Dispersion-relation-preserving finite difference schemes for computational acoustics. *J. Computat. Phys.* **107**, 262–281.

Taylor GI 1930a Strömung um einen Körper in einen kompressiblen Flüssigkeit. *ZAMM* **10**, 334–345.

Taylor GI 1930b Recent work on the flow of compressible fluids. *J. Lond. Math. Soc.* **5**, 224–240.

Terekhov AN 1979 A boundary value problem for an equation of mixed type (in Russian). In *Application of functional analysis methods in problems of math physics and comput mathematics*, pp. 128–137. Institute of Mathematics, Novosibirsk.

Terekhov AN 1980 On a unique solvability of a problem for an equation of mixed type (in Russian). In *Proc. Conf. on Differential Equations and Computational Mathematics*, pp. 62–64. Novosibirsk: Nauka.

Thiede P and Krogmann P 1989 Passive control of transonic shock/boundary layer interaction. In *Symp. Transsonicum III*, pp. 379–388. Springer.

Tijdeman H 1977 Investigation of the transonic flow around oscillation airfoils. National Aerospace Laboratory NLR TR-77090. Amsterdam, The Netherlands.

Tricomi F 1947 *On linear second-order partial differential equations of mixed type* (in Russian, translation from the Italian by FI Frankl). Gosudarstvennoe Izdatel'stvo Tekhniko-Teoreticheskoi Literaturi. Moscow, Leningrad.

Vaglio-Laurin R 1960 Transonic rotational flow over a convex corner. *J. Fluid Mech.* **9**, 81–103.

Vaigant VA 1994 An example of the nonexistence with respect to time of the global solution of Navier–Stokes equations for a compressible viscous barotropic fluid (in Russian). *Dokl. Akad. Nauk* **339**, 155–156. (Translation in Russian *Acad. Sci. Dokl. Math. 50* (1995), 397–399.)

Velichko SA, Lifshitz YuB, Neiland VM, Solntsev IA and Sorokin AM 1995 Numerical modelling of transonic flow past an airfoil in wind tunnel. *Computat. Math. Math. Phys.* **35**, 1221–1235.

Verdon JM 1989 Unsteady aerodynamics for turbomachinery aeroelastic applications. In *Unsteady transonic aerodynamics* (ed. D Nixon). Progress in Astronautics and Aeronautics, vol. 120, pp. 287–347. AIAA.

Verfürth R 1989 *A posteriori* error estimators for the Stokes equations. *Numer. Math.* **55**, 309–325.

Verhoff A 1999 Complementing computational fluid dynamics methods with classical analytical techniques. *Comput. Fluids* **28**, 389–425.

Volpe G and Melnik RE 1984 Role of constraints in inverse design for transonic airfoils. *AIAA J.* **22**, 1770–1778.

von Kármán T and Fabri J 1950 Écoulement transsonique à deux dimensions le long d'une paroi ondulée. *Comptes Rendus Acad. Sci. Sér. I Math.* **231**, 1271–1274.

Vragov VN 1977 On the theory of boundary value problems for equations of mixed type in space. *Diff. Eqns* **13**, 759–764.

Walter W 1970 *Differential and integral inequalities*. Springer.

Wang ZJ, Chen RF, Hariharan N, Przekwas AJ and Grove D 1999 A 2^N three based automated viscous Cartesian grid methodology for feature capturing. AIAA 99-3300, pp. 447–457.

Warsi ZUA 1999 *Fluid dynamics: theoretical and computational approaches*. CRC Press, LLC.

Wesseling P 2001 *Principles of computational fluid dynamics*. Springer Series in Computational Mathematics, vol. 29.

Whitlow Jr W 1988 Application of a nonisentropic full potential method to AGARD standard airfoils. AIAA Paper 88-0710.

Williams MH, Bland SR and Edwards JW 1985 Flow instabilities in transonic small-disturbance theory. *AIAA J.* **23**, 1491–1496.

Woods LC 1955 The design of two-dimensional airfoils with mixed boundary conditions. *Q. Appl. Math.* **13**, 139–146.

Woodward PR and Colella P 1984 The numerical simulation of two-dimensional fluid flow with strong shocks. *J. Computat. Phys.* **54**, 115–173.

Yang JY 1990 Uniformly second-order-accurate essentially nonoscillatory schemes for the Euler equations. *AIAA J.* **28**, 2069–2076.

Yamamoto K and Inoue O 1995 Applications of genetic algorithm to aerodynamic shape optimization. AIAA Paper-95-1650-CP. *12th AIAA CFD Conference. A Collection of Technical Papers*, Part 1, pp. 43–51.

Zingg DW, Nelson TE and Godin P 2001 A flow solwer for multi-element airfoils (http://www.utias.utoronto.ca/test/res/fd/pictures/fd/cfd-screen.gif).

Index